Technically—Write!

COMMUNICATION FOR THE TECHNICAL MAN

R. S. Blicq

RED RIVER COMMUNITY COLLEGE
Winnipeg, Manitoba, Canada

Prentice-Hall, Inc., Englewood Cliffs, New Jersey

Library of Congress Cataloging in Publication Data

Blicq, R   S
   Technically—write!

   1. Technical writing.   I. Title.
T11.B62   808.06'6'602   77-177394
ISBN 0-13-898692-4
ISBN 0-13-898676-2 (pbk.)

© 1972 by Prentice-Hall, Inc.
Englewood Cliffs, New Jersey

10 9 8

Printed in the United States of America

Prentice-Hall International, Inc., *London*
Prentice-Hall of Australia, Pty. Ltd., *Sydney*
Prentice-Hall of Canada, Ltd., *Toronto*
Prentice-Hall of India Private Limited, *New Delhi*
Prentice-Hall of Japan, Inc., *Tokyo*

Contents

1
A Technical Man's Approach to Writing

2
Meet "H. L. Winman and Associates"

3

Technical Correspondence

4

Informal Reports Describing Facts and Events

5
Informal Reports Describing Ideas and Concepts

6
Formal Reports

7
Other Technical Documents

8
Technically—Speak!

9
Illustrating Technical Documents

10
The Technique of Technical Writing

CONTENTS

11
Glossary of Technical Usage

Preface

This book presents all those aspects of technical communication that you, as a technician, technologist, engineer, or scientist, are most likely to encounter in industry. It introduces you to the employees of two technically-oriented companies, to the type of work they do, and to typical situations that call for them to communicate with clients, suppliers, and each other. At the end of each major chapter it contains a selection of detailed assignments of varying complexity, in which I assume you are employed by one of these companies, and are engaged in projects that require you to write letters and reports, and sometimes to present your findings orally.

A marking control chart (at the end of the book) offers a place for you to record your writing errors when assignments are returned to you, and to check whether you have corrected your faults in successive work. It is based on a Correction Key in "Writing to be Read," a Prentice-Hall Inc. textbook by Eleanor Newman Hutchens, with whose permission it is printed here in an altered form.

The information on technical paper presentation (Chapter 8) was presented originally as a technical paper at the IRE (now IEEE) Sixth National Symposium on Engineering Writing and Speech, Washington, D.C. It was subsequently published by Electronic Design under the title "It's Your Turn at the Rostrum—Can You Deliver a Technical Paper?", through whose courtesy it is reprinted here.

R. S. B.

Man as "Communicator"

As a communicator, man is lazy and inefficient. He is equipped with a highly sophisticated communication system, yet consistently fails to use it properly. (If our predecessors had been equally lazy, and failed to develop their communication "senses" to the level required to ensure survival, many species would be extinct today; indeed, it's even questionable that man would exist in his present form.) This communication system comprises a transmitter and receiver combined into a single package controlled by a computer. It accepts multiple inputs and transmits in three mediums: action, speech, and writing.

We spend much of our wakeful hours communicating, half the time as a transmitter, half as a receiver. If, as a receiver, we mentally switch off or permit ourselves to change channels while someone else is transmitting, we contribute to information loss. Similarly, if as a transmitter we permit our narrative to become disorganized, unconvincing, or simply uninteresting, we encourage frequency drift. Our listeners detune their receivers and let their computers think about the lunch that's imminent, or wonder if the redhead sitting under the window has a date for tonight.

As long as a person transmits clearly, efficiently, and persuasively, the person receiving keeps his receiver "locked on" to the transmitting frequency (this applies to both written and spoken transmissions). Such conditions expedite information transference, or "communication."

In the direct contact situation, in which one person is speaking directly to another, the receiver has the opportunity to ask the transmitter to clarify vaguely presented information. But in more formal speech situations, and in all forms of written communication, the receiver no longer has this advantage. He cannot stop a speaker who mumbles or uses unfamiliar terminology, and ask him to repeat or clarify what he has just said; neither can he ask a writer in another city to explain an incoherent passage of a business letter.

The results of failure to communicate efficiently soon become apparent. If a person fails to make himself clear in day-to-day communication, the consequences are likely to differ from those he anticipated, as Cam Collins has discovered to his chagrin.

Cam is a junior electrical engineer at Robertson Engineering Company, and his specialty is high voltage power generation. When he first read about a recent EHV DC power conference, he wanted urgently to attend. In a memorandum to Fred Stokes, the company's chief engineer, Cam described the conference in glowing terms which he hoped would induce Fred to approve his request. This is what he wrote:

> Fred:
> The EHV conference described in the attached brochure is just the thing we have been looking for. Only last week you and I discussed the shortage of good technical information in this area, and now here is a conference featuring papers on many of the topics we are interested in. The cost is only $75.00 for registration, which includes a visit to the Kettle Generating Station. Travel and accommodation will be about $250 extra. I'm informing you of this early so you can make a decision in time for me to arrange flight bookings and accommodation.
>
>   Cam

Fred Stokes was equally enthusiastic and wrote back:

> Cam:
> Thanks for informing me of the EHV DC conference. I certainly don't want to miss it. Please make reservations for me as suggested in your memorandum.
>
>   Fred

Cam has become the victim of his own carelessness: he has failed to communicate clearly exactly what he wanted. There may still be a chance to remedy the situation, but only at the expense of extra time and effort.

Don Ristowell, on the other hand, did not realize that he had missed the proverbial "golden opportunity" until it was much too late to do anything about it. His story stems from an incident that occurred several years ago, when he was a young engineering technician with a firm of mechanical engineers. His girl friend was a stenographer in the contracts department whose job was to type up proposals for clients. Lillian constantly complained to Don that it was impossible to make the proposals look as neat as she would like.

"The right-hand side of the typing always looks so ragged," she would say. "If only I could make the margin straight—like the one on the left."

Don told her he would look at a typewriter he had at home to see if he could modify it. It was a heavy, old-fashioned machine, but it served his purpose.

He reasoned that if the toothed bar that controls letter spacing at a fixed 10 or 12 characters to the inch could be made flexible, the length of each line could be adjusted by fractionally expanding or contracting the spacing between letters.

With the limited range of materials then available, Don could not find a practical way to stretch the toothed bar without destroying its rigidity. So he made a series of rigid toothed bars, each with a different number of teeth, which he mounted on a rotating axle that could be positioned immediately beneath the typewriter carriage. It was a clumsy contraption, too large and cumbersome for the average typewriter, but it proved that his idea was feasible.

Don felt that his employer should know about his idea—possibly the company could develop it into a marketable product, or even help him patent it. So the following day he stopped the Engineering Manager and blurted out his suggestion. This is the conversation that ensued:

Don	Mr. White
Mr. White! I've modified an old typewriter at home—to adjust the length of the typing line. . . .	
	Oh?
. . . The problem with most typewriters is the teeth are rigid, so the letter spacing is always the same. You can't get a straight right-hand margin. . . .	
	(*Mr. White appeared to be listening politely, but internally he was growing impatient*)
. . . It would help them over in Contracts if we could. . . .	
	Ah! The Contracts Department asked you to modify their typewriters?
Well—uh—in a way. It's for their typist.	
	I don't remember giving you a work order. . . .
No. I did it on my own. (*He meant he did it at home, in his own time*)	
	You took the order directly from Contracts?
It was just an idea I had. . . .	
	There's no reason why the request should not have come through the proper channels. (*His tone was now cold*

*and distant; already he was mentally
composing a strong note to the Contracts
Manager)*
You had better come into my office!

Don's simple little suggestion had become lost in a web of misunderstanding. By the time he was through explaining, he had given up trying to offer his idea to the company. It lay dormant for three years, until proportional-spacing typewriters first appeared on the scene. They served to remind him that perhaps there *had* been market potential in his modification.

If Cam Collins and Don Ristowell had paused to consider the needs of the persons who were to receive their information, they would never have launched precipitously into discourses that omitted essential facts. Cam had only to start his memorandum with a request ("May I have your approval to attend an EHV DC conference next month?"), and Don with a statement of purpose ("I have devised a gadget that I believe would make a marketable product. May I have a few moments to describe it to you?"), to command the attention of their department heads. Both Mr. Stokes and Mr. White could then have much more effectively appraised the information they offered.

Such circumstances occur daily. They are frustrating to those who fail to communicate their ideas, and costly when the consequences are carried into business and industry.

Bill Carr recently devised and installed a monitor unit for the remote control panel at the microwave relay station where he is the resident engineering technician. As his modification greatly improved operating methods, head office asked him to submit an installation drawing and an accompanying description. Here is part of his description:

> Some difficulty was experienced in finding a suitable location for the monitor unit. Eventually it was mounted on a special bracket attached to the left-hand upright of the control panel, as shown on the attached drawing.

On the strength of Bill's explicit mounting description and detailed list of hardware, head office converted his description into an installation instruction, purchased materials, assembled 21 modification kits, and shipped them to the 21 other relay stations in the microwave link.

Within a week head office received reports from the 21 resident engineering technicians that it was impossible to mount the monitor unit as instructed, because of an adjoining control unit. No one at head office had remembered that Bill Carr was located at site 22, the last relay station in the microwave link, where there was no need for an additional control unit.

Bill had assumed that head office would be aware that the equipment layout at his station was unique. As he commented afterward: "No one said why I had to describe the modification, or told me what they planned to do with it."

In business and industry it is imperative that we communicate clearly, and understand fully the implications of failing to do so. A poorly worded order that results in the wrong part being supplied to a job site, a weak report that fails to motivate the reader to take the urgent action needed to avert a costly equipment breakdown, and even an inadequate job application that fails to "sell" an employer on the right man for a prospective job, all increase the cost of doing business. Such mistakes and misunderstandings are wasteful of the country's manpower and resources. Many of them can be prevented by more effective communication—communication that is receiver-oriented rather than transmitter-oriented, and that transmits messages using the most expeditious, economical, and efficient means at our command.

1

A Technical Man's Approach to Writing

Almost every book on technical writing contains a statement that says in effect: "The key to effective writing is good organization." This rule is basically true, although many technical men who have tried to follow it too conscientiously find they have difficulty in writing clearly. Frequently their trouble is caused by overorganization, or by trying to organize too early. Organizing that starts too early is self-defeating because it stifles a person's natural ability to write creatively.

Look at it this way: Technician Dan Skinner has a report to write on an investigation he completed seven weeks ago. He works for H.L. Winman and Associates, a consulting engineering firm we will meet in Chapter 2, and he has made several half-hearted attempts to get started. But each time has never seemed to be the propitious moment: maybe he was interrupted to resolve a circuit problem, or it was too near lunchtime, or a meeting was called, or, when nothing else interfered, he "just wasn't in the mood." And now he's up against the wire and he hasn't yet set pen to paper.

Unless Dan is one of those unusual persons who cannot produce except when under pressure, he is in imminent danger of writing an inadequate, hastily prepared report that does not represent his true abilities. He knows he should not have left his report-writing project until the last moment, but he is human, like the rest of us, and constantly finds himself in situations like this. He does not realize that by leaving a writing task until it is too late to do a good job, and then frantically organizing the work, he is probably inhibiting his writing capabilities even more than necessary.

If Dan were to relax a little in the initial stages, instead of spending time trying to organize both himself and his writing task, he would find the physical process of writing reports a much more pleasant experience. But he must first change his whole approach to writing.

Every technical man, from student technician through potential scientist to practicing engineer, must recognize that he has the ability to write clearly and logically. (He may have to prove this to many responsible people in industry who continue to believe the old adage that "technical people just can't write.") He must also realize that this ability must be developed by continued practice. Only by planning and writing all types of documents will he gain the confidence that is the prerequisite to good writing.

PLANNING THE WRITING TASK

The word "planning" seems to contradict everything I have just said: it implies that a report writer must start by organizing himself and his material. However, I suggest that he do so "creatively," to allow his latent writing ability to develop naturally. I recommend that at first Dan Skinner do nothing about making an outline or taking any action that smacks of organization. All he has to do is to work through several simple planning stages that will not be the chore he probably expects.

Normally, the first stage in planning any writing project is to gather information. This means assembling all the documents, results of tests, photographs, samples, specifications, and so on, that will be needed to write the report, or that will be included with it. The writer must gather more information than he will need, because it is better to be selective and discard information than to look for additional facts and figures just when the writing is beginning to go well.

The next stage—and probably the most important—is for Dan Skinner to define his reader. He must conjure up an image of the actual person, or type of person, for whom he is writing. He must ask himself some pertinent questions: Who is this reader? What is his reading level? How much does he know about my project? Is he a technical man? Does he need to know all the details that I know? Where do his main interests lie? What will he do with my report? How will he use it? Finally, who else is likely to read it? The answers to these questions will help him not only plan a good report, but also set the right tone when he is writing it.

When he has a reader in mind, Dan can start making notes. It is at this stage that he must let his ability to generate ideas spontaneously come into full play. He must let his mind "freewheel" so that it throws out ideas and pieces of information in an uninhibited manner. He must not stop to question the relevance of this information—his role is purely to collect it in readiness for later scrutiny. First he must find a quiet place where he can work undisturbed; it's no use trying to be creative in a noisy, crowded office. (I will mention more about the need for establishing a good writing environment in the next step, when I discuss the practical aspects of settling down to write.)

Normally at this stage a technical man will take a blank sheet of paper and write down a set of arbitrary headings, such as "Introduction," "Initial Tests," and "Material Resources," arranged in logical order. These seem to be standard-type headings he has seen in other reports. But I want our report writer to be different. I want Dan Skinner to free his mind of the elementary organized headings, and even to refuse to divide his subject mentally into blocks of information. Then I want him to jot down a series of main topics that he will discuss, writing only brief headings rather than full sentences. He must do this in random order, making no attempt to force the topics into groups (although it is quite possible that this will occur naturally, since many interdependent topics are likely to occur to him sequentially). The topics that he knows best will spring most readily to mind, followed by a gradual slowdown as the flow of familiarity eases up.

When he stops his initial list, he should return to the top of it and examine each topic in turn to see if it will suggest less obvious topics. As each new or subsidiary topic comes to mind, triggered by the sight of the original topic, he must jot it down, still in random order but, if possible, keyed to the main topic. He must continue doing this until he finds he is straining to find new ideas, which should be a signal for him to stop before he becomes too objective.

It is most important that Dan not try to decide whether a topic is truly relevant during this spontaneous freewheeling session. To do so will immediately inhibit his creativeness, because it will force him to be too logical and organized. Hence all topics, regardless of their relative importance and spatial orientation in the final report, must be jotted down in this way.

At the end of this session Dan's list of topics should look like those in Figure 1–1. He can now take a break, knowing that the first details of his report have been committed to paper. What he may not yet realize is that he has almost painlessly produced his first outline.

The fourth stage calls for Dan to examine his list of headings with a critical eye, discriminating between headings that bear directly on the subject matter and those that introduce topics of only marginal interest. This is where his knowledge of the reader becomes really important, for without it he will be unable to assess whether a topic is essential to a full understanding of the report. He must delete every irrelevant topic with a stroke of his pen, as has been done in Figure 1–2.

The headings that remain he should group into "topic areas" that will be discussed together. This he can do simply by coding related topics with the same symbol or letter. In Figure 1–2, letter (A) identifies one group of related topics, letter (B) another group, and so on.

Now, at last, Dan can take his first major organizational step, which involves arranging the groups of information in the most suitable order, and at the same time sorting out the order of the headings within each group. He must consider three factors: which order of presentation will be most interesting, which will be most logical, and which will be simplest to understand. This will become his final writing plan, or report outline (see Fig. 1–3).

Building OK — needs strengthening
Elevators — too slow, too small
Talk with YoYo — elev. mfr (10% discount)
Waiting time too long — 70 sec.
Shaft too small
How enlarge shaft? — Remove stairs
Talk with fire inspector
Correspondence — other mfrs.
Talk with Merrywell — Budget $250,000
Sent out questionnaire
Tenants' preferences —

    Express elev   No 2nd floor stop
    Executive elev   Faster service
    Prestige elev   No ground floor stop
    Freight elev

Freight elev — takes too much space
Shaft only 35 x 8 (when modified)
Big freight elev — can omit Basement
Tenants OK small freight ⟵ Model "c"
YoYo — has office in Montrose   8 ft
Basement level — has loading dock
Service reputation — YoYo vs. others

Fig. 1–1   Initial list of topic headings, jotted down in random order.

(A) Building OK - needs strengthening

~~Elevators - too slow, too small~~

(B) Talk with YoYo - elev. mfr (10% discount)

(C) Waiting time too long   -   70 sec.

(A) Shaft too small

(A) How enlarge shaft?   -   Remove stairs (A)

~~Talk with fire inspector~~

(B) Correspondence - other mfrs.

(C) Talk with Merrywell   -   Budget $250,000

(C) Sent out questionnaire

Tenants' preferences -

(C)	Express elev	~~No 2nd floor stop~~
	Executive elev	Faster service
	Prestige elev	~~No ground floor stop~~
	Freight elev	

(D) Freight elev - takes too much space

(A) Shaft only 35 x 8 (when modified)

~~Big freight elev - can omit Basement~~

(C) Tenants OK small freight   ⟵ Model "C" (D)

(B) YoYo - has office in Montrose   8 ft

~~Basement level - has loading dock~~

(B) Service reputation - YoYo vs. others

Fig. 1–2 The same list of topic headings, but with irrelevant topics deleted and remaining topics coded into subject groups (A—structural implications; B—elevator manufacturers; C—tenants' preferences; D—freight elevator).

Building Condition
      OK - needs strengthening (shaft area)
      Existing elev. shaft too small
      Remove adjoining staircase
      Shaft size now 35 x 8 ft

Tenants' Needs
      Sent out questionnaire
      Identified 5 major requests
      Requests we must meet:
          Cut waiting time: < 32 sec
          Handle freight up to 7 ft 6 in (min)
      Requests we should try to meet:
          Express elev to top 4 floors
          Deluxe models (for prestige)
          Private elev (for executives)

Budget - must be within $250,000

Elevator Manufacturers
      Researched 3
      Only YoYo Co. offers discount
      Only YoYo Co. has Montrose office

Fig. 1–3   Topic headings rearranged into a writing outline. This was part of the writing plan for the formal report on elevator selection in Chapter 6; compare it with the final production on pages 206, 207, and 208.

In summary, overorganizing a report, or organizing it too early in the writing process, has an inhibiting effect on writing. The key to good report writing is to organize material in a spontaneous, creative manner, allowing one's mind to freewheel through the initial planning stages until the topics have been collected, scrutinized for relevance, sorted, grouped, and written into a logical outline that will appeal to the reader. This method will not necessarily suit everyone. If a writer already has a workable method for planning and organizing, he should continue to use it. If he does not, or if he has difficulty in starting his writing tasks, he should try doing it this way. The stages are simple, as illustrated in Figure 1–4, and apply to any major writing project.

WRITING THE FIRST DRAFT

As I sit at my desk, the heading "Writing the First Draft" at the top of a clean sheet of foolscap, I find that I am experiencing exactly the same problem that every writer encounters from time to time: an inability to find the right words—*any* words—that can be strung together to make coherent sentences and paragraphs. The ideas are there, circling around my skull, and the outline is there, so I cannot excuse myself by saying I have not prepared adequately. What, then, is wrong?

The answer is simple. In the distance I can hear the percolator in the kitchen murmuring quietly. Soon its tempo will increase to a rumbling crescendo, and then it will stop. Five minutes later my wife will appear at my study door

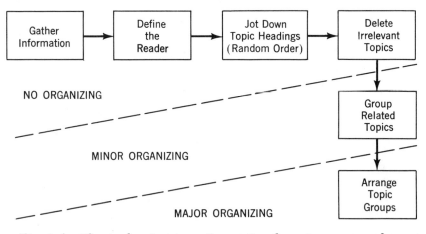

Fig. 1–4   The six planning stages. In practice, these stages can overlap.

bearing two cups of coffee and a plate of English biscuits on an old-fashioned carved tray, and I will be expected to pause for a ten-minute break. I cannot concentrate on writing when I know my continuity of thought is so soon to be broken.

Continuity is the key to getting one's writing done. In my case, this means writing at fairly long sittings during which I *know* I will not be disturbed. I must be out of reach of the telephone, visiting friends, children wanting to say good night, and even my wife, so that I can write continuously. Only when I have reached a logical break in the writing, or have temporarily exhausted an easy flow of words, can I afford to stop and enjoy that cup of coffee!

It is no easier to find a quiet place to write in the business world. The average technical man who tries to write a report in a large office cannot simply ignore his surroundings. A conversation taking place a few desks away is bound to interfere with his creative thought processes. When an office junior approaches him to collect money for the "pool" on that night's NHL game between the New York Rangers and the Toronto Maple Leafs, his train of thought will probably be completely destroyed.

The problem of finding a quiet place to write is frequently hard to solve. In business I recommend a short walk around the plant to find a hidden corner, or a small office that is temporarily unoccupied. Then perhaps, if he is lucky, the writer can just "disappear" for a while. For technical students, who frequently have to work on a tiny writing area in a crowded classroom, conditions are seldom ideal. Cooperation between students is essential if all are to obtain writing continuity. Outlining in the classroom, followed by writing in the seclusion of a library cubicle, is a possible alternative.

When Dan Skinner has found a peaceful location out of reach of the telephone, his friends, and even his boss, he can take a fresh sheet of paper and start writing. This is exactly where a second difficulty may occur. Equipped with an outline on his left, a pad of foolscap in front of him, and a stack of sharpened pencils on his right, he may find that he does not know where to start. Or he may tackle the task enthusiastically, determined to write a really effective introduction, only to find that everything he writes sounds trite, unrealistic, or downright silly. Discouraged after many false starts, he sweeps the growing pile of discarded paper into the wastebasket.

I have frequently advised technical men who have encountered this "no start" block that the best place for them to start writing is at paragraph two, or even somewhere in the middle. For example, if Dan finds that a particular part of his project interests him more than other parts, he should write about that part first. His interest and familiarity with the subject will help him to get those first few words onto paper, and keep him going once he has started. The most important thing is to *start* writing; to put any first words down, even if they are not the right words or ideas, and to let them lead naturally into the next row of ideas.

This is where continuity becomes essential. A writer who has found a quiet place to work, and has given himself sufficient time to write a sizable chunk of text, simply must not interrupt the writing process to correct a minor point of construction. If he is to maintain continuity he must not strive to write perfect grammar, find exactly the right word, insert perfect punctuation, or construct effective sentences and paragraphs of just the right length. That can be done later, during revision. The important thing is to keep building on that rough draft (no matter how "rough" it is), so that when he does stop for a break he knows that he has some words on paper that can be worked up into a presentable document.

If Dan cannot find exactly the word he wants, he should jot down a similar word and draw a circle around it as a reminder to change it when the draft is finished. It's quite likely that when he does look back, the correct word will spring to mind. Similarly, if he is not sure how to spell certain words, he should resist the temptation to turn to the dictionary, for that would disrupt the natural flow of his writing and make it exceedingly difficult to restart. Again, he should draw a circle around the word as a reminder to consult his dictionary later.

I cannot stress too strongly that a writer should not correct his work as he writes. Writing and revising are two entirely separate functions that call for different approaches; they cannot be done simultaneously without the one lessening the effectiveness of the other. Writing calls for creativeness and total immersion in the subject so that the words tumble out in a constant flow. Revision calls for lucidity and logic, which will force a writer to reason and query the suitability of the words he has written. The first requires exclusion of every thought but the subject; the second demands an objectivity that constantly challenges the material from the reader's point of view. The writer who tries to correct his work as he writes soon becomes frustrated, for creativity and objectivity are constantly fighting for control of his pen.

Most technical men find that, once they have written those first few difficult sentences, their inhibitions begin to fall away and they write more naturally. Their style begins to change from the dull, stereotyped writing we expect from the average technical man, to a relaxed narrative that sounds as though the writer is telling a co-worker about his subject. As the writer becomes interested in his topic, his speed builds up and he has difficulty in writing fast enough. The words that he chooses may not be exactly those of his final report, but when he does start checking his first draft he will be surprised at the effectiveness of many of his sentences and paragraphs.

The length of each writing session will vary, depending on the writer's experience and the complexity of his topic. If a document is reasonably short, he should try to write it all at one sitting. If it is longer, he should divide it into portions that will allow him a series of medium-length sessions that suit his particular staying power.

At the end of each session Dan should glance back over his work, note

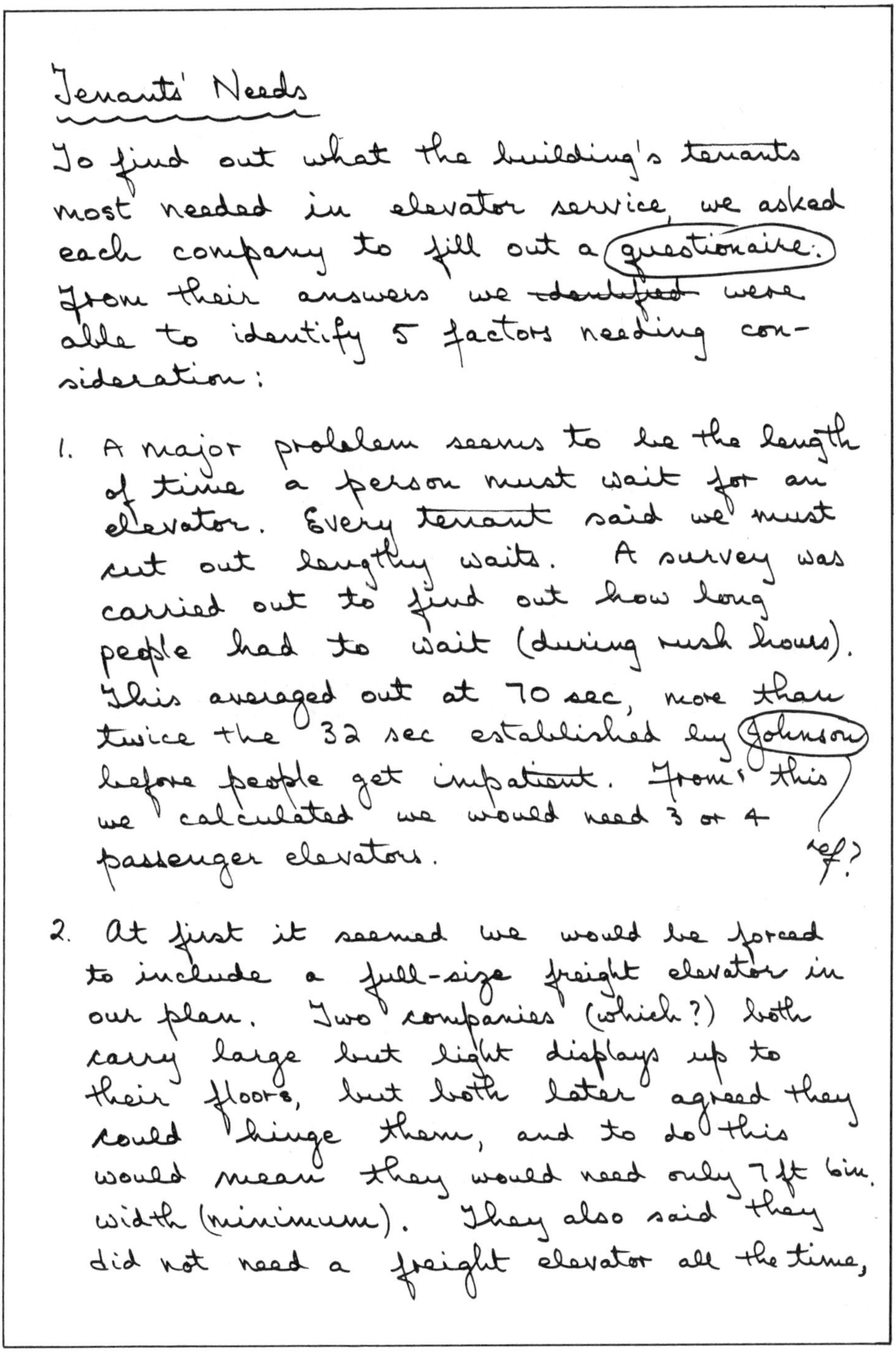

Tenants' Needs

To find out what the building's tenants most needed in elevator service, we asked each company to fill out a (questionaire). From their answers we ~~identified~~ were able to identify 5 factors needing consideration:

1. A major problem seems to be the length of time a person must wait for an elevator. Every ~~tenant~~ said we must cut out lengthy waits. A survey was carried out to find out how long people had to wait (during rush hours). This averaged out at 70 sec, more than twice the 32 sec established by (Johnson) before people get ~~impatient~~. From' this we calculated we would need 3 or 4 passenger elevators.   ref?)

2. At first it seemed we would be forced to include a full-size freight elevator in our plan. Two companies (which?) both carry large but light displays up to their floors, but both later agreed they could hinge them, and to do this would mean they would need only 7 ft 6in. width (minimum). They also said they did not need a freight elevator all the time,

Fig. 1–5 Part of the author's first draft. Note that he has not stopped to hunt up minor details. Several revisions were made between this first draft and the final product (see pages 206 and 207).

the words he has circled, and make a few necessary changes (Fig. 1-5 is a page from a typical first draft). He must not yet attempt to rewrite paragraphs and sentences for better emphasis. Such major changes must be left until later, when enough time has elapsed for him to read his work objectively. Only then can he review his work as a complete document and see the relationship among its parts. Only then can he be completely critical.

TAKING A BREAK

When the final paragraph of a long report has been written, I have to resist the temptation to start revising it immediately. There are sections that I know are weak, or passages with which I am not happy, and the desire to correct them is strong. But I know it is too soon. I must take my pages, staple them together, and set them aside while I tackle a task that is completely unrelated.

Immediate reading without a suitable waiting period encourages a writer to look at his work through rose-colored spectacles. Sentences that he would normally recognize as weak or verbose appear to be words of wisdom. Gross inaccuracies that under normal circumstances he would pounce upon go unnoticed. Passages that would be incomprehensible to the uninitiated reader are abundantly clear. His familiarity with his work blinds him to its weaknesses.

The only remedy is to wait. If he can, Dan should arrange to have a double- or triple-spaced draft typed during this time. Not only will this make his work easier to read, but he may also be fortunate enough to be assigned an efficient typist who corrects some of his punctuation and spelling errors!

READING WITH A PLAN

In practice, reading and revising are performed concurrently. I have chosen to treat them separately because I want to describe the three major checks that have to be made during the reading phase before I discuss revising the draft. The two steps must be repeated over and over until the writer is satisfied that his report says exactly what he wants it to say in as few words as possible.

Dan Skinner's first reading should take him straight through the draft without stopping to make corrections, so that he can gain an overall impression of the report. Subsequent readings should be slower and more critical, with changes written in as he goes along. As he reads he should check for clearness, correct tone, and technical and grammatical accuracy.

Checking for Clearness

Checking for clearness means searching for passages that are vague or ambiguous. If the following paragraph remained uncorrected, it would confuse and annoy a reader:

> *Muddled*   When the owners were contacted on 15 April, the assis-
> *Paragraph*   tant manager, Mr. Pierson, informed the engineer that they
> were thinking of advertising Lot 36 for sale. He however
> reiterated his inability to make a definite decision by
> requesting this company to confirm their intentions with
> regard to buying the land within two months, when his
> boss, Mr. Davidson, general manager of the company,
> will have come back from a business tour in Europe. This
> will be 8 June.

The only facts that a reader could be sure about after reading this paragraph are that the owners of the land were contacted on 15 April and that the general manager will be returning on 8 June. The important information about the possible sale of the land is effectively concealed by a plethora of words. Probably the writer was trying to say something like this:

> *Revised*   The engineer spoke to the owners on 15 April to en-
> *Paragraph*   quire if Lot 36 was for sale. He was informed by Mr.
> Pierson, the assistant manager, that the company was
> thinking of selling the lot, but that no decision would be
> made until after June 8, when the general manager re-
> turns from a business tour in Europe. Mr. Pierson sug-
> gested that the engineer submit a formal request to pur-
> chase the land by that date.

When a topic is complex it is even more important to write clear paragraphs. Although the paragraph below is quite technical, it would be generally understood even by nontechnical readers:

> *Clear*   A sound survey confirmed that the high noise level was
> *Paragraph*   caused mainly by the radar equipment blower motors,
> with a lesser contribution from the air conditioning equip-
> ment. Tests showed that with the radar equipment shut
> down the ambient noise level at the microphone positions
> dropped by 10 dB, while with the air conditioning equip-
> ment shut down the noise level dropped by 2.5 dB.
> General clatter and impact noise caused by the move-

> ment of furniture and personnel also contributed to the
> noisy working conditions, but could not be measured
> other than as sudden sporadic peaks of 2 to 5 dB.

Here paragraph unity adds much to readability. The writer has made sure that:

* The topic is clearly stated in the first sentence (the topic sentence).
* The topic is developed adequately by the remaining sentences.
* No sentence contains information that does not substantiate the topic.

If any paragraph meets these basic requirements, its writer can feel reasonably sure that he has conveyed his message clearly.

Because a writer normally knows his subject thoroughly, he may find it difficult to identify paragraphs that contain ambiguities. A passage that is abundantly clear to him may be meaningless or offer alternative interpretations to a reader unfamiliar with the subject. For example:

> Our examination indicates that the receiver requires both repair and
> recalibration, whereas the transmitter needs recalibration only, and
> the modulator requires the same.

This sentence plants a question in the reader's mind: Does the modulator require both repair and recalibration, or only recalibration? The technician who wrote it knows, because he has been working on the equipment, but the reader will never know unless he cares to write and ask. Possibly the writer indicated his true meaning in subsequent sentences, but he should not make his reader hunt for clarification.

It is possible that the technician added the little piece about the modulator as an afterthought, after he had closed off the original sentence at "only." Perhaps he even recognized that the sentence might be misinterpreted, so misguidedly inserted a couple of commas to improve it. But the only real remedy would have been to simplify it by rearranging the information:

> Our examination indicates that the receiver requires both repair and
> recalibration, whereas the transmitter and modulator need recalibra-
> tion only.

Sometimes ambiguities are so well buried that they are surprisingly difficult to identify, as in this excerpt from a chief draftsman's report to his department head:

> The Drafting Section will need three Model D7 drawing boards.
> The current price is $975 and the supplier has indicated that his

> quotation is "firm" for three months. We should therefore budget accordingly.

The department head took this message at face value and inserted $975 for drawing boards in his budget. Two months later he received an invoice for $2925. Unable by then to return the three boards he had to overshoot his budget by $1950. This financial mismanagement resulted from the chief draftsman's omitting to state whether the price he quoted applied to one drawing board or to three. If he had inserted the word "each" after $975, his message would have been clear.

Many ambiguities can be sorted out by simple deduction, although it really should not be the reader's job to interpret the author's intentions. Occasionally such ambiguities provide a humorous note, as in this extract from a field trip report:

> High grass and brush around the storage tanks impeded the technicians' progress and should be cleared before they grow too dense.

The two thoughts in this sentence should be separated:

> High grass and brush around the storage tanks impeded the technicians' progress. This undergrowth should be cleared before it grows too dense.

Checking for Correct Tone

How does a writer know when his work has the right "tone"? One of the most difficult aspects of technical writing is establishing a tone that is correct for the reader, suitable for the subject, and comfortable for the writer. If the writer tries to set a tone that is not natural for him, then his reader will sense an artificiality, an unsureness that indicates he has not felt comfortable writing at the level he has adopted.

If a writer knows his subject well and has identified his reader thoroughly, he will most likely write so confidently that he will automatically set the right tone. But if he does not know his subject well, and has not taken the time to define his reader, then his writing will lack confidence. He will make false starts, change his approach, and generally become irritable with his work because he does not know in which direction he is going. No matter how skillfully he edits his work, his readers will notice the hesitancy and become impatient.

Finding the Best Writing Level. To check that he has set the right tone, Dan Skinner must assess whether his writing is suitable for both the subject matter and the reader. If he is writing on a specific aspect of a very technical topic, and knows that his reader is an engineer with a thorough grounding in the subject, he can use all the technical terms and abbreviations that his reader will recognize. Conversely, if he is writing on the same topic for a non-technical reader who has little or no knowledge of the subject, apart from its name and a few terms that he recognizes but cannot apply, then he must tailor his approach. In this case he must generalize rather than state specific details, explain technical terms that he would normally expect to be understood, and generally write in a more informative manner.

Writing on a technical subject for readers who do not have the technical knowledge of the writer is not easy. The writer must be much more objective when checking his work, and try to think in the same way as the nontechnical reader; he must select only those technical words that he knows the reader will recognize, yet he must avoid using an oversimplified language that will irritate his audience.

In the following example from an imaginary modification report, the writer knew that his readers would be electronics engineers at radar-equipped airfields:

> We modified the AN/MPN-11 M.T.I. by installing a K-59 double-decade circuit. This brightened moving targets by 12% and reduced ground clutter by 23%.

This statement would be readily understood by the technical readers for whom it was intended, but to most readers, or to any reader not familiar with radar terminology, it would be almost incomprehensible. If the writer had wanted to report on the same subject to, say, the airport manager, he would have written it this way:

> We modified the Moving Target Indicator of radar set AN/MPN-11 by installing a special circuit known as the K-59. This increased the brightness of responses from aircraft and decreased returns from fixed objects on the ground.

For this reader he has included more descriptive detail: "AN/MPN-11" has become "radar set AN/MPN-11"; "M.T.I." has become "Moving Target Indicator"; and "moving targets" and "ground clutter" have become "responses from aircraft" and "returns from fixed objects on the ground." He has eliminated specific technical details because they might not be meaningful to the reader, and in their place made a general statement that aircraft responses were "increased" and ground returns "decreased." At the same time, he knew his reader well enough to use such terms as "Moving Target Indicator," "responses," and "returns" without defining them. He could assume that an airport manager would be familiar with these, and would use them in his own writing and conversation even if he did not understand fully their technical meaning.

Now suppose that the same writer had to write to the chairman of the local Chamber of Commerce to describe improvements in the city's air traffic control system. This time, because his reader would be nontechnical, he would avoid using any technical terms:

> We have modified the airfield radar system to improve its perform-ance, which has helped us to differentiate more clearly between low-flying aircraft and high objects on the ground.

Keeping to the Main Topic. Having established that he is writing at the correct level, Dan must now check that he has kept to the main topic. He must take each paragraph and ask: Is this truly relevant? Is it direct and to the point? Have I allowed superlatives (big, important-sounding words) to take the place of more readable words?

If Dan prepared his outline using the method I described earlier, and followed it closely as he wrote his report, he can be reasonably sure that most of his writing is relevant. To check that his subject development is con-sistent and follows his planned theme, he should identify the topic sentences of some paragraphs and check them against the headings in his outline. If they follow the outline, he has kept to the main theme; if they tend to diverge from the outline, or if he has difficulty in identifying them, he should read the para-graphs carefully to see whether they need to be rewritten, or possibly even eliminated.

Technical writing should always be as direct and specific as possible. The information that the writer conveys should be just the right amount of information his reader will need to understand the subject thoroughly—and no more. Technical writing, unlike literary writing, has no room for peripheral details that are not essential to the main theme. This is readily apparent in the following descriptions of the same equipment.

Literary description
The new cabinet has a rough-textured dove gray finish that reflects the sun's rays in varying hues. Contrary to most instruments of this type, its controls are grouped artistically in one corner, where the deep black of the knobs provides an interesting contrast with the soft gray and white background. A cover plate, hardly noticeable to the layman's inexperienced eye, conceals a cluster of unsightly adjust-ment screws that would otherwise mar the overall appearance of the cabinet, and would nullify the aesthetic appeal of its surprisingly effective design.

Technical description
The gray cabinet is extremely functional. All the controls used by the operator are grouped at the top right-hand corner, where they can be grasped easily with one hand. Subsidiary controls and adjust-ment screws used by the maintenance crews are grouped at the bottom left-hand corner, where they are hidden by a hinged cover plate.

Quick comparison between these examples shows how the technical description concentrates on details that are important to the reader (it tells *where* the controls are and why they have been so placed); it does not waste time being artistic. At the same time it observes the rules of good construction, using parallel structure to carry the reader easily through the description. Hence, it maintains an efficient, businesslike tone.

Using Simple Words. A writer who uses unnecessary superlatives sets an unnaturally pompous tone. The engineer who writes that his design "contains ultrasophisticated circuitry" seems to be trying to justify the importance and complexity of his work rather than saying that his design has a very complex circuit. The supervisor who recommends that technician Smith be "given an increase in remuneration" may be understood by the company comptroller, but he will only be considered pompous by Smith. If he had simply written that Smith should be "given a raise," he would have been understood by both men and by everyone else in between. Unnecessary use of big words, when smaller, more generally recognized, and equally effective synonyms are available, clouds technical writing and destroys the smooth flow that such writing demands.

Removing "Fat." During the reading stage Dan should be critical of sentences and paragraphs that contain an overabundance of words. A writer who is aware that his subject is new or complex will try to explain it in detail for the uninitiated reader. But if he has difficulty in finding the right level, he may stray from the narrow line between abstruseness and oversimplification and bore his reader by trying to make himself too clear.

Dan should also check that he has not inadvertently inserted words of "low information content," which means phrases and expressions that add little or no information. Their removal, or replacement by simpler, more descriptive words, can tighten up a sentence and add to its clarity. Low information content words and phrases are often hard to identify, because the sentences in which they appear seem to be satisfactory. Consider this sentence:

> For your information, we have tested your spectrum analyzer and are of the opinion that it needs calibration.

The words of low information content are "for your information" and "are of the opinion that." The first can be deleted, and the second replaced by "consider," so that the sentence now reads:

> We have tested your spectrum analyzer and consider it needs calibration.

Now try to identify the low information content words in this sentence:

> If you require further information please feel free to telephone Mr. Smith at 489-9039.

The phrase "if you require" is not entirely wrong, although it could be replaced by the single word "for," but "please feel free to" is archaic and should be eliminated. The result:

> For further information please telephone Mr. Smith at 489-9039.

Inadvertent repetition of information can also contribute to excessive length. Although repetition can be an effective way to emphasize a point, in most cases its use is accidental; Dan may explain something in one paragraph, then several paragraphs later say the same thing in different words. By welding paragraphs that repeat previously mentioned facts into a single, cohesive block of information, he will help to clarify and shorten his report.

Many writers find it difficult to pay compliments or to apologize sincerely. Frequently they set an unnatural tone by trying to say too much. A simple, brief statement seems incomplete, so they "beef it up a bit" under the false impression that they are substantiating their true feelings. The resulting tone is so false, and the words so forced, that the reader immediately senses that the compliment or apology is insincere. Such writers have to learn that sincerity is enhanced by brevity: pay your compliment or make your apology, then forget about it.

For more information on this topic see the section on "Tone" in Chapter 3.

Checking for Accuracy

Checking accuracy means examining one's work to ensure that the information is correct, and that such technicalities as grammar, punctuation, and spelling have not been overlooked.

Nothing annoys readers more than to discover that they have been presented with information that is not absolutely accurate. They automatically assume that a writer knows his facts and has ascertained that they are correctly transcribed into his report. Errors may remain undetected for a long time, possibly until an enquiring reader starts conducting further tests and making calculations based on the author's results. Suspicion of an error in the original report can lead to tedious correspondence and consumption of time until the inaccuracy is corrected. The net result is that the readers' confidence in the writer and the company he represents is downgraded.

There is no way to prevent some errors from occurring when quantities and details are being copied from one sheet of paper to another. These errors may be made by the writer himself, or by the stenographer who does his typing. Therefore, Dan must personally check that facts, figures, equations, quantities, and extracts from other documents are all copied correctly. He may feel that it is the stenographer's duty to proofread the report, but the responsibility for checking the veracity of the written work is ultimately his.

This is also the time for Dan to identify poor grammar, inadequate punctuation, and incorrect spelling. He must do this with care, because his familiarity with the subject may blind him to obvious errors. (How many of us have inadvertently written "their" when we intended to write "there," and "to" when we meant "too"?)

These examples illustrate a few of the ways that Dan can improve his report writing by careful reading and revision. Other methods are suggested in Chapter 10.

REVISING ONE'S OWN WORDS

As Dan reads his work he should make any corrections he finds necessary. Some corrections may require only minor changes to individual sentences; others may require complete revision of whole paragraphs, and even complete sections. Whenever extensive revisions are made, he should have the draft retyped, then reread and revise it. This process should be repeated until a draft emerges that is ready for final typing and printing. As he reads, Dan must continually ask questions:

Armed only with the knowledge he has now, will the intended reader be able to read right through my report without losing the thread of my argument?

Will the intended reader comprehend what I am trying to say? Will other readers?

Have I included sufficient information for the reader? Or is it more than he needs?

Is my treatment of the material suitable for the reader? Is it suitable for the subject?

Does the draft contain any ambiguities?

Have I been direct and definite?

Have I limited my discussion strictly to the topics in my outline?

Have I used overblown words when simpler words would be just as efficient?

Is the information accurate and complete?

Is the "English" good? Are there any errors of grammar, punctuation, and spelling?

Is the emphasis in the right place? Will the reader immediately recognize the important points?

Does every word, sentence, and paragraph *contribute* to the topic?

Have I eliminated all unnecessary words?

Does it contain any repetition between sections? Between paragraphs? Between sentences?

Is it too long? Have I kept it as short as the topic warrants?

By now Dan's draft should be in good shape. Any further reading and revising will be final polishing. The extent of polishing will depend on the importance of the project. If the report is for limited or in-company distribution, a workmanlike job will suffice. But if it is due for distribution outside the company, or for submission to an important client, then Dan will spend as much time as necessary to ensure that it presents the image that his company wishes to convey.

REVIEWING THE FINAL DRAFT

The final step occurs when Dan feels that he has a good report and is ready to submit it for final typing. Before charging across to the typing pool, he should pause momentarily and review what he has written, asking himself three questions:

1. Would I want to receive what I have written?
2. What reaction will it incur from the intended reader?
3. Is this the reaction I want?

In answering these questions he must be extremely self-critical, ready to doubt his ability to be truly objective. If his answers are at all hesitant, he must solicit an opinion from an independent reviewer who, ideally, should be technically and mentally equivalent to the eventual reader, but not too familiar with the project.

The reviewer must be able to criticize constructively. He should read the report completely rather than scan it, and make notes describing possible weaknesses and ambiguities. He should then take time to discuss the report with Dan, who quite likely will discover that the reviewer has identified the very paragraphs that gave him most difficulty, or did not entirely satisfy him.

.   .   .

Dan Skinner will now be able to issue his report with confidence, knowing that he has fashioned a good product. The approach described here will not have made report writing a simple task for him, but it will have assisted him through the difficult conceptual stages, and helped him to read and revise more efficiently. When Dan has to write his next report, he will be much less likely to put it off until it is so late that he has to do a rush job.

2

Meet "H. L. Winman and Associates"

One of my main objectives in writing this book is to show examples of technical documents that are based on realistic incidents and are typical of situations you are likely to face in industry. What, then, would be better than to describe a technically-oriented company, and then to base every project, whether a fully worked-out example or a short writing assignment, upon events that occur within that company?

This chapter introduces H.L. Winman and Associates, a Cleveland firm of consulting engineers, and its Canadian subsidiary, Robertson Engineering Company. To help you identify your position in this working environment, the following pages contain short histories and partial organization charts of the two companies, plus brief biographical sketches of the men who have brought them to their current position.

CORPORATE STRUCTURE

H.L. Winman and Associates was founded by Harvey Winman early in 1935. Starting with a handful of engineers working in a small office in one of the older sections of Cleveland, Ohio, Harvey gradually built up his company until it now has 570 employees, 350 at head office and 220 at a Canadian subsidiary in Toronto, Ontario. Figure 2–1 illustrates the two locations.

HEAD OFFICE: H.L. WINMAN AND ASSOCIATES

The head office of H.L. Winman and Associates is located in a modern building at 475 Reston Avenue, a major thoroughfare in the business district of

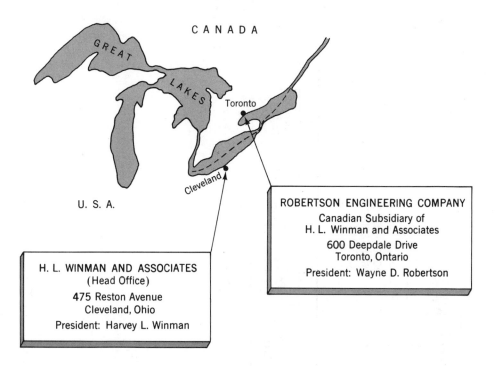

Fig. 2–1   Locations of H.L. Winman and Associates' Offices.

Cleveland, Ohio. A partial organization chart at Figure 2–2 shows Harvey Winman as president, with nine department heads, who are responsible for the day-to-day operation of the company, reporting directly to him.

Over the years H.L. Winman and Associates has developed an excellent reputation among its clients for the management of large construction projects in remote areas of the United States and Canada. It has managed the construction of whole townsites, airfields, and dams for large hydroelectric power generating stations; engineered, designed, and supervised construction of major traffic intersections, shopping complexes, bridges, and industrial parks; and designed and supervised numerous large buildings such as hotels, schools, arenas, and manufacturing plants. More recently, the company has entered the systems engineering field. Some of its current studies involve problems in air pollution, protection of the environment, and the effects of oil exploration and pipeline construction on the Alaskan arctic tundra. To handle so many diverse tasks, Harvey has built up an extremely adaptable and flexible staff with representation from many scientific and technical disciplines.

Although he is nearing retirement age, Harvey maintains an active interest in his company's business, and retains strong policy control. Short, with a round face topped by a bald head with a fringe of hair, he is often seen strolling from department to department. He stops frequently to talk to his staff

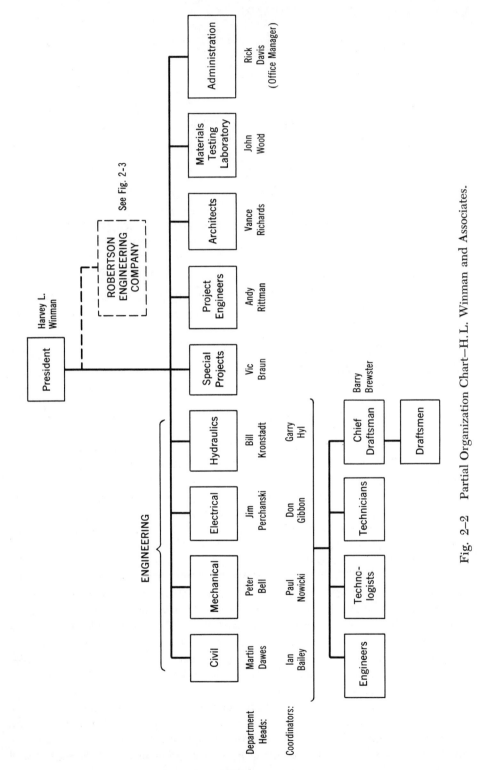

Fig. 2-2   Partial Organization Chart—H. L. Winman and Associates.

PHOTO A: Harvey L. Winman. Photo: Anthony Simmonds.

(most of whom he knows by their first name) and to enquire about both the projects they are working on and their home lives. Well-liked and thoroughly respected by all, from senior management down to the newest clerk, he is a manager of the old school, a type seldom seen in today's brusque business atmosphere.

Harvey Winman believes strongly that the written word is the means that most often conveys an "image" of his company to customers. He therefore insists that all letters, reports, and company publications be of top quality, both in construction and presentation. He even checks that quality by regularly reading a random sample of written communications, from full formal reports to interoffice memorandums. He writes well himself and expects his staff to do likewise.

CANADIAN SUBSIDIARY:

ROBERTSON ENGINEERING COMPANY

The Canadian subsidiary of H.L. Winman and Associates was acquired by Harvey Winman in 1962, when he purchased an old, established firm of consulting engineers that held a controlling interest in the Robertson Engineering Company of Toronto, Ontario. At first Harvey considered selling the Canadian company, but when he realized that Wayne D. Robertson, its president, was running a very profitable business, he decided to retain it as a subsidiary and to leave Wayne in full control.

Wayne formed Robertson Engineering Company when he left the Canadian Air Force in 1946, gathering together several ex-air force electronics specialists to set up a small equipment repair and maintenance service. He managed his business so well that his company was soon recognized for its competence and reliability. Consequently it grew steadily, with the emphasis gradually leaning toward research and development.

The company now has a well-established reputation for designing, developing, and custom-manufacturing sophisticated electronic, nucleonic, and electromechanical instruments. It also specializes in the installation and maintenance of complex communication systems across Western Canada. There are approximately 220 employees, of whom 80 are engineers, technologists, and technicians in the Engineering Department. The company occupies floors five and six of the Wilshire Building, a ten story office and light manufacturing complex owned by the Wilshire Insurance Company. A partial organization chart in Figure 2–3 shows the structure of the company and the make-up of the Engineering Department.

Wayne D. Robertson is not only an experienced electronics specialist but also an astute businessman. Recognizing the need for a strong engineering capability supported by an effective sales force, he has taken care to build a particularly good management and engineering staff to carry out his company's

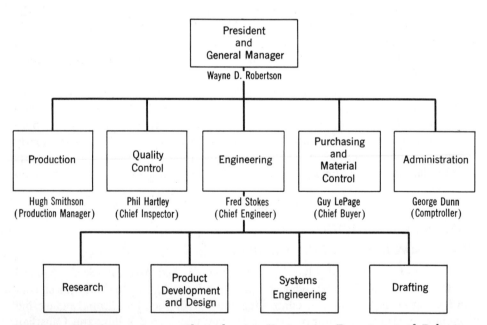

Fig. 2–3 Partial Organization Chart showing Engineering Department of Robertson Engineering Company.

PHOTO B: Wayne D. Robertson. Photo: Anthony Simmonds.

primary function. The result has been recognition by Canadian businessmen of his company's technical capability and his personal managerial ability. His design engineers have developed a specialized range of electronic and nucleonic instruments for which a small but world-wide market exists. Consequently, Wayne travels frequently to South America, Australia, South Africa, and Europe, where he is able to talk knowledgeably of the technical capabilities of his conpany's equipment.

Like Harvey Winman, he recognizes the importance of communication in his business, and particularly stresses that the operating instructions and training manuals that accompany his instruments be abundantly clear. To this end, he employs a small technical writing group that can write manuals not only in English, but also in French, German, and Spanish.

.   .   .

The letters and reports that appear throughout this book are based on work done by these two companies. Similarly, many of the end-of-chapter assignments assume that you are employed by one of the companies, and that the letter or report you are asked to write occurs naturally as part of your work.

3

Technical Correspondence

Because the letter is our most common means of written communication, we frequently overlook its importance. We scribble down a few words and hope that someone will get the message; we bury the message under a mountain of unnecessary words; or we adopt an unnatural style that conceals our character and robs the message of humanity. Many of us are so accustomed to writing letters to friends and relatives that we find it difficult to transfer from the easygoing style of the personal letter to the more formal style of the business letter. Yet this is where we err. Instead of letting our business letters be friendly, interesting, and persuasive, we force them to be stilted and dull, thinking that we are being businesslike.

CLARITY AND CONCISENESS

When we write personal letters to friends and relatives, we do so partly to be sociable and partly to convey a message. The length of our letters, and the amount of additional information we include, is unimportant. We can ramble along with very little organization, inserting interesting sidelights and comments as they occur to us. We assume our reader is pleased to have a letter from us, so we use it mainly to portray what has happened since we last wrote.

But when we write a business letter we have to be much more direct. Our readers are busy men who are interested only in facts; information they do not need irks them. For these people we must keep strictly to the point. We must check that our letters are compact, brief, and clear, and that our facts are simply but effectively presented.

Compactness

Compactness means arranging information in the smallest possible space. Like packing a small suitcase for a short trip, you have to be selective and pick out only the items you will need for the environment you will encounter, choosing only one coat, one suit, one pair of shoes, and so on. In business writing you have to be equally selective, limiting your choice to the essential elements that will convey the message. The key word for compactness, both in packing a small case and in writing a business letter, is "one."

One Main Subject to a Letter. If you have to write to a company on two different subjects at the same time, write a separate letter for each subject. To cover both subjects in the same letter is to invite one subject to be overlooked, particularly if they demand action by different departments. The first department to receive your letter may inadvertently file it after replying to you, and the second department will never know of your enquiry.

Suppose that Ian Bailey, the Civil Engineering Department Coordinator at H.L. Winman and Associates, writes to the Marvin Corporation on 6 May to request technical information regarding a new range of theodolites, and at the same time to ask for a replacement upper transit screw for his old Marvin M20 surveyor's transit. His letter goes directly to the Sales Manager who, anticipating a sale, dashes off an enthusiastic letter describing the new instruments. Ian Bailey's original letter is filed, ready to be brought forward for follow-up a month to six weeks later. In his enthusiasm to develop a sales contact, the Sales Manager has overlooked the line asking the Supply Department to send along an upper transit screw.

Only when two subjects are related, and likely to be acted upon by the same person, is it safe to write about them in the same letter. In every other case, write separate letters.

One Idea to Each Paragraph. Let the first sentence of each paragraph introduce just one idea, then make sure that subsequent sentences in that paragraph develop it adequately and do not introduce any other ideas. In technical writing it is often wise to identify the first sentence of a paragraph as the "topic sentence," so that the reader knows immediately what the writer is trying to tell him. This is particularly true of business letter writing, in which the first sentence of each paragraph should summarize the paragraph's contents. Remaining sentences must support the first sentence by providing additional information about the topic, as they do here:

> *We have tested your 12 Vancourt 60 Multimeters and find that 9 require repair and recalibration.* Only minor repairs will be necessary for 5 of these meters, which will be returned to you within one week. Of the 4 remaining meters, 3 require major repairs which will take

approximately 20 days, and 1 is so badly damaged that repairs will cost over $60.00. Since this is more than the maximum repair cost imposed by you, no work will be done on this meter.

By reading the topic sentence (italicized in this example), the reader learns immediately what he most wants to know: the number of meters that need repair. The extent of the repairs he can determine from the supporting sentences.

One Thought to a Sentence. The intent here is to keep each sentence uncomplicated. Building sentences that develop more than one thought sometimes can be done successfully by experts in literary writing, but they are confusing and out of place in the business world. Compare these two examples of the same information:

Complicated: We have had trouble with a faulty modulator, although the transmitter is operating satisfactorily, and the power supply is erratic.

Clear: We have had trouble with a faulty modulator and an erratic power supply. The transmitter is operating satisfactorily.

The first example is confusing because it jumps back and forth between trouble and satisfactory operation. The second example is clear because it uses two sentences to express the two different thoughts.

Brevity

The key word for brevity is "short": short letters, short paragraphs, short sentences, and short words.

Short Letters. A short letter introduces its topic quickly, discusses it in sufficient depth, and then closes with a concluding statement. It is impossible for me to state arbitrarily how long a short letter should be. Length should be dictated by only one factor: how many words the writer needs to convey his message adequately.

I know of a company in which the managing director has ruled that no letters issued by his organization should exceed one page. This is an effective means for forcing his staff to be brief, but it must be unbearably severe on some writers. There are occasions when one's letters have to extend to a second and even a third page. Limiting one's imagination and creativity at such times might reduce a convincing argument to a terse statement that fails to satisfy the reader's interest. Brevity at the expense of lucidity is false economy.

Many long letters can be shortened, however, by borrowing a technique from report writing. If a letter contains such supporting information as a series of test results, a cost analysis, an excerpt from another document, or similar data that is not immediately relevant, it can be placed on a separate sheet and labeled as an "attachment." A reference must be made to this data in the body of the letter, preferably by drawing a main conclusion from it, as has been done here:

> During our investigation we took a series of noise readings to determine noise level variations at night, during the day, and at weekends. These readings (see attachment) show that a maximum of 55 dB was recorded on weekdays, and 49 dB at weekends. In both cases these peaks were recorded between 5 and 6 p.m.

Short Paragraphs. Novelists can afford to write long paragraphs because they assume that they will have their readers' attention, and that their readers have the time and patience to wend their way through leisurely description. But in business and industry the reader is working against the clock: his time is limited and he wants to read no more than is necessary. He needs bite-size paragraphs of easy-to-digest information.

When I discussed compactness, I suggested that a paragraph should develop only one idea. If this paragraph is short, then the reader will be able to grasp the idea, comprehend it, and proceed quickly to the next idea. His progression from paragraph to paragraph will be rapid.

I am not suggesting that your letters should contain a series of small, evenly-sized paragraphs; then they would appear dull and stereotyped. Paragraphs should vary from quite short to medium-long to give the reader variety. How you can adjust paragraph and sentence length to suit both reader and topic, and also to place emphasis correctly, is covered in Chapter 10.

Short Sentences. If you make sure that you present only one thought in each sentence, then your sentences are likely to be reasonably short. I hesitate to stipulate a maximum sentence length, but 25 words might be a realistic one. Even this figure can vary, depending on the technical level of the reader and complexity of the information. As a general rule, the more complex the subject, the shorter one's sentences should be.

There is a danger in stringing together too many very short sentences. Each may be complete and clear in itself, but strung together they may create the effect of all starts and stops with inadequate subject development. Your objective should be to obtain rhythm by mixing short and medium-length sentences. Rhythm in sentence structure will lead the reader smoothly through each paragraph.

Short Words. To the technical man proud of his scientific environment, long words seem to convey an image of the complex world in which he works. But they can also provide a false veneer that may conceal a basic insecurity. Many extremely competent engineers feel unsure of themselves when they have to write, and hide behind large words. It is sometimes very difficult to convince them that "an error of considerable magnitude was perpetrated" is a very long-winded way to write "a large error was made."

The scientific world encompasses many long and complex technical terms that have to be used in their original form. "Diphygenic" (to have two modes of development) is an example. If we surround these terms with simple words, we will make our correspondence more readable. This does not mean using only four and five letter words, unless we want our work to read like a third grade spelling text. The criterion is to use the right word in the right place, striking a happy balance between oversimplification and ponderous heaviness.

Clarity

A clear letter conveys information simply and effectively, so that the reader readily understands its message. To write clearly demands ingenuity and attention to detail. The writer must consider not only how he will write his letter, but also how he will present it.

Create a Good Visual Impression. Experienced writers know that a nicely laid out letter impresses a reader. Subconsciously it appeals to him; it seems to be saying "my neat appearance demonstrates that I contain quality information that is logically and clearly organized."

It is a mistake to think that the person you are writing to is doing nothing but waiting for your letter to come in just so he can read it. The converse is more likely to be true. Your letter will be slipped into a pile of correspondence, each component of which is crying out for his attention. If he is an astute reader he will weed out the promotional literature (or his secretary may do it for him), so that your letter will demand his attention among a number of similar-looking letters. If yours has reader appeal he will be attracted to it and place it among those to be read first.

The appearance of a letter tells me much about the writer and the company he works for. If a letter is sloppily arranged, if it contains strikeovers, visible erasures, or spelling errors, then I imagine a careless individual working in a disorganized office. But if I am presented with a neat, tidy letter tastefully placed in the middle of the page, and carefully typed with a reasonably new ribbon, then I imagine a crisp, well-organized individual working for a forward-thinking company noted for the quality of its service. It is the latter company I want to deal with, and it is their correspondence that I will read first.

Many technical people feel that once they have scribbled out or dictated a letter their responsibility for its issue is over, apart from their signature. This is not true. It is the stenographer's job to type a neat-looking letter that presents a good image of the company. In signing the letter, the writer says that he endorses this image. In effect, he acts rather like quality control: he sets the standards of presentation that he considers correct for conveying his and the company's image. If he feels that a letter fails to meet these standards, he should reject it. A good stenographer will recognize poor workmanship and will seldom offer a letter of poor quality for signature; an inexperienced stenographer may need to learn what standards of workmanship are acceptable. Unless you explain the requirements to her, it may be a long time before she does the kind of work you want.

Develop the Subject Carefully. The key to effective subject development is to present the material logically, progressing gradually from a clear, understood point to one that is more complex. This means developing and consolidating each idea for the reader's full understanding before attempting to present the next idea. If you don't have a thorough knowledge of your intended reader, you may confuse him through inadequate development or annoy him by overstressing each point.

Break up Long Paragraphs. Ideas that need thorough development may breed overly long paragraphs. You can overcome this by using subparagraphs, which can carry the idea of a main paragraph into a series of smaller paragraphs without disturbing the main theme. Subparagraphing offers a useful way to maintain continuity through a series of points that are only partly related, and to draw attention to specific items. If there are many subparagraphs, care must be taken that the mood and tone are carried correctly from the main paragraph to all of the subparagraphs. In the following example, subparagraphing has been handled properly:

> I have analyzed our present capabilities and estimate that we can increase our commercial business from $20,000 to $30,000 per month. To meet this objective we will have to:
> 1. Shift the emphasis from purely local customers to clients in major centers. To increase business from local customers alone will require an intensive sales effort for only a small increase in revenue, whereas a similar sales effort in a major center will attract a 30–40% increase in revenue.
> 2. Increase our staff and manufacturing facilities. The cost of additional personnel and new equipment will in turn have to be offset by an even larger increase in business. Properly administered, such a program should result in an ever-increasing workload.
> 3. Create a separate department for handling commercial business. If we remove the department from the existing production organization it will carry a lower overhead, which will result in products that are more competitively priced.

Notice how each subparagraph flows naturally from the lead-in words of the main paragraph:

> . . . we will have to:
> 1. Shift the emphasis . . . .
> 2. Increase our staff . . . .
> 3. Create a separate . . . .

This is known as parallelism. There would have been neither parallelism nor continuity, and the transition would have jarred the reader, if the third paragraph had started differently:

> . . . we will have to:
> 1. Shift the emphasis . . . .
> 2. Increase our staff . . . .
> 3. The overhead could be reduced by creating a separate department for handling commercial business.

Insert Headings as Signposts. In longer letters, and particularly those discussing several aspects of a situation, a writer can help the reader by inserting headings as indicators that guide him to specific information. Such headings must be informative, summarizing clearly what is covered in the paragraphs they precede (see Chap. 10). If, for instance, I had replaced the heading to this paragraph with the single word *Headings*, I would not have summarized what this paragraph is about. The same would be true in the next paragraph if I replaced the existing heading with the single word *Language*.

Use Good Language. It hardly seems necessary to tell you to use good language, but in this case I mean language that you know the reader will understand. Use only those technical terms and abbreviations he will recognize immediately. If you are in doubt, define the term or abbreviation, or use a simple expression in its place.

Simplicity

Keep your letters as simple as possible. Say exactly what you mean, removing any extraneous or irrelevant information. Place complex technical details, drawings, and supporting data in an attachment, so that you deal only with the plain facts in the body of the letter.

Start your letters positively, particularly when replying to a letter:

> Dear Mr. Novak:
>   We have investigated the problem of paint discoloration described in your letter of 18 November, and have concluded that the paint you used may have exceeded its shelf life.

This letter assumes that Mr. Novak has an adequate filing system. There is no need to repeat much of the information contained in his original letter of enquiry, as has been done here:

> Dear Mr. Novak:
>   With reference to your letter of 18 November, in which you described discoloration of our paint color No. 177 used as a second

coat on top of No. 134 primer, we have conducted an investigation into your problem. Our conclusion is that the paint you used may have exceeded its shelf life.

The middle part of this reply tells Mr. Novak information that he already knows, hence it wastes both his and the writer's time.

. . .

The factors I have described so far are mostly manipulative details that can be learned. Armed with this knowledge and the basic letter formats illustrated later in this chapter, the inexperienced writer has some ground rules to guide him in the practical aspects of business letter writing. He has yet to acquire the more difficult technique of letting character appear in his letters without letting it become too obtrusive.

SINCERITY AND TONE

Sincerity and tone are intangible factors that defy close analysis. I can neither tell you of any quick and easy method that will make your letters sound sincere, nor suggest a check list that will tell you when you have imparted the right tone. Both are extensions of your own personality that cannot be taught. They can only be shaped and sharpened through knowledge of yourself and which of your attributes you most need to develop.

Sincerity

At one time it was considered good manners not to permit one's personality to creep into business correspondence. Today, business letters are much less formal and, as a result, much more effective. Personal qualities that experienced letter writers develop to convey their messages efficiently are:

Enthusiasm
Humanity
Directness
Definiteness

These four qualities, together with knowledge of one's subject, form the five basic ingredients that the new writer can use to inject sincerity into his correspondence.

Be Enthusiastic. Sincerity is the gift of making your reader feel that you are personally interested in him and his problems. You convey this by the words you use and the way you use them. He would be unlikely to believe you if you came straight out and said, "I am genuinely interested in your project." The secret is to be so involved in the subject, so interested by it, that you automatically convey the ring of enthusiasm that would appear in your voice if you were talking about it.

Be Human. Too many letters lack humanity. They are written from one company to another, without any indication that there is one human being at the firing end and another at the receiving end. The letters might just as well be from computer to computer.

Do not be afraid to use the personal pronouns "I," "you," "he," "we," and "they." Let your reader believe you are personally involved by using "I" or "we," and that you know he is there by using "you." Contrary to what many of us were told in school, start your letters in the first person. If you know the reader personally or have corresponded with him before, or if your topic is informal, let a personal flavor appear in your letters by using "I":

> Dear Mr. Wicks:
>   I read your report with interest and agree with all but one of your conclusions. . . .

If you do not know your reader personally, and are writing formally as a representative of your company, then use the first person plural:

> Dear Mr. Wicks:
>   We read your report with interest and agree with all but one of your conclusions. . . .

Be Direct. Very few people have acquired the knack of being naturally direct. In speech it's normal to lead up to a topic gently rather than burst out with a controversial statement. If we apply the same technique to correspondence, we will probably write weak letters that lack impact.

The key to directness is to state one's case immediately. It must be done carefully, otherwise you will sound too abrupt. The following examples compare letters with an indirect start, a direct but too abrupt start, and a naturally direct start. In each case the letter is to a client who wants to know whether a supply of steel he has in stock can be used to build a bridge in northern Alaska. He has submitted a sample to H.L. Winman and Associates for analysis, and these letters are reporting the results:

> *Indirect*
>   We conducted extensive tests on the sample of steel you sent to us for analysis. A Charpy Impact Test told us that the steel has a low transition temperature, while a Tension Test demonstrated that the steel failed in a ductile manner, with a yield point of 44,000 psi.
>   From these tests we determined the type of steel to be G40.12.

Although widely used as structural steel throughout Canada, it is less satisfactory for structures that are to be subjected to high loading in the far north. Its low temperature characteristics are only moderately good, which would make it unsuitable for a bridge at Peele Bay, Alaska.

Direct but too Abrupt

You will not be able to use the type of steel you sent to us for analysis to build a bridge at Peele Bay, Alaska. We found your steel to be type G40.12, which has poor low temperature characteristics. A better type would be G40.8.

Naturally Direct

The sample of steel you sent to us for analysis is G40.12, a type that would not be suitable for the short span bridge you propose to build at Peele Bay, Alaska. A more suitable type would be G40.8, which has better low temperature characteristics than type G40.12.

We analyzed your sample by conducting both a Charpy Impact Test and a Tension Test, details of which appear in attachment 1. These tests demonstrate that your sample has a low transition temperature and fails in a ductile manner.

The first and last letters convey the same information, but in a different order. (We can ignore the middle letter for the moment, since it is obviously too terse.) The last letter immediately tells the reader what he wants to know: whether his supply of steel can be used for the bridge. He can then read the technical details that follow more intelligently. In the first letter he has to wade through the technical details, not knowing what the result will be until he reaches the end of the second paragraph.

Unless a writer analyzes exactly what the reader wants to know, he can easily be led into writing an indirect letter. He must ask himself "What basic fact is most important to my reader?," and then introduce it as early as possible. If there are conditions or interpretations that need to be described, or additional factors that have strongly influenced the results of a project, he must bring them in afterwards. To introduce them before stating his main theme will weaken the effectiveness of his letter, and may even confuse or annoy his reader.

Be Definite. Being definite means knowing exactly what you want to say, then saying it. Persons who think better with pen in hand sometimes make decisions as they write, which results in indecisive letters that are irritating to read. Their writers seem to examine and discard points without really grappling with the problem. By the time this type of writer has finished his letter he has made his decision and knows what he wants to say, but it has been at his reader's expense. If he wants to write a definite letter that is clear and sticks strictly to the subject, he should discard the first letter and use it as a guide.

Decision-making does not come easily to many people. Those of us who hesitate before making a decision, who evaluate its implications from all possible angles and weigh its pros and cons, may allow our indecisiveness to creep into our writing. We hedge a little, explain too much, or try to say how or why we reached a decision before we tell our reader what the decision is. This is particularly true when we have to tell a reader something that is not in his favor or contrary to his expectations.

Reg Wasalusky, a field employee of H.L. Winman and Associates, has written from a project office in Alaska to ask for a raise. His supervisor, Vic Braun, writes this letter in answer to his request:

> *An Indefinite Letter*
> Dear Reg:
>
>   I could not answer your letter immediately because I wanted time to consider your request for a raise very carefully. We fully appreciate down here that you are working in rough climatic conditions and that much of your work takes place out of doors. On that count we feel that your request is justified.
>
>   On the other hand, we have to equate your position with those of all the other field staff, since to increase your salary when others are not similarly considered would be unfair. We have therefore also assessed your position and seniority, since we recognize that you have been with the company for some time and that you have a very good record.
>
>   As a result of our deliberations we find that it would not be possible to increase your salary at the moment. We would like to add, however, that your request will be reviewed again in three months, when a general salary updating program is to be inaugurated.
>
>   Vic Braun

Vic and Reg are good friends, so Vic does not like to come right out and say "No." But in trying to demonstrate that he has not treated Reg's request lightly, he has used too many wishy-washy statements that have a hollow ring:

> "We fully appreciate"   (para 1)
> "For some time"   (para 2)
> "As a result of our deliberations"   (para 3)
> "We find that it would not be possible"   (para 3)

The letter is also dangerous because the ends of both the first and second paragraphs lead Reg to believe he is being considered. Then the next paragraph contradicts this impression. This "yes-no-yes-no" attitude has a disconcerting effect that may irritate the reader more than the denial of his request. So will Vic's sudden shift from the personal "I" to the company "we," which he adopts as soon as he starts explaining why the request is going to be turned down.

   If Vic Braun wants to be definite, he must be decisive. He must state his decision immediately, then bring in his reasons. If supporting information must be included, it must be brief and to the point. Compare this rewritten version with the original letter:

A Decisive Letter

Dear Reg:

I have considered your request for a salary increase, but regret that it cannot be granted at present. All departmental salaries are to be reviewed in three months, when adjustments will be made for field staff who have seniority and who work under arduous conditions.

As soon as this salary updating program has been finalized, I will let you know if you are to receive an increase.

Vic Braun

Reg may not like hearing "No," but at least he will know exactly where he stands and what is being done. If he knows he has seniority and a good working record, then he can reasonably anticipate receiving a raise in three months. (Note that Vic Braun does not say anything that will make Reg assume his request will be granted; to do so would result in a very unhappy employee if a raise failed to come through.)

A writer will still sound indefinite, even though he approaches his subject directly and decisively, if he uses predominantly passive verbs. Passive verbs are weak, whereas active verbs are strong. Expressions such as "it was considered that" (*I consider*), "is an indication of" (*indicates*), "conducted an examination of" (*examined*), "it is recommended that" (*I recommend*), should be replaced by the much more direct expressions shown in parenthesis. For more information on this subject, see Chapter 10.

Know Your Facts. Insufficient background knowledge and inadequate preparation are the first steps to evasive, ·indefinite writing full of overblown words that avoid the issue. The big words act as camouflage to conceal uncertainty, but a discerning reader can easily penetrate them. Be absolutely convinced of your facts before you start to write, or you will find yourself filling in gaps with big words and many unnecessary adjectives and adverbs.

Tone

It is not easy for a writer to assess whether he is setting the right tone. He will probably be doing so if he has successfully developed the personal qualities that contribute to sincerity, but he cannot assume this. To achieve the right tone, his correspondence should be simple, dignified, but friendly. He should be neither ingratiating nor inadvertently insulting. He should approach his reader on a man-to-man basis, guiding his writing by following the five suggestions below.

Know Your Reader. A writer cannot hope to set the right tone unless he has satisfactorily identified his reader. Only then can he write *to* him (rather than *at* him). If the writer knows his subject well, and the reader does not, he must take care not to be condescending by implying that the subject is so complex that the reader will be able to understand only a very simple description. Even if this is true, the careful writer will select just the right terminology to hold the reader's interest and perhaps challenge him a little. By letting him feel that he is grasping some of the complexities of the subject (by using analogies within the reader's range of knowledge), a writer can often present technical information without confusing or upsetting a nontechnical reader.

Neither should an inexperienced writer attempt to "write up" to the level of a technical reader whose experience and knowledge are far greater than his own. In trying to match his reader's intellect he may be tempted to insert long, important-sounding words that are incorrect, or even faintly amusing. He should write at his own level, using sensible technical terms and expressions that fit the subject.

Take it Easy with Apologies. When apologizing in writing do so quickly and cleanly, just as you would when apologizing to a person for stepping on his foot: "I'm sorry!" is all you need say. If you have been wrong, admit it and then consider the subject closed. To keep on apologizing, to say too much, will make your apologies seem insincere. These excerpts show the difference between an overstated apology and a brief one:

Too Wordy

We were mortified to hear of damage caused to your carpeting by our repair technician's carelessly-placed soldering iron, and are most sincerely sorry that this should have happened. The person involved has been reprimanded and we can assure you that such an occurrence will not happen again. We trust that this accident will not affect our business relationship, particularly since this has never happened before and we extremely regret that it should have happened to you. Please accept our apologies and our guarantee of top-notch service in the future.

Brief

I must apologize for the inconvenience that my incorrectly-worded telegram of 20 February has caused you. The telegram should have read . . . [and so on, but with no further apologies].

Keep Compliments Simple. Some people have a natural gift for paying compliments so that they always sound sincere. They have mastered an art that most of us find difficult: the ability to tread the narrow line between understatement and overstatement.

To pay a compliment effectively in writing is equally difficult. Most of us lean toward overstatement, forgetting that the best written compliments are stated very simply. To overstate a compliment, to gush, to overwhelm your reader with intensity, is to make the compliment seem insincere. This has happened in the following example:

> We have always received excellent service from your organization in the past, so it was only natural that we should turn to you again in the hour of our need. The assistance you provided in helping us to identify a new type of transducer was overwhelming, and we would like to extend our heartfelt thanks to all concerned for their help. The prompt attention you gave to help resolve our problems was very deeply appreciated.

This exaggerated example is so badly overwritten that I automatically doubt the writer's sincerity. If he had written a simple letter of thanks, his appreciation would have been much more believable:

> We very much appreciated the prompt assistance you gave us in identifying a new type of transducer.

Criticize Effectively. A reader will usually accept criticism if it is presented "straight from the shoulder." Any attempt to soften the criticism with a lengthy preamble, or by hiding it behind big words, will reduce its effectiveness. To keep harping on the criticism once it has been stated, or to imply it without coming directly and clearly to the point, will lower the reader's opinion of you as a person who knows what he wants to say.

Implied criticism, or criticism that points a cautioning finger in the wrong direction, can so weaken a letter that the recipient will completely miss the message. Misdirected criticism caused the manager of a radar maintenance department to come running to me with a tail of woe, unaware that it was he who was at fault for failing to point out an error. On three separate occasions he had ordered a very expensive high power transmitting tube called a magnetron, and each time it arrived damaged. Three times he wrote to the manufacturer, urging him to use a better packing method, but nothing was done. "How can I get them to do something about it?" he wailed, thrusting a copy of his most recent letter into my hand. "They just ignore me!"

I examined his letter, which said (in part):

> I regret that for the third time I must write to you to report still another of your magnetrons has been damaged in shipment. It is becoming increasingly obvious that the handling methods used by the railway express department are too rough for such delicate equipment, even though it is clearly marked as such on the outside container. I have spoken to the railway about this but their only comment is that the instrument must have been packed inadequately. To prevent damage to future shipments, I therefore suggest that you use a more resilient type of packing material.

The trouble with this letter is that its writer did not criticize directly. Instead he shifted the blame onto the railway express handlers, only *implying* that the manufacturer's packing methods should be improved. If he had placed the blame where it belonged, the manufacturer would have given his letter more serious attention:

> For the third time we have received one of your magnetrons that has been damaged in shipment. This clearly indicates that your packing methods are inadequate for delicate equipment that has to be shipped by railway express. In future, will you please use a more resilient type of packing material to protect the magnetron from rough handling.

Occasionally a writer has to both criticize and pay a compliment in the same letter. If this happens, always pay the compliment first, to make the criticism easier to accept. This is not the same as softening the criticism; it simply means that the writer recognizes the good work that has been done. Placing the criticism first can so reduce the effectiveness of the compliment that the reader may not even notice that it is there. Rebuked by the criticism, he may feel that it has been added only as a sop to make him feel better. Thus neither criticism nor compliment does its job effectively.

Avoid Words that Antagonize. Any statement that implies that the reader is wrong, has not tried to understand your point of view, or has failed to make himself understood, will immediately place him on the defensive. Even though these factors may be true, they must be stated in words that will clear the air rather than electrify it. Tell a reader gently if he is wrong, and demonstrate why; reiterate your point of view in clear-cut terms, to clarify any possibility of misunderstanding; or ask for further explanation of an ambiguous statement, refraining from pointing out that his writing is vague.

BUSINESS LETTER FORMAT

There are many opinions of what comprises the "correct" format for business letters. The two methods illustrated in Figures 3–1 and 3–2 are those most frequently used by technical business organizations. The former is a conservative format known as the Modified Block, and is the more common arrangement. The latter is a modern style known as the Full Block, and is popular with organizations in which contemporary design plays an important role. Both are acceptable in today's business world.

Selection of a particular format is far less important than that a business organization be consistent and use only the one format for all correspondence. H.L. Winman and Associates, a conservative organization, uses the Modified Block form (see the letter reports in Chapters 4 and 5, and the cover letter preceding the second report at the end of Chapter 6).

The memorandum is a flexible document normally written on a prepared form similar to that shown in Figure 3–3. Formats vary according to the preference of individual companies, although the basic information at the head of

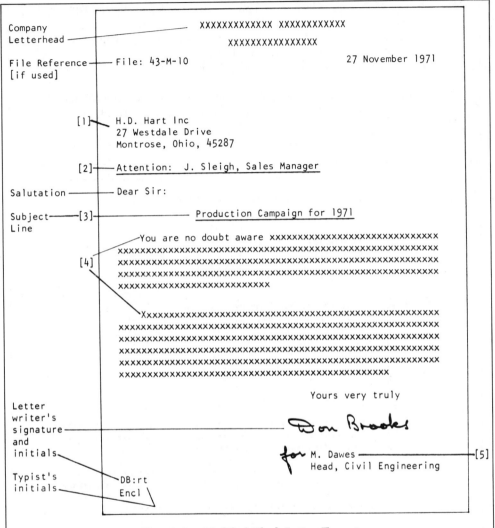

Fig. 3–1   Modified Block Letter Format.

1. Less formal alternative address and salutation does not need an attention line:

> ### Alternative
>
> Mr. J. Sleigh, Sales Manager
> H. D. Hart Inc
> 27 Westdale Drive
> Montrose, Ohio, 45287
>
> Dear Mr. Sleigh:

2. Attention line, which also appears on envelope, directs letter quickly to right department; it should be underlined.
3. Subject line is centered above text, underlined; sometimes preceded by "Subject:," "Ref:," or "Re:."
4. First word of each paragraph may be indented slightly or started flush with left margin.
5. Frequently it is company policy to direct business letters from the writer's Department Head to the reader's Department Head, rather than directly from writer to reader.

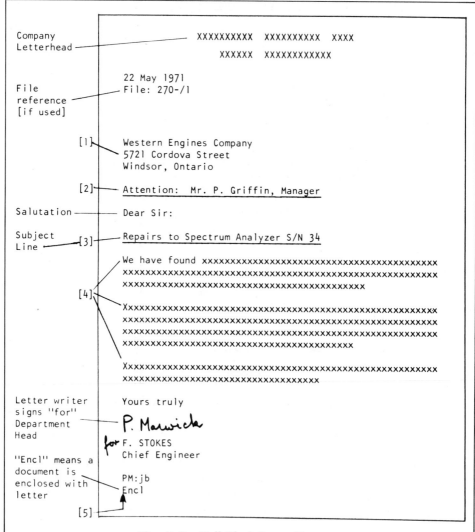

Fig. 3–2   Full Block Letter Format.

1. A less formal alternative address and salutation that does not need an attention line:

<u>Alternative</u>

Mr. P. Griffin, Manager
Western Engines Company
5721 Cordova Street
Windsor, Ontario

Dear Mr. Griffin:

2. Attention line is underlined and flush with left margin; it also appears on envelope.
3. Subject line is also flush with left margin, and underlined.
4. First word of every paragraph starts flush with left margin (never indented).
5. Key to full block letter is that *every* line starts against left-hand margin (simple for typist to set up). Parts otherwise are the same as for Modified Block.

H. L. Winman and Associates

INTER - OFFICE MEMORANDUM

(1)

From: R. Davis   Date: December 3, 1971   (4)

To: A. Rittman   Subject: <u>Early Mailing of Pay Checks</u>

                                                    <u>for Field Personnel</u>

(2)

(3)

I need to know who you will have on field assignments during the week before Christmas so that I can mail their pay checks early. Please provide me with a list of names, plus their anticipated mailing addresses, by December 8.

Checks will be mailed on Tuesday, December 14. I suggest you inform your field staff of the proposed early mailing.

RD   (5)

Fig. 3–3   Interoffice memorandum.

1. The informality of an interoffice memorandum means titles of individuals (such as Office Manager and Senior Project Engineer) often may be omitted.
2. No salutation or identification is necessary. The writer can jump straight into the subject.
3. Paragraphs and sentences are developed properly. The informality of the memorandum is not an invitation to omit words so that sentences seem like extracts from telegrams.
4. The subject line should offer the reader some information; a subject entry such as "Pay Checks" would be insufficient.
5. The writer's initials are sufficient to finish the memorandum (although some organizations repeat the name in type beneath the initials).

the form is generally similar. Examples of completed memorandums appear throughout Chapters 4 and 5.

LETTER OF APPLICATION

Knowing how to write a good letter of application is just as important to the student seeking his first job as it is to the experienced technical man looking for a better position. Both have invested time, money, and personal effort into training that qualifies them to do specialized work, and both need to display their capabilities to the best advantage. Unfortunately, I hear too often of a highly qualified person disqualifying himself by simply failing to present a good image in his letter.

Let's face it: a letter of application is a "sales" letter. It must convince a prospective employer that you are a commodity worth investigating, just as a land agent's letter must sell a prospective client the belief that the agent has a property worth looking into. The chief difference is that you, as an applicant, are trying to sell the most important commodity you own: yourself!

What happens when your letter of application first drops onto an employment manager's desk? As he reads it he forms an immediate impression, mentally classifying you as belonging to one of three groups:

1. "This man sounds interesting! I'd like to talk to him."
2. "This man *may* have something to offer. I *might* talk to him."
3. "This man has little to offer. Just a routine application—not worth following up."

You cannot afford to let him classify you in group 3, for then your dealings with this particular employer are already over. Even in group 2 you may be overlooked, particularly if you are facing competition from applicants whose letters have better reader appeal than yours. Only by writing a letter that creates enough interest for the reader to automatically classify you in group 1 can you be reasonably sure that you will reach the second stage of the hiring procedure. Your letter of application is the key that opens the door to the personal interview. If it fails to do its job properly, the door remains shut and your prospects of employment are negligible.

There are two situations that call for the job seeker to write a letter of application (ignoring for the present the many campus situations where a letter of application is not required, and also those that simply call for the applicant to drop into the office and fill out an application form). Either he has seen an advertisement in a newspaper or technical magazine that invites applications from persons having his qualifications, or he knows of an organization in which he would like to work, but does not know whether they have any job openings. In the former case his task is simplified, since he can concentrate on displaying talents required for the advertised position. In the latter situation his task is more complicated, because he does not know what the reader will most want to learn about him. Although the approach and tone are likely to differ (note the differences between the sample letters in Figs. 3–4 and 3–5), both letters follow the same basic format.

The Parts

A letter of application has three parts:

The *initial contact*, in which the applicant creates interest in himself.
The *evidence*, in which he demonstrates why his application is worth
   considering.
The *close*, in which he holds open the door to further discussion.

These are not watertight compartments of information that stand alone; rather, they carry the reader naturally and quickly through the letter so that when he has finished reading he automatically classifies it in group 1.

The Initial Contact. A good opening arrests the attention of the recipient, making him feel he wants to read the remainder of the letter. A dull, uninteresting opening can so weaken his attention that he may read the rest of the letter with only casual interest, or not even bother to read it at all.

An interesting opening will avoid the routine "I would like to apply for a position as a laboratory technician. . . ," and replace it with a direct statement of intent:

Please accept my application for the position of laboratory technician advertised in the Montrose Herald of 17 May.

or

I am applying for the position of . . . .

42 Venables Street,
Cleveland, Ohio, 44103
Tel. 777-1611

16 May 1971

Mr. J. Perchanski, P. Eng.
Head, Electric Engineering Department
H. L. Winman and Associates
475 Reston Avenue
Cleveland, Ohio, 44104

Dear Mr. Perchanski:

    In a radio interview last week you mentioned that your firm will
shortly be setting up a project group to study HV DC power transmission
for Montrose Hydro. If this project creates new openings in your
Department will you please consider me for employment., I have had
experience working in the University of Montrose high voltage laboratory
during two summer vacations, and majored in high voltage power generation during my second year at Caravelle Junior College in Cleveland.

    During periods of summer employment at the University I worked on
project Mirron (a study of special applications of DC power conducted
for the United Nations) under the direction of Dr. David C. Crawford.
Much of the time I was conducting tests of prototype systems, evaluating
results, and preparing project reports. Then at Caravelle College I
assisted in developing a HV DC Power Source for laboratory use. This
work was supervised by Mr. Samuel Wolinsky, my Electrical Instructor.
Both these persons have agreed to vouch for my experience and capability.

    If you think that a person having my qualifications could be of
value to you, may I visit your office for an interview? Any afternoon
of the week starting 27 May would be particularly suitable as I will be
writing my final examinations during the mornings and will be free
every day after 1 p.m.

                                        Yours truly,

                                        Colin Westdahl

                                        Colin Westdahl

Fig. 3–4   Unsolicited Letter of Application.

Many job applicants find the opening difficult to write because they fail to state in the first sentence why they are writing. It is relatively easy to do this when applying for a position known to be open, as in the two examples above, but can be more difficult when you do not know if a vacancy exists. Lack of knowledge can easily lead an applicant to an ineffective opening that fails to come to grips with the purpose of the letter. Colin Westdahl, the writer of the unsolicited letter of application in Figure 3–4, has clearly stated the purpose of his letter in the second sentence. But the writer below has only implied his reason for writing, and by not making a clear, specific statement of purpose has conveyed an indecisive, indefinite image:

Dear Mr. Perchanski:

A weak, indefinite contact
While working in the HV Laboratory at the University of Montrose, I became very interested in HV DC power transmission. Now I understand that you have a project going on the same topic and would like to enquire if there are likely to be any job openings.

Writing a good initial contact demands imagination, plus the ability to distinguish between a really effective approach and one that is overstated to the point that it becomes silly. Compare Colin Westdahl's approach (Fig. 3–4), and that of the following example, with the overblown approach beneath it:

Dear Mr. Wood:

A good contact
Will you please consider my application for the Programmer Analyst position advertised in Pro-Gram News of May 20, 1971? I am a graduate of the two-year Programer-Analyst course at Montrose Community College, and have three years experience as programer with Blackfriars Computers Incorporated, of Montrose.

Dear Mr. Perchanski:

An over written contact
Did you know that I majored in high voltage power generation during my last term at Montrose Community College? And that I also worked in the University of Montrose high voltage laboratory during two summer vacations? Since you shortly will be involved in high voltage engineering projects, don't you think it would be a good idea if we were to get together?

Plainly, this applicant is trying too hard. He may have qualifications for the job, but he should try proving it with a simpler approach.

The initial contact should command the reader's attention, state clearly why the letter is being written, and, if possible, demonstrate why the writer thinks he is suitable. All these requirements are met in the initial paragraph of Colin Westdahl's letter.

The Evidence. The second paragraph should present facts to support the writer's application. It should be written in firm, definite terms that emphasize his strong points, and offer specific information related to the position he is seeking. Most of all, it must never generalize. Vague statements such as "I have had many years of experience with the CanCel Chemical Corporation" make a reader suspect that the applicant is trying to conceal that he has only limited experience to draw on.

If the applicant has a lot of evidence to present, he should prepare a data sheet that describes in detail both his qualifications and his experience, and attach it to his letter. The evidence paragraph of the letter should then be no more than a brief synopsis of the more important points. (This has been done by Philip Karlowsky in his letter of application in Fig. 3–5.) If the applicant has only limited evidence, he can present it in the body of the letter. Students with little previous work experience are likely to have only a short evidence paragraph, as shown in Figure 3–4.

The Close. Most applicants have difficulty in writing an effective closing paragraph, tending to say too much because they do not know when they have said enough. The close should be short, and do the opposite of what its title suggests: it must open the door to the next step. Both sample letters do this successfully. By suggesting times when he can be available for an interview, Colin Westdahl (Fig. 3–4) has opened the door even wider than usual. If he has not heard from Jim Perchanski by, say, May 28, he can telephone and enquire whether he is to visit H.L. Winman and Associates without feeling he is being too forward in taking the initiative. Many young applicants fail to realize that a delayed reply may be caused by business pressure rather than lack of interest, and that the hiring process sometimes can be lengthy.

The Personal Biography

Many names can be applied to the personal biography, ranging from the simple Data Sheet mentioned previously, through Resume and Biography of Experience, to the Curriculum Vitae used at professional levels. But each serves the same purpose: to provide details of a person's background, experience, and qualifications that are too cumbersome to insert in the body of a letter. They may also vary in format and tone, from the rather terse approach of the Data Sheet to the well-developed narrative often seen in the Resume and Curriculum Vitae.

Philip Karlowsky's biographical details attached to his letter of application (Fig. 3–5) offer a satisfactory middle-of-the-road treatment that is neither sketchy nor pretentious. The comments below refer to the circled figures beside his biography.

1. Some applicants think that under Personal Information they should include details of their height and weight, comments on their

28-821 Bertard Street
Montrose, Ohio, 45287

March 23, 1972

Mr. R. Davis
Office Manager
H.L. Winman and Associates
475 Reston Avenue
Cleveland, Ohio, 44104

Dear Mr. Davis:

I am applying for the position of Engineering Assistant (Test Laboratory) advertised in the March 18 issue of the Montrose Herald. I have had five years experience in soils analysis and concrete testing, and am currently enrolled in a two-year Civil Engineering Technician program at Montrose Community College. I will graduate at the end of June.

My employer throughout my five years in industry was Solatest Incorporated, a firm of consultants in Reece, Minnesota. For two years I helped Mr. K.L. Pettigrew, P. Eng. set up a pavement evaluation program for the city of Reece. I then became fully responsible for the program, which involved conducting semi-annual tests on selected sections of pavement, analyzing results, and writing reports. This work continued until I resigned in 1970 to enrol at Montrose Community College. The attached biographical details provide further information on my employment background, education and experience.

I will be visiting Cleveland from May 3-5, and would appreciate having the opportunity to talk to you. May I call you after I arrive, to set up an appointment?

Yours truly,

Philip Karlowsky

Fig. 3-5   Solicited Letter of Application (with Data Sheet).

BIOGRAPHICAL DETAILS

PHILIP H. KARLOWSKY

PERSONAL INFORMATION:

① Age: 25
Marital Status: Married, one child.
Address: 22-821 Bertard Street, Montrose, Ohio.

EDUCATION:

② Graduate of Merryhirst High School, Reece, Minn. (1965).
Expect to graduate as Engineering Technician (Civil) from Montrose
Community College, May 1972.

EXPERIENCE:

③ 1970 to-date <u>Fairwarn and Merriam, Engineering Consultants</u>, Montrose, Ohio.
Part-time employment as draftsman (total 7 months), preparing
drawings of utilities and plans for a new community.

1965-1970 <u>Solatest Incorporated</u>, Reece, Minn. Junior Technician
performing soil and concrete tests for clients. For four
years took part in pavement evaluation program; under
direction of project engineer for two years, then had full
responsibility for project for two years.

1963-1965 <u>Y.M.C.A.</u>, Reece, Minn. Part-time work during high school,
assisting in program for underprivileged boys.

ADDITIONAL INFORMATION:

④ I have been an active member of the Y.M.C.A. since 1957,
having progressed through Junior, Intermediate, and
Senior Leader levels. Currently, I am a voluntary instructor
in the boys' Saturday Morning program at Central Y.M.C.A.,
Montrose.

The following persons have agreed to act as references on my behalf:

⑤ Mr. K.L. Pettigrew, P. Eng.,   Mr. David G. Friesen,
Solatest Incorporated,   Chief Draftsman,
2220 Major Drive,   Fairwarn and Merriam,
Reece, Minnesota   167 Central Street,
                                        Montrose, Ohio, 45287

Mr. G.H. Tucker,   Mr. Dan C. Hansen,
Head, Civil Engineering Department,   Director, Boys and Youth Department,
Montrose Community College,   Central Y.M.C.A.
Montrose, Ohio, 45287   Montrose, Ohio, 45287

Fig. 3-5   Solicited Letter of Application (with Data Sheet).

health and physical handicaps, and incidental items such as place of birth and ethnic origin. I do not believe such information is necessary. To include minor details seems to detract from the biography's purpose: *to offer evidence of the applicant's ability to do a good job.* If your letter of application and biography together do their work properly, they will create interest in you as a potential employee, and details will be recorded in a subsequent step in the hiring procedure (normally when you are asked to complete a formal Application for Employment).

2. There is no need to list all the schools you have attended. Simply state the name of your last school, the highest level you attained, and the year of graduation. Add to this the name of each college or university you have attended, the type of course you enrolled in, the diploma or degree you received, and the year that you graduated (or expect to graduate).

3. Experience is usually presented in reverse order, with the applicant's most recent experience appearing first and his earliest work experience last. Generally, give most coverage to recent experience, providing that the length of employment has been long enough to warrant it, and to earlier work that was similar to that of the position being sought.

4. Employers are particularly interested in an applicant's activities and interests outside of his normal work. They want to know that he is more than a routine employee who arrives at 8 a.m., works until 4:30 p.m., then drives home, eats his supper, and watches television all evening. Information on your hobbies, interests, and participation in sports and community activities tells prospective employers that you recognize your role in society, are not too rigid or too narrow, and adapt well to your environment. Employers reason that such an applicant will make an interesting, active employee who will contribute much to the company. They expect that he will take part in social and sports functions, as well as being a useful worker.

5. Try to draw your list of references from a cross-section of people you have worked for, been taught by, or served with on committees. Before including them in your list, check that all are willing to act as references. Almost everyone will be, but it is only courteous to ask them if they would mind.

The Overview

The appearance of your letter can cause an employer to form a false impression of you. The image conveyed by a poorly presented letter may lead him to classify you in group 2, or even group 3, when your qualifications warrant that you be classed in group 1. On my desk I have a letter of application from a young printer who was fully qualified technically to fill a vacancy in a scientific organization that printed its own bulletins, reports, and brochures.

But the applicant had written so carelessly, scribbling with a blunt pencil on grubby lined foolscap, that the employment manager immediately classified his letter in group 3. He reasoned that a careless applicant would be just as likely to carry this approach into his day-to-day work. Since the quality of the company's printing was important (it conveyed to customers an image of the quality of the company's scientific work), he would not even consider this applicant for an interview.

Finally, a word of warning: Try to avoid writing a letter of application that too closely resembles the stylized approach suggested by most textbooks (and I include this book!). The "textbook letter" is dangerous because its source is so obvious. Personnel managers tell me that they are more interested in an applicant who writes only an average letter, but whose personality and enthusiasm show clearly through his writing, than they are in an applicant whose letter is flawless but characterless. This by no means implies that you can be lax in your approach; rather, it means that you should try to fashion your letters of application as though *you* wrote them, without assistance from me.

Accepting or Declining a Job Offer

Regardless of whether you are accepting or declining a job offer, write to the prospective employer as soon as you have made your decision. This will enable him either to inform other applicants that the position has been filled, or to make an offer to another applicant. If you are accepting, write a brief letter stating so. Restate the important conditions of employment, such as the position you are accepting, your starting salary, and the date you will start work, to prevent any misunderstanding. For example:

> Dear Mr. Davis:
> I shall be glad to accept the position of Engineering Technician II in your Test Laboratory, offered to me in your letter of 10 May. Both the salary of $610 per month and the employment starting date of 1 July are acceptable.
> Thank you for considering me for this position.
> Yours truly,

If you are declining his offer, do so briefly and politely, if possible stating your reason:

> Dear Mr. Davis:
> Thank you for your letter of 10 May offering me employment as an Engineering Technician in your Test Laboratory. I regret that I must decline your offer, having already accepted employment elsewhere.
> Thank you for considering me.
> Yours truly,

Your tone and attitude in a letter declining a job offer are just as important as in the letter of acceptance. If the job you accept fails to work out satisfactorily, you may want to renegotiate with the employer whose offer you declined. It will help in reopening discussions if you turned down his previous offer gently.

ASSIGNMENTS

Project No. 1

Read the following letter, then complete these assignments:

1. Explain why this letter fails to convey its message clearly and effectively.
2. Identify five hackneyed phrases and expressions, or phrases of low information content, contained in the letter.
3. Rewrite the letter.

CITY OF MONTROSE
SAFETY DIVISION
AND
DRIVER IMPROVEMENT CLINIC

4 February 1972

Mr. Rick Davis, Office Manager
H.L. Winman and Associates
475 Reston Avenue
Cleveland, Ohio, 44104

Dear Sir:

We have for acknowledgment your letter of January 9th enclosing a report from Dr. K. Blake, together with your driver's license, requesting that the restriction that you wear corrective lenses while operating a motor vehicle be removed from your license. On reviewing Dr. Blake's report it is indicated that your visual acuity rating in one eye is 20/200, and 20/20 with the other eye, without the aid of lenses. On further checking this with Dr. Blake we are advised that your visual acuity rating, without glasses, in the eye that you can obtain only 20/200, is 20/20 with glasses. In short, glasses improve your visual acuity rating in the right eye to normal level. If glasses did not significantly improve your vision in your right eye, we would then not have imposed the restriction on your license. It is our policy in cases of this kind to require a driver to wear corrective lenses whenever they improve a person's vision to a degree that the driving of a motor vehicle would be made safer. Improved visual acuity rating will also improve a driver's ability to judge distances and will generally improve a person's ability to see objects clearly and distinctly, particularly under adverse weather conditions.

Therefore, in your own best interests and that of the public generally, we are of the opinion that it would be inadvisable at the present to remove the restriction from your driver's license. In the event that it can be shown that glasses make no appreciable difference in your reading, and providing that your vision in your left eye meets the minimum standard required to obtain a license, we would then be prepared to reconsider the matter.

Your driver's license is being returned herewith.

Yours truly,

H. Hall, Chairman
Safety Division

HH/ag

Project No. 2

You are employed in the Engineering Department of Robertson Engineering Company, where you are engaged in the development of electronic control equipment for a refinery. At a critical moment in the development program your Hek Model 354 Spectrum Analyzer breaks down. You take it to the foreman of the Test Equipment Repair and Calibration section, who tells you that without a Service and Parts Manual he cannot help you (a manual was supplied with the equipment when it was new, but now cannot be found). You check with other electronics companies in the city, but none of them uses the same type of Spectrum Analyzer, and so cannot lend you either an instrument or a service manual. Finally, on the fifth of the current month, you send a telegram to the Hektik Company, 315 Pennsylvania Boulevard, Pennstown, Nebraska:

> URGENT REQUIREMENT HEK 354 SERVICE AND PARTS
> MANUAL. PLEASE AIRMAIL BY RETURN.

On the eleventh you receive an air mail parcel from the Hektik Company. On opening it you find they have supplied "Operating Instructions" only, on the flyleaf of which is a reference to a special manual covering Service and Parts. You originate a second telegram:

> HEK 315 OPERATING INSTRUCTIONS RECEIVED. ALSO
> REQUIRE SERVICE AND PARTS MANUAL. URGENT.
> PLEASE AIR MAIL.

On the fifteenth of the same month you receive an air mail Special Delivery package from the Hektik Company on which they have paid $4.87 postage; the contents is a book that has obviously been specially Xeroxed for you. Its title reads: "Service and Parts Manual Model Hek 315 Oscilloscope." With dismay you look at your second telegram and realize that you have mistakenly quoted the Hektik Company's street number in place of the model number. You immediately originate a new telegram requesting the correct manual. In it you state that a letter of explanation follows.

Write the letter of explanation.

Project No. 3

You are engaged in a lake level measurement program for the State of Minnesota. At a critical moment in the program your Hektik Model 370 Water Stage Manometer breaks down. You take it apart and identify that it needs a replacement spring and drive assembly. This is the third time that the fault has occurred in the last six months, and each time the thread on the drive shaft has stripped. You previously purchased spare spring and drive assemblies from the manufacturer (Hektik Company, 315 Pennsylvania Boulevard, Pennstown, Nebraska) on 13 May and 22 August, at a cost of $78.15 each.

Today you send a telegram ordering a replacement spring and drive assembly against Purchase Order W7714.

Write to the Hektik Company to complain about the repeated failures (you may attribute the cause to any condition you wish, if you feel you need to point out the cause), and to request that this replacement part be supplied free of charge.

Project No. 4

A defense contract awarded to H.L. Winman and Associates includes preparation and printing of several lengthy technical manuals which will require the company to increase substantially its stenographic staff. Quotations for 14 identical proportional-spacing electric typewriters are obtained from all the major typewriter distributors in the area. The lowest bid is received from the Premier Sales Company at 4422 Rosedale Street, Montrose, Ohio.

At a Department meeting you are authorized to purchase 14 model GRM electric typewriters from Premier Sales. You draft a letter to the supplier announcing that he has been selected, and requesting that he arrange the shipment against Purchase Order W2117. Inadvertently, the wrong quotation gets pinned to your draft and the typist prepares the letter for Anzak Typewriters, 2820 Dundee Street, Montrose, Ohio. A form announcing that they have not been selected is mailed to the unsuccessful bidders.

You sign the letter hurriedly late in the day. Not until 10 a.m. two days later, when you receive a long-distance call from Ben Bradley at Premier Sales indicating his surprise and disappointment at not getting the order, do you realize that something is wrong. You check quickly and discover the error.

You originate a very quick telegram to Mr. B.N. Smith, Sales Representative at Anzak Typewriters, advising him that there is an error in your letter of two days ago, to disregard the letter, and that a letter of explanation follows.

Write the letter of explanation.

Project No. 5

The typical letter report illustrated in Figure 4–4 (pages 92 and 93 describes an enginering investigation conducted by H.L. Winman and Associates for television station DCMO-TV at Montrose, Ohio.

You are to assume that the TV station accepts R.W. Gray's recommendations, and on 28 October authorizes your company to monitor a step-by-step modification program. You are assigned to the project as Installation Supervisor, and by 22 November (the scheduled completion date) the following work has been done:

a. Blowers: done 6–9 November; noise reduction: 2:46 dB (average)
b. Floors: done 10–12 November; reduction (average of 12 measurements): 1.27 dB
c. Ducts: done 19–21 November; 3.9 dB reduction
d. Walls: not done; wating for results of floor carpet wear tests before going ahead

Within seven days of installing the carpet you have a problem on your hands: the carpet has started to "pill" (i.e., form tight little balls of fluff that adhere to the surface and defy vacuuming). The studio manager is dismayed: "What a mess," he says; "must be mighty cheap carpet!" You telephone Peter Bell and warn him that your client is momentarily disenchanted, and that you have decided to withhold further carpet installation until you have found the cause of the trouble. You also telephone the manufacturer, Wearwell Floor Products Inc., Chicago, Ill., who tells you that an initial surface eruption is normal and that it should stop in a few days.

By 21 November it is obvious that the pilling is growing worse, and you call him again. He agrees to send a representative to the studio on 28 November, and suggests that you and the representative study the carpet-cleaning techniques used by the maintenance staff. "It's our experience that improper cleaning techniques can be the cause of carpet wear problems," he adds. "That's No. 1 grade carpet you're installing, so you shouldn't be having problems. Anyway, pilling doesn't indicate excessive wear; it just looks untidy."

Since you have a week to wait, you return to your office in Cleveland. On Peter Bell's instructions, you write a letter to the TV studio in which you describe progress, outline the problem, and explain the delay in completing the modification program.

Project No. 6

You are a technician employed by H.L. Winman and Associates. Write a memorandum to your Department Head asking for two years leave without pay. Details are:

You want to take a computer analyst/programer course at Montrose (Ohio) Community College.

You would like to continue working part-time (Saturdays, and three months in the summer) to keep your head above water and your stomach full.

You know that in the past the company has not entertained such requests. It prefers to release employees completely, then rehire them as new employees if they eventually choose to return.

You would like the company to pay at least part of the tuition costs (assume that you plan to come back to the company for at least two years at the end of the course). There is no precedent for this; the company normally limits payment of tuition fees to night school and correspondence courses that are relevant to the person's normal duties.

The company has an IBM 360 computer which is used in many engineering applications.

You are a graduate technician of a two-year terminal course. You have worked for H.L. Winman and Associates for 4½ years. You want leave rather than a complete release so that you will not lose seniority, your group insurance plan (if necessary, you'll pay the whole premium yourself while on leave), and the pension benefits that you have accrued.

You think the computer analyst/programer course will be of value because it will help you analyze engineering problems and transcribe them onto the computer. You believe the company will also find your dual capabilities of value.

Your age is 25; you are married and have one child; your wife could work part-time to help put you through the course.

Project No. 7

Assume that "today" is Monday 3 March, and that you are employed in the Engineering Department of H.L. Winman and Associates. For the past four weeks you have been on a field assignment to Duluth, where you have been conducting an engineering study for the RamSort Corporation. You have been assisted by two competent junior technicians (Ray Hicks and Len Dominie), and you are now three days ahead of schedule. The task is to be completed by 21 March.

"Today" you receive a folder from the University of Minnesota in Minneapolis advertising a one week course. Details are:

Course title: Management for the Technical Man
Course dates: 10 to 14 March inclusive
Type of course: Maximum immersion: 8 a.m. to 5 p.m. daily plus 7 to 10 p.m. Wed. evening; approximately 20 hours of home assignments
Cost: $200; includes materials, books, and lunches, with guest speaker from industry at each lunch
Registration: No later than noon Friday 7 March; telephone registrations accepted

No. of
participants: Limited to 16

You are impressed by the technical standard of the course described in the folder and wish to attend. (Because of previous field assignments you missed a similar University Extension Department evening course held throughout last winter. Your company sponsored four engineers to attend this course, for which the fee was $75 each.)

Write an interoffice memorandum to your Project Coordinator:

1. Tell him about the course; convince him it is a good course.
2. Ask if you can attend.
3. Ask if the company will pay the tuition fee, plus travel and lodging costs.
4. Convince him that you can be spared from the RamSort task for one week.
5. Tell him to reply in a hurry (because time is so short).

Assume that your Project Coordinator can give technical approval for you to attend, but must go to the Department Head for financial approval. Also assume that he is a very busy man and notorious for placing items on which he must act immediately in an ACTION tray—with the inevitable result that they are quickly covered by other important items and overlooked.

Project No. 8

You are assigned to the DEBDEN Mobile Communications Center development program that Robertson Engineering Company is to conduct for DeBoine & DeNouvre Corporation of Montreal, Quebec. The project has a life span of two years and at peak will involve 22 engineers, technicians, and draftsmen; there is a possibility that the project may be extended for a further two years.

The Chief Engineer requires your project group to occupy an independent office area close to the Engineering Department, and assigns you to find suitable offices. You investigate space available on the fifth and sixth floors (the two floors of the Wilshire Building occupied by Robertson Engineering Company), but cannot find a large enough area.

You then look further afield and find a 2800 square foot self-contained office on the fourth floor which, although presently occupied by the Premier Sales Company, will become vacant at the end of the month. This office is immediately below the present Engineering Department and would be ideal for your project group, particularly if a hole could be cut in the floor/ceiling between the fifth and fourth floors so that a spiral staircase could be installed to connect the two areas.

You discuss the idea with the Chief Engineer, who suggests you write a letter to the landlord asking if the area is available for rent. He adds that you should indicate that Robertson Engineering Company will want a lease for at least two years, and possibly four, and that you should request a firm rental price for the area. You should also request the landlord to construct a communicating stairwell between the two floors.

The landlord is:

Wilshire Insurance Company
1554 Tuscan St.
Burlington, Ontario

The office on the fourth floor that you wish to rent is identified on the building plan as Area 4J27.

Project No. 9

While on a surveying project in a remote area north of Weekaskasing Falls for H.L. Winman and Associates, one of your crewmen is stricken with acute appendicitis on Sunday morning. Remick Airlines hears of the incident and flies a special helicopter to the area to transport the sick man (Patrick O'Riordan) to the nearest hospital. The date is 27 October.

On returning to Cleveland on 6 November you learn that the airline has waived all charges for the flight, considering it an essential emergency. Write to the Flight Operations Manager (Mr. Carl Nickerson), to thank him and the pilot (Bill Badger). Remick Airlines' head office is in Room 301 of the Arrival Building at Montrose Municipal Airport.

Project No. 10

This project assumes that you are seeking permanent employment at the end of a technical training program. You are to reply to any one of the following advertisements, using your present situation and actual background. If it is still early in your training program, you may update the time and assume that it is now two months before graduation date.

ROPER CORPORATION (OHIO DIVISION)
requires a
CHEMICAL TECHNICIAN

to join a project group conducting research and development into the organic polymers associated with the coating industry. He will also assist in the development of control techniques for producing automated color tinting. Apply in writing, stating salary expected to:

Mr. P. Rassmusen,
Personnel Manager,
P.O. Box 1728,
Montrose, Ohio, 45287

(Advertisement in Montrose Herald, 17 April)

We require a
CIVIL ENGINEERING TECHNICIAN

with an interest in building construction to prepare estimates, do quantity takeoffs, and assume responsibility for company sales and promotion activities.

Apply in writing to:

Personnel Department
PRECASCON CONCRETE COMPANY
227 Dryden Avenue
Montrose, Ohio, 45287

(Advertisement in Montrose Herald, 26 February)

DESIGN DRAFTSMAN

Duties:

Under the general direction of the Chief Engineer, to be responsible for detailed product design of prototype machines. Also will investigate and recommend improvements to existing product lines.

Qualifications:

Graduate of a recognized Design and Drafting course with emphasis on Mechanical Drawing. Experience with or knowledge of agricultural equipment would be an asset.

Please apply to:

Chief Engineer
Agricultural Manufacturing Industries
3720 Harvard Avenue
Montrose, Ohio 45287

(Advertisement in Montrose Herald, 26 February)

ARCHITECTURAL DRAFTSMAN
required by
a well-established firm of architects and planners.

Applicants should have completed a Design and Drafting course at Montrose Community College or similar training institution, in which strength of materials and structural design were a curriculum requirement. Experience in preparation of reports and proposals will be considered in selection of successful applicant.

Write to: Box 821, Montrose Herald.

(Advertisement of 28 March on College Notice Board)

MONTROSE PAPER COMPANY

Offers excellent opportunities for recent graduates to join an expanding manufacturing organization in the pulp and paper industry.

ENGINEERING ASSISTANTS

Two positions are available for mechanical technicians to assist in the design, installation, and testing of prototype production equipment. Previous experience in a manufacturing plant would be helpful. Innovative ability will be a decided asset.

ELECTRICAL TECHNICIAN

This person will assist the Plant Engineer in the maintenance of power distribution systems. He should be a graduate of a two-year course in Electrical Technology who has good knowledge of automatic controls and machine application. Ability to read blueprints and working drawings is essential.

COMPUTER TECHNICIAN

This position will suit either a graduate of a Computer Technology course, or an Electronics Technician who has specialized in Computer Electronics.

Duties will consist of installation, maintenance, and troubleshooting of existing and future computer equipment.

PROGRAMER ANALYST

Duties will consist of program maintenance and development of computer systems on a Honeywell Model 120–16K tape system. Knowledge of Cobol and Fortran will be an asset. This position has excellent potential for growth and progress.

Salaries for the above positions will be commensurate with experience and qualifications. Excellent fringe benefit program available. Reply in confidence to:

Manager of Industrial Relations
MONTROSE PAPER COMPANY
Montrose, Ohio, 45287

(Advertisement in Montrose Herald, 10 March)

H. L. WINMAN AND ASSOCIATES
475 Reston Avenue
Cleveland, Ohio, 44104

CIVIL TECHNICIANS

This established firm of consulting engineers needs three Civil Technicians to assist in the construction supervision of several grade separation structures to be built over a two-year period. Graduation from a recognized Civil Technology course is a prerequisite. Successful applicants will be hired as term employees. Opportunities are excellent for transfer to permanent employment before the project ends.

Apply in writing to:

Mr. A. Rittman
Senior Project Engineer

(Advertisement on College Notice Board)

TECHNICIAN

required by manufacturer of
automated car wash equipment

Duties: Assist plant engineer design and test mechanical and electrical components; supervise installations in cities across U.S. and Canada; troubleshoot problems at existing installations.

Applicants must be free to travel extensively. Excellent salary and promotion possibilities.

Apply in writing to:

Personnel Manager,
DIAL-A-WASH Incorporated
2020 Waskeka Drive
Montrose, Ohio, 45287

(Advertisement in Montrose Herald, 23 April)

ROBERTSON ENGINEERING COMPANY
invites applications from
ENGINEERS, TECHNOLOGISTS, AND TECHNICIANS
interested in Metrology

We are building a first class Standards Laboratory that will be a calibration center for precision electrical, electronic, and mechanical measuring instruments used in our military and commercial equipment maintenance programs.

Applicants for senior positions should hold a B.S. in Electrical Engineering. Laboratory Technicians should be graduates of a recognized electronics course who have specialized in precision measurement.

Applications are invited from forthcoming graduates, who will work under the direction of our Standards Engineer. All applicants must be able to start work on 1 July.

Apply in writing to:

Mr. F. Stokes
Chief Engineer
Robertson Engineering Company
600 Deepdale Drive
Toronto, Ontario

(Advertisement in Toronto Globe and Mail, 18 March)

Project No. 11

This project assumes that you are seeking permanent employment at the end of a technical training program, but few job openings have been advertised in your field. Write to H.L. Winman and Associates (for the attention of one of the department heads) applying for employment. Use your knowledge of the company and your actual background. If it is still early in your training program, you may update the time and assume that it is now two months before graduation date.

Project No. 12

Assume that you are looking for summer employment. Reply to any one of the following advertisements:

CITY OF MONTROSE, OHIO
TRAFFIC DIVISION

The city of Montrose is carrying out an Origination-Destination (OD) Study throughout June and July, with computation and analysis to be carried out in August.

Approximately 40 students will be required for the OD Study, and 20 for the computation and analysis. For the latter work, some knowledge of statistical analysis will be an advantage.

Students interested in this work are to apply in writing to Mr. G. Braithwaite, Room 310, Civic Center, City of Montrose, Ohio, 45287.

(Notice on College Notice Board)

NORTHERN POWER COMPANY
BOX 1760, FAIRBANKS, ALASKA

Applications are invited from male students in two- or three-year technical programs who are seeking three months of well-paid summer employment. The work will include construction of living quarters and associated facilities for a mining camp 30 miles south of Prudhoe Bay, Alaska. Terms of employment will be:

    13 full weeks—10 June to 10 September
    Excellent pay: $180 per week
    Transportation: Paid
    Accommodation: Paid
    Health: Applicants should be in topnotch physical condition.

Students interested in this work are to write to Mr. G.R. Cardinal at the above address. Although previous experience in construction work is not essential, knowledge of general construction methods will be considered an asset for some positions.

(From *Construction—North*, 20 February)

REMICK AIRLINES
offers
SUMMER EMPLOYMENT
to a limited number of male and female students.

Duties: To work as part-time dispatchers, baggage handlers, aircraft cleaners, cafeteria assistants, etc., during the peak summer period (June 15—September 10) at Remick Airlines Terminals at:

        Fort Wiwchar
        Weekaskasing Falls
        Lake Lawlong

Ample free time
Free transportation to destination and return
Good pay
Write, describing previous work experience (if any), to:

        Carl Nickerson
        Flight Operations Manager
        Room 301
        Montrose Municipal Airport

Project No. 13

Assume that you are looking for summer employment, but that few summer jobs have been advertised. Write to H.L. Winman and Associates (Attention: R. Davis, Office Manager), asking for a summer job. Use your knowledge of the company, plus your actual background, to write an interesting letter.

4

Informal Reports
Describing Facts and Events

When someone tells me that he has just finished writing a technical report, I tend to imagine a nicely-bound formal document, tastefully typed and printed. In some cases I would be correct, but most of the time I would be wrong. Far fewer formal reports are issued than are informal reports, which reach their readers as letters and memorandums.

The Memorandum Report

The memorandum report is the least formal of all technical reports. Normally an interoffice or interdepartmental communication that enjoys the informality of presentation and tone discussed in the previous chapter, it may be almost terse in sticking strictly to the point. At other times it may develop its subject in considerable detail to present a convincing argument to its readers.

When Pete Small writes a memorandum to Ian Bailey to tell him that he has finished phase 1 of a pavement evaluation project on a stretch of highway one mile north of Weyburn, and that he is heading 160 miles west to Hadashville to start phase 2, he is sending an informal report. He may also comment on the weather, road conditions, difficulties he has experienced in carrying out his tests, that he has had car trouble, and that somewhere along the route he has lost his watch. Perhaps he may request some equipment to be shipped to him, and for Ian Bailey to arrange that his next paycheck be mailed to Hadashville. These little bits of information seem unrelated, but if organized properly and arranged in logical sequence, they form a report. Pete Small, a member of the Engineering Department staff, is reporting his progress to Ian Bailey, his Department Coordinator. The type of report he is writing is known as a progress report.

Hank Williams is alone in the office, working late one Tuesday evening, when the telephone rings. The caller is Bob Walton, a member of the electrical engineering staff who is on a field trip to Fort Albert. He tells Hank that he has had an automobile accident near Humbrol, and some of his equipment is damaged. He wants Jim Perchanski, his department head, to send out replacement items by air express, and to confirm officially to the operations manager at his destination that he will arrive two days late.

Hank jots down notes while Bob talks. Since he plans to be out of town the following day, Hank knows he will have to write a transcript of the phone call and leave it on Jim Perchanski's desk. He then organizes his notes, goes to a typewriter, and pecks out a summary of the conversation. He prepares it as a memorandum, but basically it is a routine occurrence report. It tells Jim Perchanski what has happened to a member of his staff, that he is not hurt, where he is now, and how soon he will be able to move on. It also tells him that equipment is damaged and spare parts are needed.

If Bob Walton has communicated his information and needs effectively to Hank Williams, and if Hank has written a clear, concise report, then when Jim Perchanski walks in on Wednesday morning he can quickly appreciate the situation and know immediately what action he has to take. He does not need to ask questions because he has been placed in the picture. The memorandum in Figure 4–1 is an example of an informal report that might have been prepared in this hypothetical situation.

The Letter Report

The letter report, although still basically informal, can vary in formality according to its purpose, the type of reader, and the subject being discussed. Some letter reports may be as informal as a memorandum report, particularly if they are conveying information between organizations whose members know each other well or have corresponded frequently. Others may be almost severely formal, although examples of such reports are increasingly rare. Most often they are simple, dignified business letters that convey technical information from one company to another. As such, they serve their purpose by avoiding the rigidity of style and format demanded by the full formal report described in Chapter 6.

The letter in Figure 4–4 is a typical informal letter report written to a client by Bob Gray, an engineering assistant in the Mechanical Engineering Department of H.L. Winman and Associates. In the first paragraph he summarizes his findings to give the reader a quick understanding of the results of his investigation. Then in subsequent paragraphs he describes what he has found out, what methods can be used to alleviate the problem, and which methods he recommends. He has found the key to writing an effective letter report because he has organized his material into a logical, coherent order that will be readily understood.

Although there are many types of informal reports, most appear in either the memorandum or letter form illustrated here. There are variations in format, content, and writing approach, but all conform to a basic pattern: (1) a short introduction of the topic or problem; (2) a discussion of the report writer's ideas, in which he presents the data he has amassed and describes how he has used it; and (3) a conclusion in which he sums up the results of his endeavors and possibly recommends what should be done next. This pattern is evident in almost every report, whether it is a simple occurrence report, a field trip report, a project progress report, an investigation report similar to that in Figure 4–4, a feasibility study, or a technical proposal.

Writing Style

The reports described in this chapter deal with specific information. They are written in a direct, informative style that is crisp and to the point; their writers have something to say and they waste no time in saying it. Because they describe events that have already occurred, they are written mostly in the past tense, which helps the report writer to be consistent. The only time that he needs to shift gear into the present or future tense is when he is describing something that is presently occurring because of a previous happening, outlining what will happen in the future, or suggesting what needs to be done. All three tenses occur in the following extract from a project progress report:

> We installed indoor/outdoor carpet on the floor of the satellite studio control room during the off-air hours of the night of 8–9 January. Noise readings taken at 12 different locations on 10 January showed that the ambient noise level had decreased an average of 1.27 dB. The greatest decrease was 2.1 dB near the east wall, with the smallest decrease 0.8 dB close to the west wall. Favorable comments were received from operating staff, who immediately noticed a marked decrease in clatter caused by impact noises.
>
> After two weeks of use we now notice that the carpet has a tendency to "pill." The manufacturer's representative states that this is normal and does not indicate that the carpet is wearing too quickly. He suggests that it may be caused by improper carpet cleaning techniques, and probably can be easily corrected. However, it is unsightly and our client is not pleased with the carpet's appearance.
>
> The manufacturer's representative and I will return to the control room between midnight and 2 a.m. on 28 January to study the carpet cleaning techniques used by the maintenance staff. I will report our findings to you by telephone later in the day.

The first paragraph is written entirely in the past tense because it deals only with what has already been done. In the second paragraph the writer shifts to the present tense to describe an event currently occurring. Then the final paragraph jumps into the future to outline what he plans to do. This past-present-future arrangement is natural and logical; the reader feels comfortable making the transitions from one tense to the next. The writer has taken care not to jump back and forth between tenses, because this would have destroyed the continuity.

There are occasions when it is permissible to shift tense several times in a short document. In the accident report in Figure 4–1, the writer does so fairly frequently. In his first paragraph he uses the past tense to state what has happened. He then uses the present tense to describe what is being done to remedy the situation, and occasionally the future tense to state what else will be done. Because the reader knows from the first paragraph—indeed, from the first sentence—exactly what has happened, he does not feel uncomfortable shifting from one tense to another; he expects the person involved in the accident to be taking action and planning what is to be done next. Thus the tone and style of the memorandum suit the situation.

OCCURRENCE REPORT

As we have seen in the example in Figure 4–1, an occurrence report simply describes an event or events that have occurred. In this instance, Hank Williams will not be available to answer questions the following morning, so he has taken care to describe the situation clearly. The comments below are keyed to the memorandum by numbers:

1. Hank knows that a subject line must be informative, telling what the memorandum is about and stressing its importance to the reader. If he had simply entered a vague statement, such as "Transcript of Telephone Call from R. Walton," he would not have captured Jim Perchanski's attention nearly as quickly.
2. This brief summary gets right to the point by immediately telling Jim Perchanski what he wants to know:

   Why was this memo written?
   What happened?
   When did it happen?
   What are the implications?

3. Hank knows that Perchanski must act quickly to ship the replacement parts. He has already mentioned that equipment was damaged, so he can introduce the needed items quite naturally into his second paragraph. He intentionally uses a list to help Perchanski identify the items. If he had simply described them in paragraph form, they would have been much more difficult to recognize:

H. L. Winman and Associates

INTER - OFFICE MEMORANDUM

From: H. Williams   Date: 18 November 1971

To: J. Perchanski   Subject: Preliminary Accident Report –   (1)
                                           (Telephone call from R. Walton)

Bob Walton telephoned from Humbrol at 7:35 this evening to report he was involved in a highway accident at 5:15 p.m. He was not hurt but his panel   (2)
truck and some of his equipment were damaged.

He will need replacements for the following items, which he wants you to ship
by air express to the Remick Airlines terminal at Montrose and to mark the
shipment "HOLD FOR PICK UP BY R. WALTON 20 NOV":

        1 Modification Kit MOD-57

        5 Circuit breakers, heavy duty, Temron Type CB-3   (3)

        1 set Burndy Extraction Tools

The panel truck will be repaired by noon on Thursday. Bob plans to drive to
Montrose that afternoon, and will pick up the spare parts in the early
evening. He will drive on to Fort Albert on Friday morning and will arrive   (4)
there about noon. He has informed the night crew that he will be two days
late, but he would like you to confirm this officially by telephoning the
Operations Manager at Fort Albert Wednesday morning.

While at Humbrol Bob is staying at the Mayfield Motel (Tel 287-4166), where
he is preparing a complete accident report which he will mail to you.   (5)

HW

Fig. 4–1   An Occurrence Report Written as a Memorandum.

He will need replacements for a MOD-57 Modification Kit, 5 Temron CB-3 heavy duty circuit breakers, and a set of Burndy extraction tools. He wants you to ship these items air express to the Remick Airlines terminal at Montrose, and to mark them. . . .

4. It is important that instructions and movement details be explicit, otherwise the equipment and Bob Walton may not meet at Montrose. Hank Williams has taken care to identify specific days, and once even the date, to make sure that no misunderstandings occur. To have stated "tomorrow" or "the day after tomorrow" would have been simple, but might have caused Jim Perchanski to assume a wrong date, since he will be reading the memorandum one day later than it was written.
5. In this brief closing paragraph Hank indicates what further action is being taken and where Bob Walton can be contacted if more information is needed from him, or if a message needs to be relayed to him.

When the writer's role is solely to report the event without adding supporting information or analyzing details, he should confine his report to the bare facts. If Hank had not had to write about spare parts and future movement details, he could have conveyed the essential facts in a single paragraph:

Bob Walton telephoned from Humbrol at 7:35 p.m. to report he was involved in a highway accident at 5:15 p.m. He was not hurt but his panel truck and some of his equipment were damaged. He is preparing a detailed report for you and will mail it at noon tomorrow (Wednesday). If you wish to contact him he is staying at the Mayfield Motel (Tel. 287-4166).

There are occasions when a writer is able to analyze an event and the factors that may have caused it. In the following report the writer describes what has occurred, analyzes how it may have happened, and suggests remedial measures to prevent it from happening again:

Fire Damage in Mobile Unit No. 2

During our annual overhaul of your No. 2 Mobile TV Unit we discovered that fire has caused extensive damage to the Sony Videotape Recorder. From your operating log we determined that the VTR was last used on 5 November, so that the fire could have occurred at any time between then and 9 December. The fire was confined to the rear of the VTR and the wall of the vehicle.

The fire probably was caused by careless smoking habits. Although "No Smoking" signs are posted in the mobile unit, multiple scorch marks on the top of the VTR indicate that technicians habitually rest burning cigarettes against the raised lip at the back of the lid. We believe that in this instance a cigarette dropped behind the VTR and fell into an assortment of paper scraps that had accumulated there.

To prevent recurrence of this type of damage we suggest that you update your mobile unit housekeeping procedures to include weekly vacuuming behind the VTR, and that you enforce more stringent control of the no smoking regulation.

An occurrence report that has been developed to this extent is somewhat similar to, although much simpler than, the investigation report described later in this chapter.

FIELD TRIP REPORT

When an engineer or technologist returns from a field assignment, he is expected to write a field trip report describing what he has done. He may have been absent only a few hours, inspecting cracks in a local water reservoir; he may have spent several days installing and testing a prototype pump at a power station in a nearby community; or he may have been away for two months, overhauling communications equipment at a remote defense site. Regardless of the length and complexity of his assignment, he will be expected to remember many details and to transcribe them into a logical, coherent, and factual report. To help him, he should carry a project notebook in which to jot down daily occurrences. Without such a record to rely on, he will be likely to write a disorganized report that omits many details and emphasizes the wrong parts of the project.

The simplest way to write a trip report is to answer three basic questions:

1. *Who went where?*
   Why and when did he go?
2. *What did he do?*
   What did he set out to do and was he able to do all of it?
   What problems did he run into? What additional work did he do?
3. *What remains to be done?*
   What work could he not complete, and who should now complete it? When and how should it be done?

In technical language, these questions would be replaced by more formal headings, such as:

1. *Assignment Details*
2. *Work Accomplished and Problems Encountered* (often treated as two separate headings)
3. *Follow-up Action Required*

Long trip reports require headings like these to help their readers find specific information. They should also contain subheadings and a simple paragraph numbering system that the writer will find easy to use and the reader easy to follow. The example shown in Figure 4–2 demonstrates how a standard format can help a writer organize his material, describes the type of information that would follow each heading, and includes excerpts and sample paragraphs that might appear in a typical field trip report.

Short trip reports do not need headings. A brief narrative following the three question pattern normally carries the story quite effectively:

> Dave Makepiece and I visited site RJ-17 from 15 to 17 January to install a prototype automatic alarm system for the Roper Corporation.
>
> The installation was completed without difficulty and no major problems were encountered. Installation instruction W27 was followed throughout with the exception of step 23, which called for connections to be made to the remote control panel. This step was omitted since the remote panel has been permanently disconnected.
>
> The prototype alarm will be field-evaluated for one month. It will be removed in mid-March by M. Tutanne, who will be visiting the site to discuss summer survey plans.

With the exception of the section describing follow-up action, trip reports should be written entirely in the past tense.

PERIODIC PROGRESS REPORT

Progress reports are management's means for keeping in touch with what its project groups are doing. Whenever a project is likely to continue over several months, management normally will specify that progress reports be submitted at regular intervals so that it can gain an overall picture of project progress and identify problems as they develop. Management also needs the information in these reports to write progress reports to its clients.

As initial purveyors of progress information, project group leaders must include everything that might be of interest to management. They must be able to differentiate between routine details that need not be reported, and facts that management should know. If there have been labor problems that have affected work output and slowed progress, they should include such information in their reports, even though they are aware that it is not the type of information that should reach a client. Management wants to know when problems are rumbling beneath the surface, and will decide whether the client should receive such information.

A periodic progress report may be no more than a one-paragraph state-

H. L. Winman and Associates

INTER - OFFICE MEMORANDUM

From: K. Young

To: I. Bailey

Date: 20 February 1972

Subject: Trip Report - Installation of Prototype Alarm System at Site RJ-17

This guide has been written in the format for a typical field trip report, using a system of headings, subheadings and paragraph-numbers that will help you to organize your trip reports along easy-to-follow lines. Some of the information contained below comprises writing instructions, while some represents typical report entries.

1. Assignment Details

   Under this main heading state the purpose of the trip and include any supporting information that the reader may want to know. The four sub-headings below represent typical entries.

   1.1 Purpose

   The prototype automatic alarm system developed by the Component Development Group was to be installed at a typical monitoring station to evaluate its ability to perform in a low-temperature environment. Copies of the installation instructions and test specifications are attached as appendixes to this report.

   1.2 Authority

   I. Bailey Memorandum 27/J/1656 dated 5 December 1971, and RamSort Corporation purchase order 2727W dated 28 August 1971.

   1.3 Personnel assigned:

K. Young	Project Leader (Electronics)
K. Makepiece	Mechanical Technologist
M. Baslett	Installation Technician

   1.4 Dates Involved

   21 January to 15 February 1972

2. Work Accomplished

   2.1 In this section of the report describe the work that was done. Normally present it in chronological order unless more than one project is involved, in which case describe each project separately. Keep it short: it is not necessary to describe at great length routine work that ran smoothly. Frequently it

Fig. 4–2   Format for a Field Trip Report.

is possible to refer simply to the work instructions, which can be attached to the report for reader reference:

> The manually-controlled alarm system was disconnected as described in steps 6 to 13 of installation instruction W27.

2.2   Only if difficulty is encountered, or if work is necessary beyond that anticipated by the instructions, should you launch into more detail:

> At the request of the site maintenance staff the manual alarm was installed in the power house as a temporary replacement for a defective GG20 alarm. This entailed modifying the control panel to permit the larger alarm to be installed. Parts removed from the panel, together with instructions for returning the panel to its original configuration, were left with the Senior Power House Engineer.

Note that this extract demonstrates to the reader that the report writer has anticipated that the temporary modification will have to be replaced, and has shown how this is to be accomplished.

2.3   If parts of the assignment could not be completed, identify them and explain why the work was not done:

> Test No. 23 was omitted because the remote control panel for the RamSort equipment has been permanently disconnected.

In this case no entry under "Follow-up Action" (see para 4 below) is necessary because the words "permanently disconnected" have made it clear that the remote control panel will not be used in future.

3.   Problems Encountered

3.1   In contrast to the conciseness recommended for the Work Accomplished section of the report, this section should describe problems in detail. Knowledge of problems encountered, and the methods used to overcome them, can be invaluable to the engineering department, which may be able to prevent similar problems from occurring during other field trips, or to anticipate related problems and forewarn field teams of difficulties they may encounter.

3.2   An entry that does not tell the reader what the problem was or how it was overcome is virtually useless:

> Considerable time was spent in trying to mount the miniature control panel. Only by fabricating extra

Fig. 4–2   Format for a Field Trip Report.

parts were we able to complete step 17.

3.3   If the information is to be used by the engineering department, it must be much more specific:

> Considerable time was spent in trying to mount the miniature control panel according to the instructions in step 1. We found that the main frame has additional equipment mounted on it, which prevented us from using most of the parts supplied. We overcame the problem by fabricating a small sheetmetal extension to the main frame and mounting it, with the miniature control panel, as shown in the attached drawing.

4.   Follow-up Action Required

4.1   This section of the trip report ties up any loose ends by telling the reader what still remains to be done. If any work has not been completed, draw attention to it here even though you may already have mentioned it in para 2. Identify what needs to be done, if possible indicate how and when it should be done, and state whose responsibility it now becomes. (The responsibility to complete the work does not always rest with the original writer, who may be transferred to other assignments soon after his return from a field trip.) This follow-up action entry refers to the work described in para 2.2:

> The manual alarm mounted as a temporary replacement in the power house is to be removed when a new GG20 alarm is received. This will be done by the Senior Power House Engineer, who will use special instructions and spare parts left with him for this purpose.

The Power House Engineer probably will assign the task to a member of his staff. Who actually performs the work interests neither you nor the reader; your responsibility ends when you inform the reader that the task has not been left unassigned.

4.2   In some cases you may direct follow-up action to someone else in your department:

> The prototype alarm is to be removed after one month. This will be done by M. Tutanne, who will be visiting site RJ-17 in mid-March.

In his trip report at the end of March, Mike Tutanne will state that he has removed the alarm (he will write this in the Work Accomplished section of his report). Until he makes this entry your assignment remains incomplete, and the file for your trip report cannot be closed.

Fig. 4–2   Format for a Field Trip Report.

ment describing progress of a simple design task, or it may be a multipage document covering many facets of a large construction project. (I am not including form-type progress reports, which call for simple entries of quantities consumed, yards of concrete poured, and so on, with cryptic comments.) Regardless of its size, it should answer three main questions that the reader is likely to ask:

1. *What progress have you made?*
   How much work has been done, and what period does it cover?
2. *Have you had any problems?*
   What were they and how have they affected project progress?
   Can you see any problems peeping over the horizon?
3. *Is the project on schedule?*
   If not, how do you propose to get back on schedule?

There is a pattern to the answers to these questions which results in a natural flow of information:

THIS IS THE WORK DONE DURING THE REPORT PERIOD	WHILE DOING THIS WORK WE ENCOUNTERED PROBLEMS	AS A RESULT, WE ARE NOW AHEAD OF, ON, OR BEHIND SCHEDULE

In lengthy progress reports this pattern can be broken down even further to form the six major headings in Table 4–1. The headings in the first column are in almost the same order as the basic pattern. The second column represents one of many alternative orders that are equally suitable, and may be preferred by report writers who like to state whether or not they are on schedule early in their reports.

TABLE 4–1—PROGRESS REPORT HEADINGS

Suggested Order	*Alternative Order*
Period Covered	Period Covered
Progress	Adherence to Schedule
Problems Encountered	Progress
Adherence to Schedule	Problems Encountered
Future Problems	Future Problems
Additional Information	Additional Information

In appearance the memorandum-type periodic progress report is very similar to the field trip report illustrated in Figure 4–2. Although the information that appears under each main heading listed in Table 4-1 will depend on the size and complexity of the project, the arrangement should be similar to that suggested below:

Period Covered

This need be no more than a one-sentence statement, such as:

This report covers the month of February 1972.

Progress

Progress and schedule information are frequently confused by the new report writer. True, whether or not a project is on schedule depends primarily on how much progress has been made, but the two are not inextricably linked. In the alternative order shown in Table 4-1 they are placed next to each other so that the writer who does not like to separate them can write:

Heavy rains in the middle of the month slowed progress, so that we are now three days behind schedule.

I prefer to separate them because they convey different information. The first describes the *amount* of work done (progress), while the second discusses the *rate* at which it has been done in relation to the allotted time span (schedule).

The Progress section describes the amount of work that has been accomplished during the report period. For a straightforward project it may involve only one or two sentences:

During the period 492 test samples were received, 306 of which were rated as acceptable and were retained in the laboratory for analysis. This represents a 12% increase over the number of samples analyzed the previous period, and 3% over the number projected for this period.

For large projects it is wiser to start with a summarizing statement describing general progress (para 2.1 below) and to follow it with a series of short paragraphs discussing different aspects of the project (paras 2.2 to 2.4):

2. *Progress*

    2.1 Construction progress this month has been only fair. Outside work was hampered by heavy rains that fell for three days during the third week, and sunny weather at the end of the month enabled us to retrieve only some of the lost time. Except in the north wing, inside work continued at the planned rate.

2.2 Work on the west wing was accelerated to enable the wing to be opened by month-end. This was done by transferring approximately 60% of the work force from the north wing. Exterior and interior decoration were completed and electrical fixtures were installed and tested.

2.3 Very little construction progress was made in the north wing. Subcontractors laid only 20% of the floor tiles and started installing air conditioning pipes and vents. The electricians wired all the 400 volt outlets and installed 15% of the lighting fixtures.

2.4 Landscaping started on 16 September but was abandoned three days later when rain turned the earth into a quagmire. Work restarted on 25 September, but by month-end only the parking area had been completed.

This report generalizes because the task is a large one. Small details are intentionally omitted unless they have significantly affected progress. Specific quantities, if included, would appear as tabular data in an attachment to the report.

Specific information should appear in the progress section only when it is an integral part of the project, or when the reader would find the information of particular interest. In this example, the writer considered that the procedure he used to mount and align the antenna, and the results he obtained, were sufficiently important to warrant inclusion:

The height finder antenna was installed and aligned. A gin pole procured from a local supplier was erected to raise the antenna parts to the tower roof. The pedestal, rotary coupler, and 50 ft parabolic dish were assembled and installed without difficulty, following a revised procedure developed at the previous site (see progress report No. 24, pp 6 to 9). Alignment in azimuth was made by reference to the north mark painted on the tower roof. Alignment in elevation was accomplished during the evening and early morning of 17–18 August, using sun strobe technique SST-20/1. Ten sightings were made and compared against the true elevation. From these, an antenna elevation error of −11 seconds was calculated and marked on the inside of the left-hand vertical support, one foot below the hub.

Note that entries in the progress section are concise, written in the past tense, and held strictly to the subject.

Problems Encountered

Management wants to know if problems are encountered during a project, and to be told how they were overcome or what steps have been taken to overcome them. As I mentioned in the section on the Trip Report, such knowledge alerts management to difficulties that might be averted on future projects (see Fig. 4–2, para 3). It also indicates project trends. A series of problems encountered during a design project can indicate that insufficient time has been allowed for it, that more engineering skills are needed, or that funds budgeted for the project are inadequate. While the Progress section shows how a project is getting along generally, the Problems section can frequently act as a warning that progress may slow down unless immediate steps are taken.

Problems should be described in depth. The writer should state clearly what the problem was, how it affected his project, what measures he took to overcome it, and whether his remedial measures were successful. This paragraph discusses several incidents that can be lumped together as a single problem:

> Juvenile vandalism has proved to be a petty but time-consuming problem. On 3 September (Labor Day), several youths scaled the fence around the materials compound and made off with about $170 worth of building supplies. On 16 September they managed to start up a front end loader, drove it into the excavation, then got it stuck in the mud and burned out the clutch while attempting to move it. From 18 September the night watch has been doubled and the site has been policed by a patrol dog. There have been no further attempts at vandalism.

If there are several unrelated problems, each should be discussed in a separate paragraph.

Adherence to Schedule

If the project is on schedule at the end of the reporting period, a single sentence will suffice for this section:

> The project is on schedule.

The writer has used the present tense because he assumes the reader will understand that he is reporting project status as of the last day of the reporting period (not the actual date the report was written, which probably was several days later). If he wants to keep to the past tense for this section, he must write it this way:

> At the end of the reporting period the project was on schedule.

If the project is not on schedule, the entry should state the number of weeks, days, or hours that it has dropped behind:

> At the end of the reporting period the project was six working days behind schedule.

This statement is factual, but poses two questions in the reader's mind: Is the schedule continuing to slip? What is being done to bring the project back on schedule? The report writer must anticipate these questions and answer them with a positive statement of *action* taken. A vague reference to remedial measures is insufficient, as the following examples demonstrate:

> *Positive:* At the end of the reporting period the project was six working days behind schedule. To accelerate progress, 10 plasterers and 5 carpenters were transferred to the project from the Northern Inn construction site on 22 September. They will be retained on the project until we are back on schedule, which will be approximately 24 October.
>
> *Vague:* At the end of the reporting period the project was six working days behind schedule. If the good weather continues and we can hold the extra labor brought in temporarily from the Northern Inn construction site, we hope to be back on schedule in three weeks.

The positive explanation will convince the reader that the report writer has grasped the problem and taken effective steps to rectify it. The vague statement only implies that something is being done and creates doubt in the reader's mind that the remedial measures will be sufficiently effective.

Future Problems

The man responsible for a project can frequently recognize problems before they develop and, with prompt action, sometimes avert them. But he must not keep this information to himself. By forewarning management of an impending problem he may be able to prevent costly work stoppages or equipment breakdowns. Management has more resources to draw on, plus experience in coping with previous problems, and he should make use of them.

In describing an anticipated problem he should also indicate that he has tried to work out a resolution. It is not sufficient to drop the problem in someone else's lap and trust that a resolution will be handed down. This extract from a construction project progress report describes an existing situation and warns management of its implications should it develop into a problem:

> 5.1 Unless the strike at Vulcan Steel Works ends shortly, it will soon curtail our construction program. Our present supply of reinforcing barmats will last until mid-October, by which time an alternative source of supply must be found. I have re-

searched other suppliers, but have been unofficially warned by
union representatives that any attempt to obtain steel from elsewhere may result in a walkout at other plants.

This writer has tried to resolve the problem on his own, but now can do no
more; he must depend on management assistance to resolve the problem. In
the following example the writer is able to recommend definite alternative
action:

If we have not finished pouring concrete by 1 November, we will
have to start providing heat on very cold nights. Since our own heating units will not be available until after 15 January, I have explored
other sources of supply and have discovered that the U-Rent-It
Equipment Company will rent portable space heaters to us for $6.00
per unit per week. I have therefore made arrangements to rent 12
of their space heaters from 1 November.

Additional Information

Items of information that do not naturally fall into the other sections
can be entered here. They may include comments on the work and the contract,
details of accidents affecting staff or equipment, changes in personnel, and
similar minor activities.

Continuity Between Reports

The need to maintain continuity of information between successive
progress reports can be easily overlooked. If a writer introduces a problem that
has not been resolved in one report, he must refer to it in the next report even
though it may have been solved only a day or two later. He must never drop a
problem simply because it no longer applies.

In the first example quoted under Future Problems, the report writer
mentions in his September report that a strike at Vulcan Steel Works might
affect construction in mid-October. His October report must say what happened, otherwise the reader may remember the previous report and ask: Did
he run out of steel? Did he find an alternative source of supply? Or is the strike
over? If the problem did not develop, the follow-up entry need be no more
than a brief explanation:

The possibility of a shortage of steel barmats mentioned in paragraph 5.1 of my September report was averted when the strike
at Vulcan Steel Works ended on 6 October.

Where does this entry go? Since it is no longer a future problem, it must be transferred either to Problems Encountered or Additional Information. In this case the latter would be the best choice, because the occurrence did not become a problem. If the strike had continued and a shortage of steel became a reality, then the continuing entry would naturally appear under Problems Encountered.

OCCASIONAL PROGRESS REPORT

Management does not normally ask for progress reports when an assignment is short, but sometimes events call for the man on the job to write one. Jack Binscarth, one of Robertson Engineering Company's laboratory technicians, has been assigned to Cantor Petroleums in Edmonton to analyze oil samples. The job is expected to take five weeks, but problems develop that prevent Jack from completing the work on time. To let his chief know what is happening, he writes the brief progress report in Figure 4–3.

Occasional progress reports like this are similar to occurrence reports. They have an event or events to relate, and they do this best by adopting a past-present-future pattern:

Summary	A brief description of the overall situation
Progress	The work that has been done, the problems that have been encountered, and the effect these problems have had on progress
Situation Now	What is being done at present
Future Plans	What will be done to complete the project, and when it will be done

INVESTIGATION REPORT

I have used the term "Investigation Report" to cover several types of reports which, broadly speaking, can be grouped together. These include laboratory reports, project reports, and any other report in which the writer describes how he has had to perform tests, examine data, or conduct an investigation using tangible evidence. Basically, the writer starts with known data which he analyzes and examines so that the reader can see how he conducted his investigation and reached his final results. The report he produces may be issued as a memorandum or a letter, or on a special form.

Most investigation reports are prepared as interoffice memorandums between departments within the same company (their format is the same as

robertson | engineering   company

INTER - OFFICE MEMORANDUM

From: J. Binscarth [at Cantor Petroleums]   Date: 14 October 1971

To: F. Stokes, Chief Engineer,
Head Office

Subject: Delay in Analysis of
Oil Samples

My analysis of oil samples for Cantor Petroleums has been delayed by problems
at the refinery. I now expect to complete the project on 25 October, nine
days later than planned.

The first problem occurred on 23 September, when a strike of refinery
personnel set the project back four working days. I had hoped to recover all
of this lost time by working a partial overtime schedule, but failure of the
refinery's spectrophotometer on 13 October again stopped my work. To date,
I have analyzed 111 samples and have 21 more to do.

The spectrophotometer is being repaired by the manufacturer, who has promised
to return it to the refinery on 19 October. I have informed the Refinery
Manager of the delay, and he has agreed to an increase in the project price to
offset the additional time. He will confirm this to you in writing.

Providing there are no further delays I will analyze the remaining samples
between 20 and 24 October, then submit my report to the client the following
morning. This means I should be back in the office on 26 October.

Jack.

Fig. 4–3   An Occasional Progress Report.

that for the evaluation report in Fig. 5–2). Sometimes they appear first as such a memorandum, or as a memorandum between a project group and a department head, and then the essential details are rewritten as a letter report to a client. The investigation report in Figure 4–4 illustrates the informal yet businesslike tone to be found in such a letter.

Although they are not always readily identifiable, there are standard parts to a well-written investigation report that help shape the narrative and guide the reader to a full understanding of the topic. These parts are relatively easy to recognize in long investigation reports, where headings act as signposts introducing each parcel of information. They are more difficult to identify in short reports which use a continuous narrative, as in the investigation report in Figure 4–4. These standard parts are described below.

Summary. A very brief description of the whole report, stated in as few words as possible. It gives busy readers a quick understanding of the investigation from which they can learn the results and assess whether they should read the whole report.

Background. Events that led up to the investigation, knowledge of which will help the reader place the report in the proper perspective. This part is sometimes referred to as the Introduction.

Investigation Details. A complete narrative description of the investigation, carefully organized so that the reader can easily follow the report writer's line of reasoning. Although every topic will be different, and the shape of each report will vary, the writer should follow a basic pattern when presenting his data. He may not use all the following parts, but those that he does use should contain information similar to that described here, and should appear roughly in the same order:

> *Start with Known Facts.* Introduce the reader to all the technical data and tangible evidence available at the start of the investigation, or used during the investigation.
>
> *Introduce Guiding Factors.* These are the requirements or limitations that controlled the direction of the investigation. They may be as diverse as a major specification stipulating definite results that must be attained, or a minor limitation such as price, size, weight, complexity, or operating speed of a recommended prototype or modification.
>
> *Outline Investigation Method.* Tell the reader the planned approach for the investigation, so that he can understand why certain steps were taken.
>
> *Describe Equipment Used.* If tests were performed that called for special instruments, describe the test set-up and, if possible, include a sketch of it. In long reports this may have to be done several times, to keep details of each test set-up close to its description.
>
> *Narrate Investigation Steps.* Many investigation reports lend themselves to a chronological presentation of information. In most cases

H. L. Winman and Associates

PROFESSIONAL CONSULTING ENGINEERS
475 Reston Avenue-Cleveland, Ohio, 44104

21 October 1971

Television Station DCMO-TV,
P.O. Box 890,
Montrose, Ohio, 45287

Attention: Mr. D.R. Carlisle,
          Technical Operations Manager

Dear Sir:

        Reduction in Noise Level - Satellite Studio Control Room

We have investigated the high background noise reported in the control room
of station DCMO-TV's satellite studio at 21 Union Road, and have traced it to
the building's air-conditioning equipment. The noise could be reduced to an
acceptable level by soundproofing the air-conditioning ducts and blowers, and
by carpeting the control room.

Our investigation was authorized by your letter of 28 August 1971. Tests
conducted with a noise meter at various locations in the control room established that the average ambient noise level is 36 dB, with occasional peaks
of 39 dB near the west wall. This is approximately 8 to 10 dB higher than
the noise levels measured in the control room of your No. 1 studio on
Westover Road.

The unusually high noise level is caused by the air conditioning equipment,
which is located in an annex adjacent to the west wall of the control room.
Air conditioner rumble and blower fan noise are carried easily into the
control room because the short air ducts permit little noise dissipation between the equipment and the work area. The flat hardboard surface of the
west wall also acts as a sounding board that amplifies the noise.

We have considered three methods that could be used to reduce the ambient
noise to an acceptable level:

1.   The most effective method would be to move the air conditioning equip-
      ment to a more remote location. This, however, would require major
      structural alterations which would make the approach impractical except
      as a last resort.

Fig. 4–4   An Investigation Report in Letter Form.

2.   A noise reduction of approximately 6 dB could be effected by replacing the existing blower fan assembly with a Model TL-1 blower manufactured by the Quietaire Corporation of Detroit, and by lining the ducts with Agrafoam, a new soundproofing product developed by the automobile industry in West Germany.

3.   A further reduction of 1.5 dB could be achieved by replacing the vinyl floor tiles with indoor/outdoor carpet, a practice that has been tried with some success at Air Traffic Control centers. Mounting carpet on the whole of the control room's west wall, and possibly along the first five feet of the north wall, would also reduce the noise by an additional 1 dB.

Since relocation of the air conditioning equipment would be unduly expensive, we recommend that methods 2 and 3 be adopted to reduce overall ambient noise by approximately 8.5 dB for a total cost of $2080.00. We suggest that the modifications be made progressively so that actual noise reductions can be measured on completion of each step, which will permit the final step to be eliminated if earlier modifications achieve better results than anticipated. The four steps would be:

Modification	Anticipated Ambient Noise Reduction	Approximate Cost
(a) Replace blower fan assembly	2.5 dB	$ 560.00
(b) Line ducts with Agrafoam	3.5 dB	$ 600.00
(c) Install floor carpet	1.5 dB	$ 580.00
(d) Install wall carpet	1.0 dB	$ 340.00

We believe that these modifications will provide the quieter working environment desired for your operating staff at moderate cost.

Yours truly,

RW Gray

for Peter Bell
    Head, Mechanical Engineering

PB:bls

Fig. 4–4   An Investigation Report in Letter Form.

a step-by-step description will be both logical and easy to follow, particularly if the investigation has progressed naturally through a series of steps involving the introduction to a problem, development of a means to overcome it, application of corrective measures, and tabulation of results. But when an investigation has comprised an analysis of processes, equipment, or methods, chronological presentation cannot be used. In such cases divide the information into suitable subject areas (e.g. equipment, processes, application) and examine the advantages and disadvantages of each. For more information on the chronological and subject method of report development, refer to the "Organization of Discussion" section in Chapter 6.

Discuss Test Results. Analyze the results of each test and discuss whether or not they satisfy the needs of the investigation. (It is not sufficient simply to tabulate the test results and assume that the reader will infer their meaning and implications.) Such an analysis after a test or series of tests can demonstrate that a specific trend influenced or altered the course of the investigation.

Develop Ideas and Concepts. The preceding parts of the report have all dealt with facts. At some point in his narrative the writer may want to introduce ideas and concepts that he has evolved as a result of his investigation. He may introduce them periodically throughout the report to demonstrate what he had in mind before taking the next investigative step, or he may group them together near the end. They should develop naturally from the flow of the report, and should be persuasive: the writer may need to use them to influence the reader's acceptance of a new technique or unusual method he wants to suggest.

Conclusions. This is a brief summing-up in which the writer draws the main conclusions that evolved from his investigation. He may also recommend what steps he considers need to be taken next, providing that they develop naturally from his conclusions.

These standard parts can be visualized as a basic flow diagram (see Fig. 4-5) that can be applied to almost any engineering investigation. As an

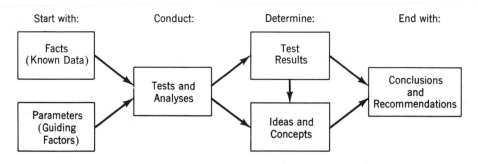

Fig. 4–5   Basic Flow Diagram—Investigation Report.

example, suppose that Harvey Winman, as president of his company, decides to purchase an executive jet aircraft because he travels a lot, but cannot decide which aircraft would be most useful. He might assign one of his engineers to investigate a selected range of aircraft and to recommend the most suitable model. Starting with facts (specifications of the types of aircraft Harvey is known to favor) and parameters (probable route mileage, maximum operating costs, maximum acceptable purchase price), the engineer would analyze the different aircraft and determine which offers the most advantages under the proposed operating conditions. He might also conceive the idea of leasing an executive aircraft which could be rented, complete with crew, on an as-required basis from a local aviation company. Armed with all this information he would prepare a report of his findings, ending with a conclusion that identifies the most suitable aircraft to purchase, but which also points out the economic advantage of leasing. The report he would submit to Harvey Winman would be an investigation report that starts with known information but ends by developing an idea.

Some companies preface their investigation reports with a standard title and summary page similar to the H.L. Winman and Associates' design illustrated in Figure 4–6. This page saves a reader the trouble of searching for the summary and the report's identification details. Subsequent pages, which have not been included with the example, contain the report narrative, starting with Background Information. Reports written in this way tend to adopt a slightly more formal tone than memorandum and letter reports, particularly in the use of the first person. Sometimes they are called Form Reports, although I prefer to identify them as Semiformal Investigation Reports.

ASSIGNMENTS

There are 15 writing assignments of varying length and complexity included with this chapter. The first nine are individual projects that comprise:

2 Occurrence Reports (p. 97)
3 Trip Reports (p. 99)
2 Progress Reports (p. 104)
2 Investigation Reports (p. 111)

The remaining six are assignments that grow naturally out of a single project that I trace from its inception (p. 116) through to its completion.

. . .

I have classified the major assignments in this and the next two chapters into those calling for informal reports (Chaps. 4 and 5) and those calling for

H. L. Winman and Associates

INVESTIGATION REPORT

Report No. __70/26__   File Ref. __53-Civ-26__   Date __20 March 1970__

Prepared for __City of Montrose, Ohio__

Authority __City of Montrose letter Hwy/69/38; 7 Nov 69__

Report Prepared by _~~G. Winman~~_   Approved by _~~M. Davies~~_

SUBJECT OR TITLE

INVESTIGATION OF STORMWATER DRAINAGE PROBLEM --

Proposed Interchange at Intersection of Highways 6 and 54

SUMMARY OF INVESTIGATION

1. The proposed interchange to be constructed at the intersections
    of Highways 6 and 54, on the northern perimeter of Montrose, Ohio,
    incorporates an underpass that will depress part of Highway 54
    and some of its approach roads below the average surface level
    of the surrounding area. A special method for draining the storm-
    water from the depressed roads will have to be developed.

2. Two methods were investigated that could contend with the
    anticipated peak runoff of 77 cfs. The standard method of
    direct pumping would be feasible but would demand installation
    of four heavy duty pumps, plus enlargement of the three-quarter
    mile drainage ditch between the interchange and Lake McKing.
    An alternative method of storage-pumping would allow the runoff
    to collect quickly in a deep storage pond that would be excavated
    beside the interchange; after each storm is over the pond would
    be pumped slowly into the existing drainage ditch to Lake McKing.

3. Although both methods would be equally effective, the storage-
    pumping method is recommended because it would be the most
    economical to construct. Construction cost of a storage-
    pumping stormwater drainage system would be $184,000, whereas
    that of a direct pumping system would be $267,000.

Fig. 4–6   Title and Summary Page for a Semiformal Investigation Report.

formal reports (Chap. 6). Many of these reports, however, can be presented in either form, depending on specific writing situations.

Many of the assignments in these chapters also interlock with assignments suggested for Chapters 8 and 9. Your instructor will inform you of specific writing projects that are to be combined with speaking or illustrating assignments.

OCCURRENCE REPORTS

Project No. 1—An Accident Near Hadashville

You are employed by H.L. Winman and Associates. During a field assignment you are a passenger in a vehicle that is involved in a traffic accident. Your Department Head requests a report of the accident for the department's files. Use the following information for writing the report.

Whose vehicle was it?	Belongs to Multiple Industries, 2820 Rosser Road, Montrose
Who was driving the vehicle?	Pete Crandell, Project Supervisor for Multiple Industries
Why were you there?	On way to visit project site RJ-17 at Tangwell
When did this happen?	10:45 a.m. yesterday
What type of vehicle was it?	Land Rover T2711
How many passengers was it carrying?	Only 2 (you and the driver)
Where did this happen?	Highway 4, 2 miles N-E of Hadashville (see illustration)
Was anyone injured?	Only your driver: broken left arm, left kneecap; suspected concussion. You were only shaken about, sustained one or two bruises—no more than that
What happened next?	Only vehicle that could still be driven was the truck; used to take your driver to hospital. You stayed at scene, dealt with police, then later visited Pete in hospital. In evening returned to Cleveland by bus

Assume that you attach a sketch of the intersection showing the positions of the vehicles (you need not draw this sketch; just state it's there). Also assume that the only written entries apart from highway identification are the name and license number of each vehicle, e.g. Pontiac MD 372, Truck T 485.

SITUATION JUST BEFORE ACCIDENT   N

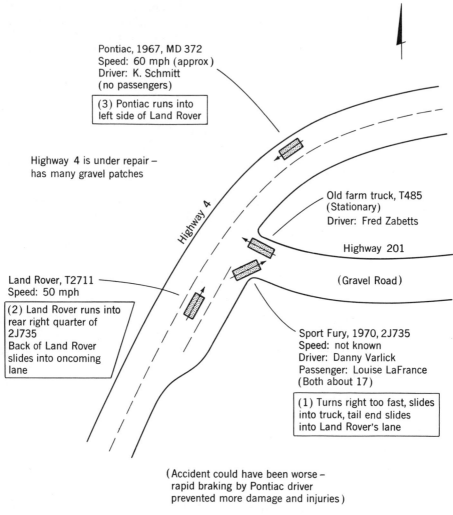

Pontiac, 1967, MD 372
Speed: 60 mph (approx)
Driver: K. Schmitt
(no passengers)

(3) Pontiac runs into
left side of Land Rover

Highway 4 is under repair –
has many gravel patches

Highway 4

Old farm truck, T485
(Stationary)
Driver: Fred Zabetts

Highway 201

(Gravel Road)

Land Rover, T2711
Speed: 50 mph

(2) Land Rover runs into
rear right quarter of
2J735
Back of Land Rover
slides into oncoming
lane

Sport Fury, 1970, 2J735
Speed: not known
Driver: Danny Varlick
Passenger: Louise LaFrance
(Both about 17)

(1) Turns right too fast, slides
into truck, tail end slides
into Land Rover's lane

(Accident could have been worse –
rapid braking by Pontiac driver
prevented more damage and injuries)

Project No. 2—Our Surveyor's Transit Has Been Damaged

You are the Team Leader of a four-man H.L. Winman crew en route to
an assignment at Minnowin Point in Alaska. On Sunday evening you take a
cab to Cleveland Airport to catch Remick Airlines Flight 679 to Winnipeg,
Canada, the first stop on your route. As one of the crew's suitcases is lifted by
an airline employee from the weigh scale to the conveyor belt, its shipping
tag tangles with the strap of the case containing a Surveyor's Transit, which
has been placed carelessly on the counter with its strap hanging over the
edge. The moving suitcase pulls the Transit to the floor, then drags it beside

the conveyer belt and bumps it against obstacles along the way. You try to rescue the Transit before it disappears through the swinging doors to the baggage room, but it crashes through them and disappears from view. You eventually retrieve the Transit and note that its carrying case is badly battered.

You consider that the Transit cannot be used in its present condition and decide to leave it with a Remick Airlines' representative to be picked up by your company early in the week. You obtain a receipt for it and board the aircraft to Winnipeg.

While in flight you write a report of the incident for your Department Coordinator (Ian Bailey), asking him to pick up the damaged Transit and ship it to the manufacturer's service office. You also ask him to borrow or rent a replacement Transit from the local manufacturer's representative, and to ship it on Monday night's flight to Winnipeg, for routing to you at Minnowin Point. You hand your report to the flight crew and ask them to send it by messenger to your office on their return to Cleveland on Monday morning.

NOTES:

Type of Transit: Fennel "Tropl" 5¼" with optical plummet, Model
   A0150
Manufacturer: Otto Fennel Co., Kassel, W. Germany
North American Service Office: 35 Kringle St., Peterborough,
                        Ontario, Canada
Local Sales Representative: Engineers' Supply Company,
                        409 Cumberland Ave., Cleveland, Ohio
Which member of your crew placed the Transit carelessly on the
   counter?—Doug Rickerson

TRIP REPORTS

Project No. 3—Substandard Split-bolt Connectors

H.L. Winman and Associates is management consultant for Montrose Hydro during construction and start-up of a power transmission line between Wabagoon Falls and Montrose, Ohio. Coordinator for the project is Don Gibbon, and you are an engineering assistant in his project group. The contractor is Welland and Welland Inc., also of Montrose. The supervisor at site M9 has called in to report that he has found three faulty KS-28 split-bolt connectors (SBC's) in the site's stock, and asks if this is a common occurrence. Don Gibbon decides to investigate whether the fault has occurred elsewhere, and sends you to check on the connectors at site M17, one mile east of Wabagoon Falls. Your authority to proceed on the assignment is Travel Order W67.

At M17 you test all the split-bolt connectors on site, and find four that are faulty (three in stock and one that has been installed). You also notice that there appear to be two types of connectors, one a slightly darker gray than the other.

On returning to the office you write a report of your trip for Don Gibbon. In addition to informing him of your findings, you suggest that all the SBC's be tested, including those in the contractor's stock. You also try to draw a conclusion from the differences in shading.

NOTES:

1. The method you used to test the SBC's is specified in company procedure PR27–7.
2. There are 18 installation sites.
3. All the faulty connectors were in the darker gray group.
4. The fault is visible as a fine hairline crack when the connector is placed in tension.

Project No. 4—Selecting a Site for an Auto-Mart

H.L. Winman and Associates has been commissioned by the president of Auto-Marts Inc. to recommend new sites for Auto-Marts. The project coordinator is Ian Bailey, and he has assigned you to find a good location in Montrose, Ohio. (Auto-Marts are drive-in convenience stores.)

Before you leave on your assignment, Ian Bailey hands you the following guidelines prepared by the client:

> Aside from exterior facade and foundation details, Auto-Marts are built to a standard 80 by 30 foot pattern with an order window at one end of the 80 foot wall, and a delivery window at the other. Auto-Marts carry only a limited selection of grocery items, but boast 90-second service from the time of placing an order to its delivery at the other end of the building. For this reason, Auto-Marts attract the householder hurrying homeward from work, the impulse buyer, and the late night shopper, rather than the selective shopper. The store depends more on volume of customers than on the volume of goods sold to each individual.
>
> The chief consideration in selecting a site must therefore be a location on the homeward-bound side of a main trunk road into a large residential area. The site must have quick and easy entry onto and exit from this road, even during peak rush hour traffic. The residential area should be occupied mainly by upper and middle class people who own cars and have money to spend. There should be little competition from walk-in grocery stores.

The site you select is a vacant lot at the southwest corner of the intersection of Alexander Street and Evergreen Drive (see sketch).

On returning to head office the following day, you prepare a trip report for Ian Bailey, in which you:

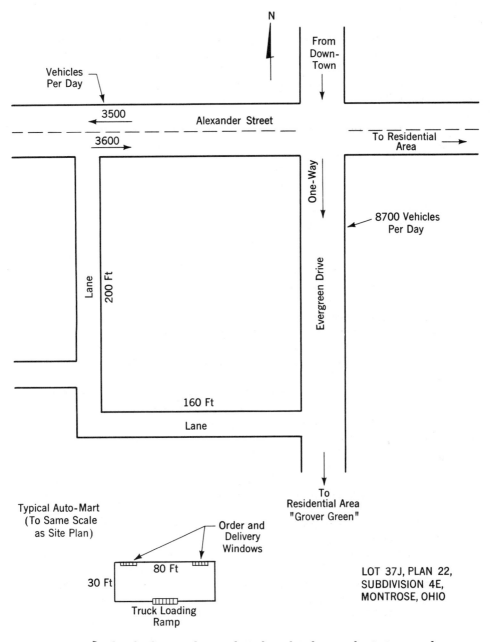

1. Describe the lot you have selected, and indicate why it is a good site for an Auto-Mart.
2. Outline how the Auto-Mart and approaches will be arranged on the site (may be substantiated by a sketch showing the building, entry and exit roads, loading ramp, parking area for three employees' cars, and a large advertising sign).

NOTE:

City of Montrose By-law 261 stipulates that no exits may be placed
within 100 feet of a major intersection, that separate cuts must be
used for both entry and exit, that entries must turn off the road at a
30° angle, and that exits must approach the road at a 60° angle, as
shown below:

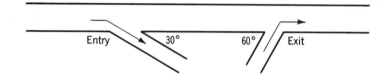

The nearest walk-in grocery stores are one third of a mile west on
Alexander Street, and half a mile south on Evergreen Drive.

Project No. 5—A Visit to Wakeling Processors Inc.

Wakeling Processors Inc. has its head office in Cleveland, Ohio, and numerous manufacturing plants throughout the surrounding area. Oliver R.
Wakeling, President and General Manager, calls in H.L. Winman and Associates to help him resolve a management problem with technical implications.
The problem concerns the Wakeling Processors Inc. plant at 2045 Seventh
Avenue West, Montrose, Ohio, and centers around the operation of the power
house. This subsidiary is engaged in vegetable oil processing, for which the
power house has to produce large supplies of hot water and a moderate
supply of compressed air.

Mr. Wakeling is concerned because there has been a continuous upward
increase in power house costs at the Montrose plant over the past two years.
Fuel costs have risen by 18% and numerous breakdowns have occurred which
have interfered with production, with a consequent increase in overhead costs.
Yet management visits have revealed little that could be attributed to poor
operation; in fact, the power house has always been immaculate.

Mr. Wakeling believes that another approach at management level will not
produce the answers he requires. He suggests to Peter Bell that H.L. Winman and Associates send someone at technician or technologist level who will
be able to talk on technical subjects to all the people in the power house,
and at the same time evaluate production problems. They call you into the
office, and Mr. Wakeling gives you a brief outline of the situation:

The present chief engineer at the Montrose plant is David Skyla,
and he is to retire in three months. Management has to decide
whether to promote Barry Bishop, the existing Senior Shift Engineer,
or to bring in a new Chief Engineer from outside the plant. On
paper, Bishop is ideal for the job: He has worked in the plant for 15

years (he is now 36) and always under Skyla, so his knowledge of the power house and its operating conditions cannot be challenged. Yet the rising costs indicate that all is not correct in the plant, and management wants to be sure that the new Chief Engineer does not perpetuate the present conditions.

Mr. Wakeling says that he will advise the Montrose plant that he has engaged H.L. Winman and Associates to study the hot water and steam generating system at Montrose, and to expect your arrival.

You drive to Montrose one week later, planning to spend two days in the plant.

During your talks to plant staff and tours of the power house you make the following notes:

1. Housekeeping excellent—whole place shines (but is this only surface polish for impression of visitors?).
2. Maintenance logs are inadequately kept—need to be done more often. Need more detail. Equipment files not up to date and not properly filed.
3. Boiler cleaning badly neglected. Firm instructions re boiler cleaning need to be issued by head office.
4. Flow meters are of doubtful accuracy. May be overreading. Not serviced for three years. Manufacturer's service department should be contacted (these are Vancourt meters). Manufacturer needs to be called in to do a complete check and then recalibrate meters.
5. Overreading of meters could give false flow figures—make plant seem to produce more steam than is actually produced.
6. Good housekeeping obviously achieved by neglecting maintenance. Incorrectly placed emphasis probably caused by frequent visits from company president, who likes to bring in important visitors and impress them. Skyla likes reflected glory (so does Bishop).
7. Shift engineers are responsible for maintenance of pumps and vacuum equipment. Not enough time given over to this. They seem to prefer straight replacement of whole units on failure rather than preventive maintenance. Costly method! Obviously more breakdowns: they wait for a failure before taking action. A preventive maintenance plan is needed.
8. Bishop seems O.K. Genial type; obviously knows his power house. Proud of it! But seems to resist change. Definitely resents suggestions. Does he lack all-round knowledge? Is he limited only to what goes on in his plant? Is he afraid of new ideas because he doesn't understand them? Young staff hinted at this: too loyal to say it outright, but I felt they were restive, hampered by his insistence that they use old techniques that are known to work but are slow. Nothing concrete was said—I just "felt" it.
9. Skyla's done a good job training Bishop. Made him a carbon copy. Skyla doesn't do much now. Bishop runs the show, and has for

over a year. He *expects* to get the job when Skyla retires. It'll be
a real blow to him if he doesn't! Wakeling might even lose a good
company man.
10. Discussed RAMSORT 2300 power panel with staff. Young engineers had read about it in "Plant Maintenance"—eager to have
one installed (I described the one I'd seen at Abotinam Pulp and
Paper). But Skyla and Bishop knew nothing about it—weren't interested. Obviously not keeping up-to-date with technical magazines.

When you return to head office you inform Don Gibbon verbally of your
findings. He asks you to prepare a routine memorandum-type field trip report,
which he can use to prepare a letter report for the client.

PROGRESS REPORTS

Project No. 6—Installing EDP Equipment

Robertson Engineering Company is converting to Electronic Data Processing (EDP) equipment for use in inventory control, accounting, and other
departmental operations. Coordination of the installation phase has been
assigned to you. The installation contractor is the Milward Corporation of
Detroit.

The EDP equipment is scheduled to be fully installed and ready for handover by the contractor on the last working day of the current month. The
changeover date from manual accounting and inventory control to the EDP
system has been set for the last working day of next month.

You already know that the contractor will not meet his target date. Today
he is 11 working days behind schedule.

He has had problems:

1. There has been a ten-day delay in delivery of some major components for the system (they were shipped to Montreal in error).
2. The main control panel was damaged in transit; the contractor
wasted five days trying to repair it; a replacement ordered by
Telex five days ago is due at Robertson Engineering one week
from today.
3. The contractor's crew normally comprises four men and a supervisor. But one man quit four days ago (a replacement being flown
in from Detroit is due here tomorrow), and another is sick (absent
six days to date, due back in about a week).

Write a progress report to George Dunn, Robertson Engineering's Comptroller. Advise him of the problems and delays. Forecast a new installation
completion and handover date, and a revised system changeover date.

Project No. 7—Building an Airfield at Lac le Roulet

When you report to work at H.L. Winman and Associates on June 12, Project Engineer Andy Rittman informs you that you are assigned temporarily to an airfield construction project at Lac le Roulet in Northern Quebec. He hands you your field assignment instructions:

Proceed to Lac le Roulet, Quebec, arriving at approximately noon on June 14. You are to:

1. Take over the duties of Resident Engineer L. Martineau.
2. Supervise a six-man construction crew erecting temporary campsite buildings.
3. Supervise a four-man crew surveying the sites for a proposed airstrip, radio transmission tower, and permanent campsite.
4. Ensure that the work performed by the two crews is complete, and that all construction materials and facilities are on hand, in readiness for the fly-in of the main construction crews on July 2.
5. Take photographs showing the general topography of the area, using the Polaroid camera carried by P. Lamonde. (These photographs will be displayed at a management conference to be held in Toronto on June 19.)
6. Shortly after arrival, prepare a report of the progress at Lac le Roulet, with details of problems encountered. Have this information, plus the photographs, ready for pick-up by helicopter at noon on June 15.

Before leaving for Lac le Roulet you visit Larry Martineau in the hospital, where he is recovering from pneumonia. He briefs you on the personalities of the two crew foremen, intimating that Pete Lamonde (the survey crew foreman) is very reliable, but saying that Dave Grogann (the construction crew foreman) is "a slippery character who needs constant watching."

You arrive at Lac le Roulet at 12:15 on June 14 and immediately start gathering facts. The two crew foremen show you their Crew Check Lists (see Attachments 1 and 2), from which you establish that the survey work is pretty well on schedule but that the construction work, which was on schedule up to June 11 (the day Martineau was taken ill), now is two days behind schedule. Grogann vaguely attributes the delay to "unsatisfactory" weather, his search for local labor, and trouble with equipment. He has had trouble with the gasoline-operated generator because the wrong types of plugs and gasoline were supplied to him, and he needs tar paper to complete the roofs of the temporary buildings he has erected. Both foremen give you a list of tools and materials they require (page 106).

Later, you discuss topography with Lamonde and ask him for the Polaroid camera and any photographs he has taken. With some embarrassment he admits that he can offer you neither. Reluctantly, he informs you that the

Items required – Survey Crew

1. A Wild Tripod with wooden legs to fit our Wild N-3 Surveyor's level.

2. Steel Surveyor's chains to replace 3 broken chains. Require 2 100-ft chains and 1 200-ft chain.

Construction Crew

We can't use the gas-operated generator because head office sent us unleaded gas and the wrong type of plugs! We need leaded gasoline (8 50-gallon drums) and <7 plugs – about a dozen.

Also need six auger bits Type S for soil sampling, 4 inch diameter.

15 gallons latex paint, half white and half olive green.

Tar paper for roofs (comes in 50 yard rolls, 36" wide). 6 rolls.

---

morning after Martineau was flown to hospital, Grogann borrowed the camera and disappeared with it and his crew. They returned 36 hours later, with what seemed to be massive hangovers but minus the camera, having visited a village three miles to the north ostensibly to recruit labor for the main construction program. You question Grogann about the camera, but he answers evasively about its loss: "Just one of those things," he says with a shrug.

That evening you write a memorandum-type progress report to Andy Rittman, as requested in paragraph 6 of the field assignment instructions. (Refer to p. 246 of Chapter 7 for instructions on how to prepare a request for parts and materials.)

H. L. Winman and Associates

LAC LE ROULET INSTALLATION -- CONTRACT 77-Mon-72

SURVEY CREW CHECK LIST

Item	Description	Scheduled Completion Date	Comments; date complete
1.	Run a closed traverse for reference points	21 May	Complete - 22 May
2.	Prepare a contour map showing: Water areas Swamp & Muskeg Safe travel routes	23 May	Complete - 25 May
3.	Prepare a topographic map of area showing: Proposed locations of runway, radio transmission tower, and main campsite Locations of hills and rock outcrops	10 June	Complete - 9 June
4.	Locate water table and determine depth of frost penetration	15 June	Complete - 14 June W. Table - 4 ft 8 in. Frost Pen. - 6 ft.
5.	Conduct an accurate cadastral survey of the location for the radio transmission tower	18 June	In process
6.	Visually inspect type of vegetation in the area and determine whether any building materials (e.g. gravel) are obtainable locally	21 June	

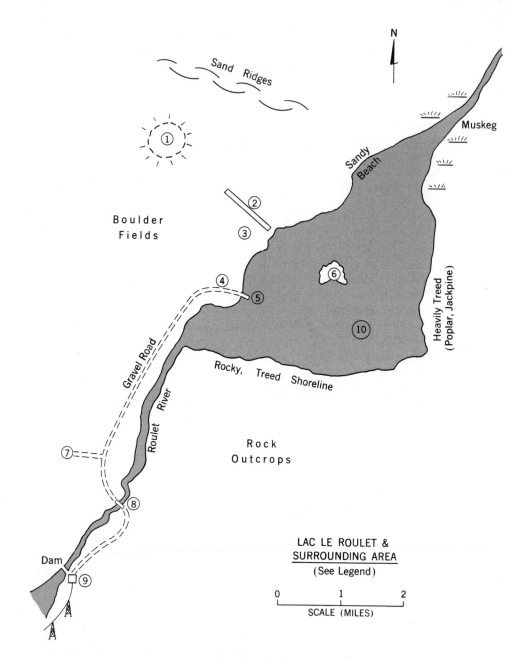

N

Sand Ridges

Muskeg

①

Sandy Beach

Boulder Fields

②

③

⑥

Heavily Treed (Poplar, Jackpine)

④

⑤

⑩

Gravel Road

Roulet River

Rocky, Treed Shoreline

⑦

Rock Outcrops

⑧

Dam

⑨

LAC LE ROULET &
SURROUNDING AREA
(See Legend)

0   1   2

SCALE (MILES)

LEGEND

for

MAP OF LAC LE ROULET AND SURROUNDING AREA

1.   Hillock; top 837 ft Above Mean Sea Level (AMSL); has a gravel base and
      a park-like appearance with scattered evergreens growing to a height of
      40 ft; also has low underbrush.

2.   Proposed airstrip; 4000 ft long; 780 ft AMSL.

3.   Site for Permanent Camp.

4.   Site of Temporary Camp.

5.   Existing seaplane dock; deteriorating badly; needs complete rebuilding.

6.   Island; top 791 ft AMSL plus height of trees (approx 30 ft).

7.   Site for Radio Transmission Tower.

8.   Bailey Bridge over rapids; water level drops 8 ft through rapids.

9.   Roulet Hydro-electric power generating station; water levels: 806 ft
      AMSL above dam; 774 ft AMSL below dam.

10.   Lac le Roulet; water level 765 ft AMSL; level varies no more than ±2 ft;
      contains excellent stock of whitefish and northern pike.

LAC LE ROULET INSTALLATION -- CONTRACT 77-Mon-72

CONSTRUCTION CREW CHECK LIST

Item	Description	Scheduled Completion Date	Comments; date complete
1.	Clear bush from areas designated for temporary buildings and storage sites	21 May	*Done - May 22nd*
2.	Erect: Toilets Kitchen/dining shack Bunkhouse Shed for portable power plant Equipment shed Wash-up shack Radio shack Hut for construction office	1 June	*All done by June 2nd except for tar paper on roofs and painting*
3.	Select and fence suitable storage area to house tools and construction materials supplied for construction of permanent site buildings	9 June	*OK - June 8th*
4.	Select and fence fuel storage area (in accordance with Safety Specification S177)	11 June	*We'll finish this job today*
5.	Select area and erect building for remote powder magazine (in accordance with Safety Specification S216)	14 June	
6.	Assist Staff Technologist to conduct a soil survey at proposed airstrip location. Submit soil samples to Cleveland office	20 June	*Need auger bits*
7.	Determine whether local labor available near Lac le Roulet; if so, assess how many men are willing to work as laborers on main construction project	22 June	*Small village 3 miles north. 7 men and 2 youths will work. agreed to come for $1.40 per hour.*

INVESTIGATION REPORTS

Project No. 8—Choosing the Right Vehicles

H.L. Winman and Associates has two vehicles that are used for field assignments. One is a nine year old Jeep, used principally by the survey crews. The other is a seven year old station wagon which is used mainly for highway inspection, construction supervision, and other field assignments. There is always a demand for these vehicles, and many people feel that there should be at least one more company-owned vehicle to reduce the demand in summer months. Staff use their own cars, and are reimbursed for doing so, when very little equipment has to be carried and their assignments require only highway travel.

At a management meeting, Harvey Winman authorizes the purchase of three new vehicles, placing a budget ceiling of $10,000 for all three. The two present vehicles are considered to have a total sale or trade-in value of $950, which can be added to this figure.

Discussion of the proposed purchase results in a wide divergence of opinions. Andy Rittman strongly recommends purchase of at least two, and preferably three, Willys Jeeps. "They're ideal for our survey crews," he says. "They're tough, and they've got good road clearance plus a four-wheel drive."

"So have a lot of other makes!" counters Ian Bailey. "Take the Land Rover. In the army we. . . ."

"Keep away from imports," Jim Perchanski interrupts, "stick to domestic makes; you won't have parts problems that way."

"Personally, I think we should be looking for vehicles with some degree of driver comfort," Martin Dawes suggests.

"No good in the bush," Andy argues. "They're not built for it."

Harvey Winman suggests that three different vehicles be purchased, with a different purpose in mind for each. But Peter Bell, Head of Mechanical Engineering, disagrees. "Better to have a fleet," he says, "so we can stock common spare parts."

The disagreement grows in scope, some wanting easy interchangeability of vehicles, some suggesting commercial panel trucks, and one even wanting a 13-seater bus, until Harvey Winman decides to halt the argument. "Get one of your technicians to investigate our needs and the types of vehicles available, and then report back to us," he says to your department head.

Ten minutes later your department head calls you into his office and outlines the problem. He suggests that you make a fairly comprehensive survey and prepare a report, complete with recommendations, that he can hand out at the next meeting. (He adds that firm recommendations will show the management team that you have researched the problem, and will counter personal preference and opinions that have not been fully developed.)

Before starting your investigation you establish that:

1. The survey crew needs a vehicle that can carry four men (instrument man, chainman, rodman, and one cutter) plus equipment. Equipment consists of 100 lb of surveying instruments, plus 400 lb of camping equipment (tents, bedrolls, stoves, axes, food) when operating in remote areas. Personal effects for each crewman can be calculated at 40 lb. A vehicle with high ground clearance and four-wheel drive is essential. Utilization: 13,000 miles annually.
2. The Civil Engineering Department needs a vehicle for building and highway inspection projects, where travel generally will be on prepared surfaces. Up to five passengers (plus the driver) and 300 lb of equipment would be the maximum payload. Utilization: 11,000 miles annually.
3. The Electrical Engineering Department needs a vehicle for occasional transmission line inspection, which will mean travel over rough terrain where four-wheel drive would be an advantage. Utilization of 5000 miles annually.
4. The Mechanical and Electrical Engineering Departments both would like to have a small station wagon or light pick-up in which to transport their instruments to project sites and clients' offices. Estimated utilization: 14,000 miles annually (Mech Eng, 8000 miles; Elec Eng, 6000 miles).
5. The project engineers need a station wagon to carry themselves and a limited amount of equipment and drawings to construction and installation sites. Approximate utilization: 8000 miles annually.
6. The Materials Testing Laboratory would like a light runabout for transporting small quantities of materials and equipment. It would have limited use, but could also be utilized by the Special Projects Group, which would like an import-size station wagon to carry charts and special equipment to management meetings and project sites in the local area. Estimated utilization: 10,000 miles annually (Mat. Test Lab, 4000 miles; Spec Proj Grp, 6000 miles).

You find that the existing Willys Jeep traveled 17,000 miles per year over the past three years, and the station wagon traveled 24,000 miles per year.

You discuss vehicles with local car and truck suppliers, and list the vehicles you consider would best suit the company's needs. Major factors you take into account for each vehicle are:

Load capacity	Price
Seating capacity	Service and parts availability
Gross vehicle weight	Performance
Road clearance	Robustness
Gas consumption	Mechanical details (two- or four-wheel drive, etc.)

You prepare an informal investigation report in which you make specific recommendations and substantiate them with sound evidence.

NOTE:

The range of vehicles to be investigated is so large that a means must be found for grouping them. For example, you could evaluate:

The vehicles available from a specific manufacturer; or

The types of vehicles available within a specific range, such as payload or price; or

The range of vehicles suitable for a specific purpose (e.g., for the survey crew, for local deliveries, for construction supervision).

Alternatively, the evaluation could be divided between individuals or small project teams, each of whom would evaluate one group of vehicles. The results of their efforts could then be discussed at a meeting and presented in a briefing, as described in the Assignments section of Chapter 8.

Project No. 9—Relocating the Drafting Department

You are employed in the Engineering Department of Robertson Engineering Company, H.L. Winman and Associates' Canadian subsidiary. When Fred Stokes, the Chief Engineer, calls you into his office on 26 June, he assigns a department relocation project to you. In explanation he hands you an interoffice memorandum from the head of the company (see figure).

Mr. Stokes instructs you to assess the three recommended areas and to select one of them; to investigate the types of drafting equipment available, and to recommend instruments and furniture that should be purchased; to make a plan of the selected area, with furniture and equipment in position; and to prepare an informal report for him to submit to Mr. Robertson. He adds that Mr. Robertson will want to know the total cost for furnishing the new area, and a cost breakdown should be included in your report.

You then talk briefly to the Chief Draftsman, who furnishes you with the following information:

1. The blueprint machine will have to be replaced. The present one is an Ozalid Streamliner, about 14 years old. It might have a small trade-in value (no more than $150).
2. There are now eight draftsmen in the department, plus the Chief Draftsman. In your planning make provision for another two draftsmen who may be hired next year.
3. There are six ancient four-drawer wooden filing cabinets. The drawing cabinets are also very old and overcrowded. The current number of drawings is:

engineering   company

INTER - OFFICE MEMORANDUM

From: ......W.D. Robertson, President........ Date: .....22 June 1971......

To: ....F. Stokes, Chief Engineer.......... Subject: Relocation of Drafting
.... Department

As we discussed in last Thursday's management meeting, the area on the fifth
floor now occupied by the Drafting Department will become the location of a
Calibration Laboratory due to start operation in November. The Drafting
Department therefore will have to move out on 15 September.

I can offer you three locations to choose from [see rough sketch attached]:

"A" -- An L-shaped air-conditioned area of 1564 sq ft on the sixth floor,
         adjacent to the Engineering Department. It's in the center of the
         building, hence has no windows.

"B" -- A nearly square area of 1800 sq ft on the fifth floor, close to
         where the Drafting Department is now. It has windows on the west
         and south sides, which make it a little warm in the summer, and has
         a magnificent view of the river, Civic Buildings, and Memorial
         Park. It is not air-conditioned.

"C" -- An 80 ft long area of 1600 sq ft on the north wall of the sixth
         floor. Although not air-conditioned it remains reasonably cool in
         the summer. It has the best windows and natural light in the
         building, and is a very quiet area.

Since it will have to be a complete move for Drafting, I think this will be
an excellent opportunity to entirely re-equip the department. As you
mentioned at the meeting, most of the major drafting equipment was acquired
when we purchased Longman's Drafting Service eight years ago; it was old then
and has since been written off.

Will you therefore prepare a report recommending your preferred location for
the Drafting Department and the equipment it will require. I would like to
have your report on my desk before 15 August.

WDR:rk

Size A—8½ in. × 11 in.—8400
Size B—17 in. × 11 in.—7100
Size C—17 in. × 22 in.—4500
Size D—34 in. × 22 in.—2800
Size E—34 in. × 44 in.—1600

Number of drawings prepared annually: 1400

4. Existing drawing tables are trestles with drawers. Only the Chief Draftsman has a drawing machine (a Vard).

ROBERTSON ENGINEERING CO. - PROPOSED SPACE FOR DRAFTING DEPARTMENT

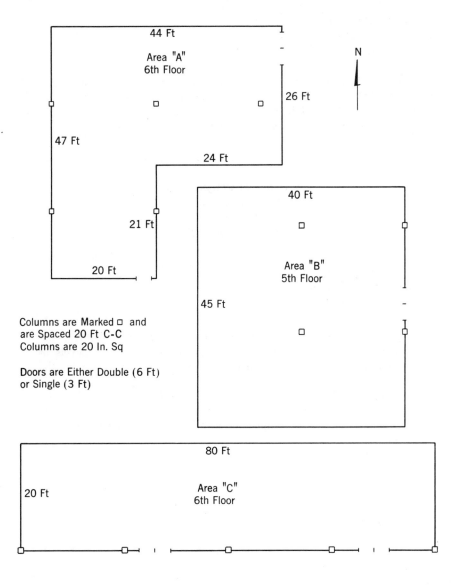

MULTIPLE-ASSIGNMENT PROJECT

Project No. 10—Rectifying Audio Level Problems

This multiple writing assignment traces a technical project from its inception through to its completion. It is divided into six parts, which will require you to write one each of the following:

1. A field trip report
2. An informal "thank you" letter
3. A letter requesting a price quotation
4. A letter-form investigation report
5. A project progress report
6. A project completion report

Introduction. H.L. Winman and Associates receives the following letter from Mr. D.C. Scrivener, Operations Manager of Radio Station DMON, Montrose, Ohio. Don Gibbon is appointed Project Coordinator, and you are assigned technical responsibility for the project.

RADIO STATION DMON
"The Voice of Montrose"
217 Vermont Avenue
Montrose, Ohio, 45287

18 October 197__

H.L. Winman and Associates
475 Reston Avenue
Cleveland, Ohio, 44104

Dear Sirs:

Since moving into our new office in the basement of the Waskeka Building, we have experienced sudden changes in audio levels at infrequent intervals and of short duration. These interfere with the technical quality of our programs.

Will you please investigate this problem, identify its cause, suggest possible remedies and recommend the most suitable, and provide me with a cost estimate and approximate time schedule for its rectification.

Yours truly,

D.C. Scrivener
Operations Manager

DCS:al

Part 1—Field Trip Report. You drive to Montrose on 21 October, where you find that Radio Station DMON rents the entire lower level of the Was- keka Building. It is a two-level structure, with the main floor occupied by Antrim Insurance Corporation. You introduce yourself to Mr. Scrivener and then start your investigation, making notes as you go along.

INVESTIGATION NOTES

1. Audio Level Measurements:
   a. Normal audio levels reasonably stable: fluctuations no more than ±2 dB (dB = decibel, a unit of sound measurement).
   b. Sudden drops of approx 7 to 10 dB (see table).
   c. Drops occur at random intervals (5 to 21 minutes apart).
   d. Drops occur night and day, more frequently during day; par- ticularly a problem last summer; less noticeable now.
   e. Measurements made with Tektronix Model 549 Storage Oscil- loscope.
2. Cause:
   a. Sudden temporary drops of approx 10–12 V in line voltage that exceed limitations of ±5% of 120 V.
   b. Checked incoming three-phase power line voltage supplied by Montrose Hydro: stable—only minor fluctuations (within 117– 123 V).
   c. Checked equipment in building; traced cause to compressors in air-conditioning equipment, which draws heavy load on start-up.
3. Possible Remedies:
   a. Power company install separate lines for air-conditioning equip- ment; or
   b. Install line voltage regulator (LVR) to smooth out bumps.
4. Will Remedies Work?
   a. Remedy (a): Probably; can't be guaranteed—depends on inde- pendence of lines and future loads.
   b. Remedy (b): Definitely; needs high quality regulator with fast response (suggest Vancourt 6 kVA solid state LVR with 5 cycle response time).
5. Cost:
   a. Remedy (a): $950 (estimate from Montrose Hydro).
   b. Remedy (b): $1400 (about $800 for regulator, plus $600 for installation).

You decide that you will recommend the LVR because its performance is guaranteed. It cannot be affected by sources beyond DMON's control, as might happen with a separate line.

Before leaving Montrose on 23 October you tell Mr. Scrivener that it will be relatively easy to resolve his problem, and that you will obtain firm price quotations before sending in your report.

Upon returning to Cleveland you write a field trip report addressed to Don Gibbon.

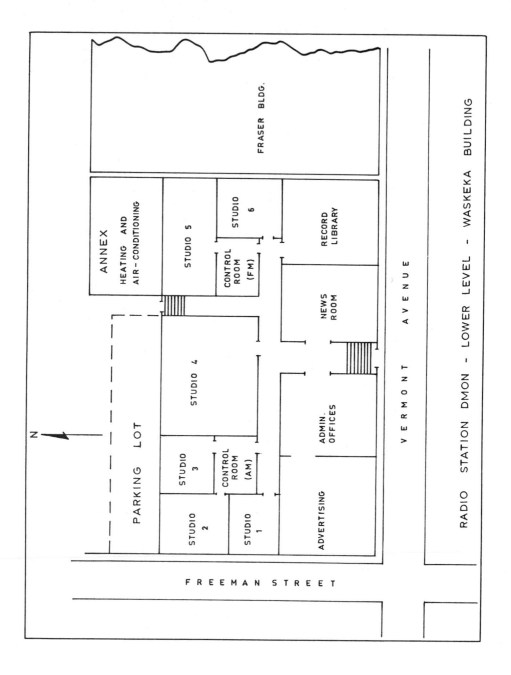

RADIO STATION DMON – LOWER LEVEL – WASKEKA BUILDING

H.L. WINMAN AND ASSOCIATES *CHECKS OF AUDIO AND LINE VOLTAGE DROPS* *RADIO STATION DMON—22–23 October 197__*					
	Time	*Audio Drop*	*Line Voltage Drop*	*Outside Air Temp. (F)*	*No. of Persons in Building*
Morning	0912	6 dB	8.6 V		
	0923	8 dB	10.7 V		
	0938	9 dB	11.4 V	48°	66
(0900–1000)	0947	8 dB	10.5 V	(partly cloudy)	
	0959	8 dB	10.4 V		
	1431	9 dB	11.7 V		
	1438	7 dB	9.7 V		
	1444	7 dB	9.5 V		
	1449	8 dB	10.2 V		
Afternoon	1455	7 dB	9.3 V	64°	71
(1430–1530)	1501	8 dB	10.6 V	(sunny)	
	1508	8 dB	10.4 V		
	1514	9 dB	11.3 V		
	1520	10 dB	12.1 V		
	1526	8 dB	10.3 V		
	2012	7 dB	9.8 V		
Evening	2027	8 dB	10.1 V	56°	12
(2000–2100)	2043	8 dB	10.6 V		
	2056	8 dB	10.5 V		
Night	0113	8 dB	10.3 V		
(0100–0200)	0131	7 dB	9.9 V	44°	4
	0148	7 dB	9.6 V		

Part 2—Letter of Thanks. On your second evening in Montrose, the Program Manager at DMON invites you to dine with him and his family at a local night club known as The Red Barn. His name is Jimmy Doyle, his wife's name is Nannette, and his 20 year old daughter's name is Barbara. Write an informal thank you letter to Jimmy Doyle. In addition to thanking him for the evening out, ask him to extend your thanks to Don Smythe, a technician at the radio station who came back at night to assist you, and rounded up additional equipment you needed.

Part 3—Request for Quotation. You write a letter requesting a price quotation for installing the LVR at Radio Station DMON and send it to three electrical contractors in Montrose. You specify the exact type of LVR and inform the contractors that they have to include it and the price of all materials in their quotation. You also stipulate that the work must be done in three nights between midnight and 6 a.m., and that maximum off-air time allowed for hookup is one hour. Closing date for bids is two weeks from the date of your letter. The LVR has to be installed on the main control panel, which is in the building annex (see diagram).

You need write only one letter, and address it to CanForm Electric Company, 212 Wall Street, Montrose, Ohio, 45287. Sign the letter "for" D. Gibbon.

Part 4—Investigation Report. The lowest quotation received is from CanForm Electric Company, whose letter appears below:

CanForm Electric Company
212 Wall Street
Montrose, Ohio

November 12, 197__

H.L. Winman and Associates
475 Reston Ave.
Cleveland, Ohio, 44104

Attention: Mr. D. Gibbon
                    Coordinator.

Dear Sir:

Installation of LVR—Radio Station DMON

In response to your invitation to bid on the installation of a 6 kVA Vancourt Line Voltage Regulator with 5 cycle response at Radio Station DMON, we submit the following quotation:

Materials	$	835.00
Installation		580.00
	$	1,415.00

This price will remain firm for three months from the above date.

Yours truly,

L. Minsky
Manager

LM/bls

You can now write the letter-form investigation report to Mr. D.C. Scrivener at DMON. Suggest that your company would also like to supervise the installation if he accepts your recommendation. If you need to mention H.L. Winman and Associates' fees, they are:

For carrying out the investigation:	$ 525.00
For supervising the modification (if so requested):	$ 420.00

Again, sign the investigation report "for" D. Gibbon.

Part 5—Project Progress Report. On 26 November you receive a letter from D.C. Scrivener authorizing H.L. Winman and Associates to supervise the installation of the LVR specified in your report. The dates stipulated for the project are 12 to 14 December. Mr. Scrivener ends his letter with this statement:

> On completion of the project will you kindly provide me with a report describing the reduction in line voltage fluctuations achieved by the modification.

You advise the contractor on 26 November, by a telegram in which you refer to your request for quotation and his bid, to proceed with the installation.

On 29 November Lou Minsky, the manager at CanForm, telephones to say that Vancourt cannot deliver the specified 6 kVA LVR until 18 December. He has a Vancourt 7.5 kVA LVR in stock which he can supply for the same price. Will you accept the change? You advise him that the 7.5 kVA LVR is acceptable providing it meets all other specifications. He acknowledges this.

During the installation you keep a project diary:

INSTALLATION PROJECT NOTES
(RADIO STATION DMON)

11 Dec. 1. Arrived Montrose by car at noon
2. Checked with contractor: ready to start at midnight
3. Checked with Scrivener at DMON—OK to start

12 Dec. 1. Contractor arrived at midnight
2. Problems getting to power junction box. Key to power control panel left at telephone switchboard, but key to annex not available. Obtained from Maintenance Engineer's home: woke him up at 0130 to get it!
3. Rechecked audio and voltage fluctuations (a before-modification check):

Time	Audio Drop	Voltage Drop
0203	7 dB	9.4 V
0222	8 dB	10.3 V

Equipment used: Tektronix Model 549 Oscilloscope
4. Contractor installed mounting rack for LVR

13 Dec. 1. Complaint from Maintenance Engineer, routed through D.C. Scrivener: "Contractor awfully messy—couldn't he clean up area after night's work?" Passed message on to contractor's two-man team: Ken Waldo (Electrician) and Danny Belinsuk (helper)
2. Contractor installed LVR on rack, plus cabling and new junction box

3. Left note for Scrivener to anticipate brief power cut about 3 a.m. tomorrow (14 Dec)

14 Dec. (night)

1. Arranged with night staff to schedule 30-min off-air break from 0300 to 0330
2. Contractor completed all but final hookup by 0220
3. Arranged for temporary generator to start up
4. 0300—off air, started hookup
5. 0318—completed hookup
6. 0320—Contractor made voltage and continuity checks—OK
7. 0330—Back on air
8. 0415—Conducted postmodification audio and voltage level checks with Tektronix 549 Oscilloscope:
   - Audio: Less than 1.5 dB variation from nominal level
   - Voltage: Less than 2.5 V variation from nominal level (120 V)
9. 0447—Noted sudden momentary drop in audio level. Decided to monitor levels between 0500 and 0600:

Time	Audio Drop	Voltage Drop
0508	5 dB	7.7 V
0523	4 dB	6.8 V
0544	4 dB	6.5 V
0558	4 dB	6.2 V

   From 0515 checked time air conditioning compressor cut in; corresponded with audio and voltage drops at 0523, 0544, and 0558
10. Initial reaction—LVR not functioning properly; providing only partial regulation.

At 8 a.m. on 14 December you call Lou Minsky and together you check the specifications for the 7.5 kVA LVR. You discover that it is an older model (not solid state) and that its response time is too slow to react to the sudden drops caused by the air-conditioning compressor. Minsky calls the supplier (Vancourt), who advises him that the original model ordered has just arrived in Cleveland and that he can have it shipped to Montrose by 6 p.m. the following evening (15 December).

Minsky tells you he will need two additional nights to change over the LVR (mounting details are different), which means the job will not be finished until 6 a.m. on 17 December, three days later than scheduled. He also says that the labor for the two additional nights will be an extra charge to the contract price. You deny this, insisting that the LVR he supplied failed to meet your specifications. He counters this by replying that you approved the change in a telephone call on 29 November. You agree to meet with him after the project is complete to hash out the details of extra payment.

You inform Scrivener of the delay, and the reason for it, then return to your motel to write a progress report to D. Gibbon.

Part 6—Completion Report. The changeover goes faster than planned and the job is finished at 2:45 a.m. on 17 December. You measure audio and voltage levels and find that fluctuations are no greater than ±1.9 V. These are fully within tolerances acceptable to Mr. Scrivener and Jimmy Doyle. The audio drops are now no greater than 1.3dB.

At a meeting between you and the contractor on the afternoon of 17 December, the contractor agrees to accept the original contract price because both your request for quotation and his bid specify a 5 cycle response time.

You drive back to Cleveland on 18 December. The following day you write the project completion report requested by Mr. Scrivener.

5

Informal Reports
Describing Ideas and Concepts

In the previous chapter I discussed reports that deal almost entirely with facts, known data, or tangible evidence. Only at the end of any of these reports does the writer have the opportunity to introduce new ideas or concepts. The reports in this chapter have a reverse flow: they start with an idea or a concept and then develop it, using both tangible and intangible evidence to reach a conclusion based on facts and probabilities. They are written in a fluent narrative style that is both persuasive and convincing; their writers have new concepts and techniques to present, and they want to help their readers understand their line of reasoning.

EVALUATION REPORT

The similarity between evaluation reports and investigation reports frequently causes technical writers to confuse one with the other. As a comparison of the flow diagrams in Figures 4–5 and 5–1 indicates, the approach is quite different, although the basic report form is similar.

In an evaluation report, the writer starts with an idea or concept that he wants to develop. He establishes guidelines that will keep its development within prescribed bounds, and researches known data to analyze the idea. He then evaluates his concept in the light of the parameters he has set and the data he has amassed, conducting tests and analyses to prove or disprove his theories. At the end of his study he is able to draw the conclusion that his concept either is or is not feasible, or perhaps feasible in a modified form.

The Robertson Engineering Company report on training methods in Figure 5–2 is an evaluation report because it deals with ideas and concepts

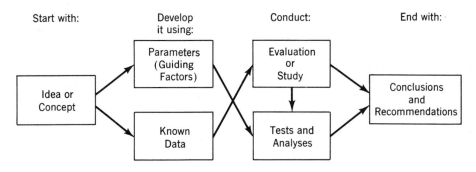

Fig. 5–1   Basic Flow Diagram—Evaluation Report.

rather than with facts and events. The *idea* is management's belief that new training methods will increase production; the *known data* are the four training methods discussed by Fred Stokes; the *evaluation* is his assessment of each method, or combination of methods, in the light of the company's training needs. The standard parts that shape the narrative of this report are, however, the same as those for the investigation report (see Chap. 4, p. 92 and 93).

The comments below identify the standard parts that Fred Stokes has used, and discuss briefly how they have helped him fashion an effective report. His use of paragraph numbers and headings not only serves as a useful model for us when we want to do the same, but also provides me with an easy means of reference.

Paragraph 1 is a brief *summary* of the evaluation and main conclusions.

> Rather than write a summary of the main points of the whole report, Fred homes in on the one point of immediate concern to management: How much is all this training going to cost? He makes it clear that improved training methods are mandatory, and that it is going to cost the company X or Y dollars. He has followed the first rule of summary writing: Identify what is most important to your readers *before you write.*

Paragraphs 2 and 3 provide *background information* and details of events that led up to the evaluation.

> Fred knows that the person to whom the report is directed will already be aware of much of this information, but he also recognizes that there may be other readers to whom it may be new.

Paragraph 4 is a *guiding factor* that limits the extent of the evaluation.

> Fred probably discussed these limitations with Wayne Robertson when he first established his terms of reference. No engineer or tech-

| robertson | engineering company | INTER - OFFICE MEMORANDUM |

From: F. Stokes, Chief Engineer   Date: 20 November 1971

To: W.D. Robertson, President   Subject: Updating Our Training
                                                 Methods

1. If we are to meet our production schedules with the semi-skilled labor currently available, during the next year we shall have to spend $26,500 for adequate but stereotyped training methods, or invest $51,000 in a unique system that will meet all our future needs and have many additional uses. After the first year operating costs will be either $20,000 or $23,000 per year.

2. Over the past five years the scientific instruments we manufacture have become increasingly complex and have demanded highly sophisticated manufacturing techniques. At the same time our source of skilled labor has decreased, so that we have had to use semi-skilled and unskilled labor for many positions on the assembly lines. This has resulted in a gradual slowdown of production, an increase in the number of manufactured units rejected by Quality Control, and a consequent increase in manufacturing costs.

3. At the management meeting held on 20 September 1971 it was recognized that a better training program than we now have would do much to improve the manufacturing skills of our employees. I was requested to evaluate alternative training methods and to recommend a training technique that might increase productivity.

4. I have confined my report to an analysis of training methods that could be used in our plant. Development of a specific training program I will discuss in a separate report to be prepared after the training method has been selected.

Analysis of Industrial Training Methods

5. There are four basic training methods, ranging from inexpensive established methods to costly new techniques, that we could utilize to train our staff. The suitability of each is discussed generally here and is supported by a comparison chart attached as an appendix.

6. On-the-Job Training [OJT]

   6.1 This time-honored training method is used extensively in industry and is the method currently used on our production lines. It calls for the trainee to learn a manipulative task in which a sequence of movements has to be mastered. The trainee "learns by doing" under the watchful eye of an instructor, who often is another fully-trained employee. When he has learned the task the trainee is given a proficiency test to assess whether he can work on his own.

Fig. 5–2   An Evaluation Report.

6.2　Every time we hire a new employee, or transfer an existing employee from one production line to another, he is given OJT which continues until he is fully able to perform his new task. It is an effective training method, but is slow and has many hidden costs. The trainees are poor producers for a long time, and a higher-than-normal proportion of their work is below standard and has to be rejected. To continue relying solely on OJT would be to perpetuate a training method that has already been proved to be inadequate for our needs.

7.　Classroom Instruction [C.I.]

7.1　The classroom situation is the form of training most of us know best. Employees recognize this type of teaching and, even though some may resent the schoolroom atmosphere it evokes, they usually respond quite readily to it. Courses follow a basic pattern, with the instructor setting objectives, preparing instructional material to meet them, then presenting lectures. An end-of-course test demonstrates whether the trainees have learned enough to be able to perform their tasks under supervision.

7.2　To be effective, C.I. demands the proper atmosphere, a good instructor, and receptive trainees. This means providing space for a properly equipped training room rather than just a corner of the workshop; it means keeping an instructor who can teach specialist subjects on the payroll, or drawing instructors from the supervisory line staff [this must be done carefully because not all supervisors make good instructors]; and it means making sufficient employees available for training at the same time.

7.3　The cost of classroom instruction is high in relation to the degree of training it provides. It is particularly suitable for teaching theory, but of less value when purely manipulative skills must be learned. Hence it must be supplemented by practical training such as OJT.

8.　Programmed Instruction [P.I.]

8.1　Programmed instruction is a relatively new technique that only for the past 12 years has been generally accepted as a useful training aid. Because it permits students to work alone and at their own speed, it is invaluable for teaching routine tasks to very small groups or to individuals. The type of programmed instruction most of us recognize is the programmed textbook that teaches a small piece of information at a time, tests the trainee on what he has just learned, provides correction if he still has not understood, and then proceeds to the next piece of information. Trainees progress quickly or slowly through the program, depending on their previous knowledge and ability to learn rapidly. They are tested periodically, and at the end they are given a proficiency test. Except for monitoring the trainee's progress and marking the final test, an instructor is not needed.

8.2   P.I. textbooks are ideal for teaching specialist subjects and new techniques, but seldom can be used to teach manipulative skills. Therefore, any P.I. manuals that we produce would have to be supplemented by some verbal instruction and practical training. Preparation of P.I. materials is also expensive and very slow -- too slow to meet the rapid training needs of many of our production lines.

9.   Closed Circuit Television [CCTV]

9.1   As an educational training aid, CCTV is rapidly gaining popularity with departments of education across the country. Though its use until now has been limited primarily to large schools [because its initial cost is high], there is no reason why the new, less expensive systems should not be used in industry. Training lessons showing manufacturing processes, assembly techniques, packaging methods, etc, could be prepared as TV "programs" and stored on videotape ready for playback to trainees whenever the need arises.

9.2   The initial cost of setting up CCTV would be much greater than for the other three methods, and operating costs would remain moderately high. However, the advantages of CCTV would out-weigh many of the objections to its high cost. The familiarity of TV, combined with its uniqueness as an inplant teaching method, would promote learning and employee acceptance of training courses. The image of a progressive company that it would convey would be invaluable for building employee morale. From a training standpoint we would be able to show programs repeatedly, either to large classes or just to individuals, at virtually no cost; we could demonstrate the right and wrong ways of doing things; and we could look at tiny objects and en-large them for all to see simultaneously. It would also give us the ability to prepare and record programs quickly, so that teaching of new assembly methods would be able to start very early in future programs.

9.3   However, CCTV would not be able to stand alone as a training medium. We could use it to teach theory and to demonstrate manipulative skills, but it would not replace the final step: practice. This would still have to be provided by OJT.

Selection of a Training Method

10.   The training method that we select must teach both theory and manipulative skills, and also provide opportunity for practice. Since none of the training methods I have discussed meet all of these require-ments, we will have to use a combination of methods that provides the best training for the least cost.

- 4 -

11. OJT must be retained because it is the only method that provides the practice that is essential before a trainee can become fully productive. But because OJT has so many hidden costs, it must be reduced to the minimum time possible by combining it with the most effective teaching method.

12. P.I. can be eliminated because of its high costs, slowness of preparation, and inability to teach complicated manipulative skills.

13. Either C.I. or CCTV could be combined successfully with OJT to provide a comprehensive training program. Both would significantly reduce the time required for OJT. The C.I./OJT combination would cost $26,500 the first year, and $20,000 per year thereafter. The CCTV/OJT combination would cost $51,000 the first year, and $23,000 per year thereafter. CCTV, however, offers many peripheral advantages that C.I. does not.

14. I consider that the advantages of the CCTV/OJT combination outweigh those of the C.I./OJT combination. The image that CCTV would convey and the enthusiasm that it would spark are intangible factors that cannot be measured in dollars. CCTV equipment could also be used for activities such as sales seminars, management training, time and motion studies, and advertising, so that maximum value would be obtained from our investment.

15. I recommend that we update our training methods by installing a CCTV system at a cost of $28,000, and budget for an annual operating cost of $23,000. If this plan is adopted I will prepare specifications for purchasing the equipment, and will develop a detailed training implementation program.

F. Stokes

COMPARISON OF FOUR TRAINING METHODS

	OJT	CLASSROOM (C.I.)	P.I.	CCTV
What is initial purchase and set up cost?	Nil	$6,500	$1,000	$28,000
What are annual operating costs (including instructors' salaries)?	$2,000	$18,000	$46,000	$21,000
How effective is training method?	Very effective	Quite effective	Very effective	Extremely effective
How quickly do trainees learn?	Slowly	At moderate speed	Varies; depends on student	Very quickly
How readily do trainees accept training method?	Readily	Very readily	Varies; resisted by some	Very readily (promotes learning)
Do trainees have to be removed from production line to take training?	No	Yes	No	Yes
How quickly can a specific training program be set up?	Very rapidly	Fairly rapidly	Very slowly	Fairly rapidly
Can students work independently or must they work in groups?	Independently	Groups	Independently	Either
How adaptable is training method?	Quite adaptable	Fairly adaptable	Rather rigid	Very adaptable
Can it be used for teaching most types of skills?	No; mainly practical skills	No; mainly theory training	No; depends on subject	Yes
Are there any limitations to using method?	No	No	Yes; trainee must be able to read English	No; equipment can be used for other than training purposes

- 5 -

Fig. 5-2   An Evaluation Report.

nician should attempt to write a report without first clearly establishing his objectives.

Paragraphs 5 through 9 comprise *investigation details*. Paragraph 5 outlines the *investigation method*, while paragraphs 6, 7, 8, and 9 each evaluate one of the training aids.

> Fred's analysis is interesting and logical. He uses a three-step approach that makes the evaluation of each training method a small report in itself: (1) a brief description of the training method and how it is used; (2) a discussion of its application to industry and the implications of using it; and (3) a conclusion that sums up the value of the training method to Robertson Engineering Company.

Paragraphs 10 through 13 *discuss evaluation results.*

> By first introducing the guiding factor that will most influence his selection, Fred is able to eliminate one training method quickly, identify another that has to be used, and establish that either of the two remaining methods can be used with it. His discussion is unquestionably logical, and follows naturally from the conclusions drawn at the end of the four analyses.

Paragraphs 14 and 15 state the main *conclusion* and make a *recommendation.*

> Fred now identifies which method he thinks is most suitable, explains why, and recommends in the first person what action he thinks management should take.

Note that Fred Stokes does more than simply describe each training method in the investigation details section. By also discussing the advantages and disadvantages of each method, and then skillfully inserting a persuasive concluding statement at the end of each description, he conditions you to agree with its acceptance or rejection *before* you read his conclusions. This approach is used frequently in reports dealing with ideas and concepts, particularly when the writer wants to influence the reader's acceptance of his report. Fred Stokes, for example, wants to convince Wayne Robertson to invest in the more costly training method rather than the cheaper one. (This does not mean that a report writer can allow bias to creep into his work. His evaluation must always be objective, even though it may seem to be persuasive.)

The difference between an evaluation report and an investigation report can be seen more readily if I compare two similar situations that call for different types of reports. In the previous chapter I described a situation in which Harvey Winman *has decided* to purchase an executive jet aircraft, and assigns an engineer to *investigate* which type of aircraft would be most suitable.

Now suppose that Harvey only *has an idea* that an executive jet might be useful to him and his company. He assigns an engineer to *evaluate* whether purchase of a private aircraft would be practicable, and possibly to recommend what action should be taken.

This engineer's first step is to establish parameters: How often would an aircraft be used? How many persons would use it? What sort of budget allocation can be made available for its purchase and operation? What is the likely route length? He then identifies the types of aircraft available and collects data on each, such as purchase price, operating costs, and performance. He also looks at leasing as an alternative to purchase. Armed with all this data, he evaluates the advantages of having an executive jet under varying conditions, developing cost analyses for each situation. (In practice this would be done by computer, with the results possibly expressed as a series of benefit-cost ratios.) From the evaluation he establishes whether the idea is economically feasible; if it is, whether it would be better to purchase or rent an aircraft; and which type of executive jet would be most suitable for the company's operations. The report he submits to Harvey Winman would be an evaluation report because it starts by outlining an idea, develops it by using known information, and ends by drawing definite conclusions based on an objective analysis of concepts, data, and parameters.

FEASIBILITY STUDY

A feasibility study is very similar to an evaluation report. It also starts with an idea or concept, and then develops and analyzes it to assess whether it is technically or economically feasible. The chief difference lies in the name and application of each document. An evaluation report is generally based on an idea that is originated and evaluated within the same company; hence it is nearly always informal. A feasibility study is normally prepared at a slightly higher level: the management of company A requests company B to conduct a feasibility study for it, because company A does not have staff experienced in a specific technical field. It is unlikely, for instance, that a company engaged solely in wholesale distribution of dry goods would have the capability to assess the feasibility of purchasing an executive jet. The company would seek advice from a firm of management consultants, who would submit their findings in a report called a feasibility study.

When a feasibility study is prepared for a client, the document that results may be either a letter or a formal report. A letter is used when the project is relatively small and there is no need to present the information formally. If the evaluation of training methods presented as a memorandum report in Figure 5–2 had been prepared for a client, it would have been called a feasibility study and written as a letter report. There would have been a few minor changes in narrative and format, but the basic information would have been the same. A formal report is used for more comprehensive projects and for special clients, as in the case of the "Evaluation of Sites for Montrose Residential Teachers College" at the end of Chapter 6. The margins of difference among the feasibility study, the evaluation report, and the technical brief (described in the next section) are so narrow that it is frequently personal preference that dictates which label is applied to any one of them.

TECHNICAL BRIEF

The purpose of a technical brief is to introduce a new idea to a reader and to develop it in sufficient depth to enable him to assess its practicability. The idea should be innovative, logical, and practical, and the brief should indicate that it is worth the reader's time to pursue the subject further. It can be originated by almost anyone and cover almost any topic or situation: a foreman with a novel scheme for calibrating test instruments, a salesman with ideas for strengthening a product line, a technician with a new method for testing mechanical components, and an engineer with plans for a unique product can all present their ideas to management in a technical brief.

Scientists, supervisors, technicians, engineers, foremen, and technologists write memorandums every day to their managers suggesting new ways of doing things. These memorandums become technical briefs if their writers have developed their ideas sufficiently to convince management that their proposals are sound. A memorandum that introduces a new idea without demonstrating fully how it can be applied is no more than a suggestion. A memorandum that introduces a new idea, discusses its advantages and disadvantages, demonstrates the effect it will have on present methods, calculates cost and time savings, and finally suggests what further developments might accrue, proves that its originator has done a thorough research job before attempting to put pen to paper. Only then does a suggestion become a technical brief.

The technical brief in Figure 5–3 promotes development of an audiovisual training aid called the APL System. Although its originator, Ron Brophy, could have written his brief entirely as an interoffice memorandum, he has chosen to prepare it as a separate document and to preface it with a short covering memo to Jim Perchanski, his department head. This arrangement allows additional copies to be printed without the memo for distribution to readers within and outside H.L. Winman and Associates. The memorandum opens the door for further development by suggesting where the APL System can be tested, and how much money would be involved in initial development.

Within the brief, Ron describes why there is a need for such a system and establishes parameters for its development. Since he is dealing with a concept, he discusses how the system can be applied, and describes clearly

H. L. Winman and Associates

INTER - OFFICE MEMORANDUM

From: R. S. Brophy   Date: 13 January 1971

To: J. Perchanski   Subject: New Product Design for the
                                                     Education field.

The attached technical brief describes "The APL System", a method of Audiovisual Programed Learning that I have devised. I believe it could be devel -
oped into a marketable product for the education field.

Its development grew out of a conversation I had with Les Walters, head of
the Training Aids Department of Montrose Community College, when we were
designing modifications for their remote control room. He would be interested
in evaluating a prototype system in a classroom situation, which would give
us a chance to assess its potential without excessive development costs. An
initial system for up to 16 students could be built for $2300.

RdB.

Fig. 5-3   A Technical Brief.

H. L. Winman and Associates

PROFESSIONAL CONSULTING ENGINEERS
475 Reston Avenue-Cleveland, Ohio, 44104

Technical Brief

DESIGN FOR AN INEXPENSIVE AUTOMATED TEACHING SYSTEM

Prepared by

Ron S. Brophy
Development Engineer
Electrical Engineering Department

A study of the automated teaching machines currently available to the educational field or still under development indicates that they are divided into two general groups. The first comprises a fairly wide variety of simple, fairly cheap machines that can be used by only one person at a time and that normally require a written response. The second is a smaller range of very sophisticated, highly expensive systems that simulate complex conditions and situations, often presenting information on a television-type screen; these vary from one-operator units to complete classroom installations.

Between these groups there is a need for an inexpensive, basically simple system that can be used by a whole class simultaneously. "The APL System" described below falls into this category.

General Description of the APL System

Audiovisual Programed Learning is a system that combines the simplicity of audiovisual training aids with existing programed learning techniques to provide an effective method for automatic teaching of groups of students. Its cost for a 50-student system would be less than $3500. Significant features are:

1. It is fully automatic; once set in operation it continues through a complete lesson of up to one hour's duration without attention.

2. It demands continual student participation.

3. It monitors each student's progress separately, checking on student response as the lesson progresses and providing a quantitative assessment of the student's total subject assimilation at the end of the lesson.

4. It corrects errors as they occur; if a student's response to a question is incorrect, he is immediately told of his error and given further instruction to correct the fault before proceeding to the next step.

System operation consists of a series of 35-mm slides projected onto a screen in turn and accompanied by a spoken commentary, heard by each student through a pair of headphones. Periodically the slides and commentary contain multiple-choice questions that the students have to answer by depressing one of three pushbuttons corresponding to three possible answers. A multiple-track commentary then informs each student separately whether or not his answer is correct and, if incorrect, provides further instruction to correct his knowledge. Simultaneously, a counter registers his total number of correct responses.

Equipment Description

A complete APL System comprises:

* At the teacher's position: a tape recorder reproducer (wired for "Playback" only and modified for simultaneous reproduction of all four channels of 4-track quarter-inch tape), an automatic slide projector, and a screen.

* At each student position: a pair of headphones with an individual volume control, three pushbuttons for selecting answers to questions, and a small 3-digit counter for recording the number of questions answered correctly.

These components are itemized on page 4 and shown as part of an overall system in the "APL System Block Diagram" on page 5.

System Operation

The slide tray and the 4-track tape are loaded into the projector and the tape deck. At each student position the counter is reset to "000" (counters are the pushbutton "total reset" type, therefore cannot be advanced by the students). The output from each of the four tape recorder tracks is applied as follows:

1. The output from track 4 is fed only to the slide projector. This track carries "beeps" (low frequency tone signals, 100 to 200 Hz) which cause the projector to change slides at the correct points in the spoken commentary.

2. The audio outputs from tracks 1, 2 and 3 are fed to pushbuttons "A", "B" and "C" at each student position. Individual students hear only the commentary carried by the track they have selected by depressing one of the pushbuttons.

3. The counters at each student position are triggered by inaudible low frequency tone signals inserted immediately after each question but placed on only one of the three audio tracks (always the track carrying the commentary signifying that a "correct" answer has been selected). Thus counters advance one digit only at those student positions where the depressed pushbutton corresponds to a correct answer.

Initially, the same commentary is carried on all three tracks. Slides projected onto the screen change as the commentary progresses, until a point is

- 2 -

reached when the program needs to test student assimilation of the instruction.

The next slide projected onto the screen poses a question and provides three possible answers, coded "A", "B" and "C". The commentary asks the students to read the question and to depress the pushbutton corresponding to the answer they consider correct. (For the example below, assume answer "B" is correct).

Just before the answer commentaries start, a tone signal inserted on track 2 (answer "B") causes the counters of all students who have selected the correct answer to advance one digit. Different commentaries are then carried simultaneously on the three tracks:

Tracks 1 and 3 (Answers "A" and "C"	The commentary informs the students that the answer they selected is not correct. It then describes why it is incorrect and why they should have selected answer "B".
Track 2 (Answer "B")	The commentary informs the students that they have selected the correct answer. It then states briefly why the answer is correct in case a student was unsure of the correct answer and hopefully selected "B".

Further instructional slides and a general commentary common to all three tracks continue until subject assimilation again needs to be checked. This procedure continues throughout the lesson, instructional material alternating with test questions to provide positive feedback of student response. At the end of the lesson individual results are read from the counter at each student position.

Conclusion

The foregoing represents a first assessment of a simple, practical method of whole-class automated teaching. Because it uses equipment and methods that are already accepted as valuable classroom training aids it should not meet with the resistance usually accorded by the teaching profession to automated teaching systems. The combination of relatively low price and simplicity of operation should make it attractive to both teachers and to school boards. Thus I believe it has the prerequisites of a marketable product.

List of Major Components

4-Track Tape Recorder   Any commercial 4-track tape recorder
                                     modified for simultaneous reproduction
                                     of all 4 channels; alternatively, an
                                     industrial 4-channel tape recorder.

Slide Projector   Kodak Carousel AV-900, Ident. No. B914,
                                     with Ektanon 4-inch f/3.5 lens.

Detector Circuit for   Local design and manufacture; al-
Slide Projector   ternatively, Kodak Programer B65.

Pushbuttons   Switchcraft 3-button multiswitch Part No.
                                     7300 (or Part No. 17300 if illuminated
                                     pushbuttons are used); alternatively,
                                     Microswitch 3-position coordinated
                                     Manual Control-Honeywell Catalog 69.

Counters   Veeder Root Series 1770, 3 figures, dc
                                     operation, pushbutton reset.

Counter Detector Circuit   Local design and manufacture (built as
and Headphone amplifier/   one unit, integrally with pushbuttons).
volume control

Instructional Material   For each program (lesson) a 'package'
                                     comprising a Kodak Carousel Universal
                                     Slide Tray containing up to 80 slides,
                                     plus tape commentary on a plastic-
                                     cartoned reel.

- 4 -

Fig. 5-3   A Technical Brief.

130

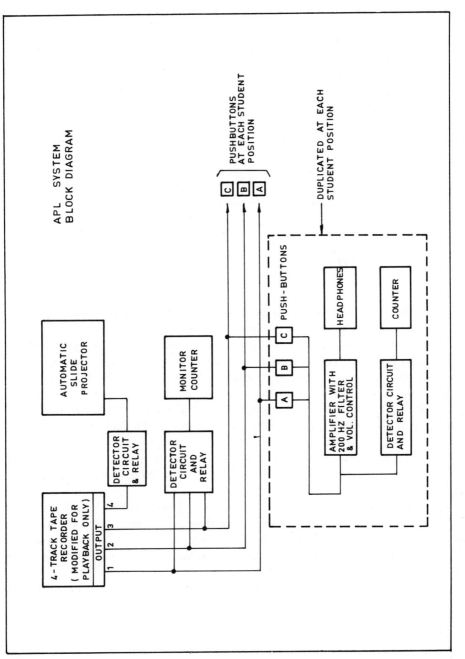

Fig. 5-3 A Technical Brief.

how it works. Then, when he has presented enough evidence to convince the reader of its capabilities and practicability, he draws a brief conclusion encouraging further development. His approach is generally similar to that used for an investigation report (Chap. 4).

Not all technical briefs are generated through the writer's own initiative. An idea may spring unexpectedly out of a conversation between engineers and technologists, resulting in one person being asked to prepare a brief for management's perusal. Or an innovative technician may construct a test jig simply to speed up part of his work, and it proves so effective that he is prompted by fellow technicians to "write it up" for the boss. His write-up, if he develops it properly, becomes a technical brief.

Formal technical briefs occasionally are written at intercompany level, and sometimes between companies and government departments, but these are far less common and not likely to be encountered by the newcomer to industry.

TECHNICAL PROPOSAL

A technical proposal is prepared by a company that seeks to impress or convince another company of its technical capability to perform a specific task. The best known technical proposals are the impressive documents prepared for multimillion dollar contracts. Less known but equally important are the many short, informal proposals covering smaller tasks. A proposal is classified as informal when it is written as a letter that simply describes how a task would be tackled and establishes the proposing company's ability to do the work. Letter proposals are used when the subject matter is relatively thin, when the work is routine, or when the dollar value of a project is low. The importance attached to a technical proposal is often geared to the price tag of the project that may result from it, and the possibility of additional work that it may engender.

I do not intend to cover preparation of high level technical proposals in this book. They are normally impressive documents prepared under extreme pressure, and call for techniques beyond the scope of most courses you are likely to encounter before you enter industry. If you eventually become seriously involved in technical proposal writing, I recommend you read Chapter 2 of Ehrlich and Murphy's short textbook.[1] Until then, I suggest you prepare only informal technical proposals using the letter form suggested for the investigation report.

[1]Ehrlich, Eugene and Daniel Murphy, "The Technical Proposal," in *The Art of Technical Writing* (New York: Thomas Y. Crowell Company, 1964), p. 10. (Also available in paperback: New York, Apollo Editions, A–226, 1969.)

ASSIGNMENTS
EVALUATION REPORTS AND FEASIBILITY STUDIES

Project No. 1—An Electrostatic Copier for Engineering

You are employed in the Engineering Department of H.L. Winman and Associates. Ian Bailey (your department coordinator) comes to your desk and asks you to join him for a short meeting in Mr. Dawes' office.

Martin Dawes (your department head) outlines the reason for the meeting: Because the office copying equipment that serves the whole company is badly overloaded, he has decided to install an additional office copier in the Engineering Department. He asks you to investigate the electrostatic-type copiers currently available, to prepare a report of your findings, and to recommend the make of copier you consider most suitable. As he speaks, you jot down the following factors that will guide you in making your selection:

1. The new copier will probably make between 8000 and 10,000 copies per month.
2. The total monthly cost for rental, labor, operating expenses, and maintenance should be no more than $350.
3. The new copier should be rented rather than purchased.
4. It should be installed "fairly soon," preferably within a month to six weeks.

As the first step in your assignment you walk down to the front office to examine the existing office copier. You talk to Rick Davis (the Office Manager), who tells you that his machine is the "Little Gem" manufactured by the Bel Copying Company. It has been in use for four years and has given consistently good service. But as the workload has gradually increased, its comparatively slow speed of only seven copies per minute has proved to be a bottleneck, particularly on Friday afternoons, when everyone wants his work processed and mailed before the weekend.

At first glance the Little Gem seems to be a cheap machine. There is no rental charge, but the manufacturer bills H.L. Winman and Associates at a basic rate of 3½¢ for every copy made (a counter on the machine counts the copies). When costs for materials and labor of 1½¢ per copy are added to the basic rate, the total cost per copy amounts to 5¢.

Rick Davis tells you that almost two-thirds of the work done by the Little Gem comes from Engineering. If Engineering rents its own machine, his workload will drop to a manageable 6000 copies per month.

You then contact the manufacturers of electrostatic office copiers and examine 30 different machines, 27 of which you discard as unsuitable for various reasons: too expensive, too slow, unable to copy pages from bound books, and so on. The remaining three machines you compare in a "Table of Comparative Data." Each is built by a different manufacturer, and each has a basic rental charge varying from $50 to $125 per month. There is an additional cost per copy of 2¢ or 3¢, which is made up of:

	Model A (per copy)	Models B & C (per copy)
Paper	½¢	½¢
Printing "powder"	½¢	½¢
Labor	1¢	½¢
Counter charge per copy	—	1½¢
Total	2¢	3¢

All three equipment manufacturers tell you that practically everyone underestimates the quantity of copies that will be made, and advise you to plan on making at least 20% more than you have estimated for the first year. They add that you can expect this quantity to increase by 15% annually during the next three years.

You also learn how an electrostatic copier works. The original is photographed by a camera which projects the image onto a rotating drum. This drum is coated with selenium and charged electrostatically. Where the image projected on the drum is white, the electrostatic charge is dissipated; where the image is black or colored, the electrostatic charge remains on the drum. This electrostatic image is held briefly by the drum which, as it turns, is coated by a black powder that adheres to the charged areas. A piece of paper then passes over the drum and the electrostatic charge is reversed, causing the black powder to detach itself from the drum and cling to the paper. Finally, so that the image cannot simply be brushed off, the paper is passed over a small heater which partially melts the powder and bakes it onto the surface of the paper.

Before making your final selection you check on the power supplies in the department. You discover that 120 volt power outlets rated up to 30 amperes are readily available, and that a 208 volt power source is available in the basement. It will cost $180 to bring the 208 volt power up to your floor.

Using all this information, you select the most suitable copier for the department and write your report.

NOTES:

1. It makes little difference which copier you select: all are suitable for the department, and each has advantages over the other two. One is cheap but slow, another is fast but expensive, while the third seems to strike a happy medium between speed and price. The most important factor in making your selection is to know *why* you chose one particular model. If *you* are convinced, then you will be much more likely to write a convincing report.
2. Start your report by outlining the parameters, or factors, that will guide you in your assessment of the copiers. Also indicate which parameters will most influence your final selection.
3. Before describing the copiers decide in which order you will present them:
   a. In the order listed in the Table of Comparative Data.
   b. In descending order of preference (from most suitable to least suitable).

ELECTROSTATIC OFFICE COPIERS
TABLE OF COMPARATIVE DATA

	MODEL A "CLEARCOPY"	MODEL B "MULTICOPY"	MODEL C "AUTOCOPY"
Basic rental charge	$50/month	$90/month	$125/month
Additional cost per copy (includes materials and labor)	2¢	3¢	3¢
Speed	Slow; up to 4 copies/minute	Fairly fast; up to 12 copies/minute	Very fast; up to 20 copies/minute
Convenience (degree of automation)	Semiautomatic; requires manual restart for each copy required	Fully automatic; operator feeds original into machine, selects number of copies required	Fully automatic; operator places original on glass plate, dials number of copies required
Copy quality	Fair–Good	Good–Very good	Very good
Limitations	Does not copy light blue; copies any original up to 8 in. × 13 in.	Copies all colors; copies any original up to 8½ in. × 14 in.	Copies all colors; copies any original up to 10 in. × 15 in.; can reduce image 15%, to fit standard 8½ in. × 14 in. paper
Floor space required	Table-top model 24 in. W × 20 in. D × 10 in. H	Floor model 72 in. × 48 in. D × 36 in. H	Floor model 84 in. W × 50 in. D × 42 in. H
Power requirements	3-pin plug; 120v, 15 amp	3-pin plug; 120v, 25 amp	Junction box; 208v, 30 amp
Manufacturer	Rostock Corporation, Montrose, Ohio	Valiant Office Products Inc., Tampa, Florida	The Print-All Company, Reece, Minnesota
Manufacturer's reputation for providing service	Only fair	Reasonably good	Excellent
Delivery from date of order	Immediate; available from stock	1 month	2 months

c. In ascending order of preference (from least suitable to most suitable).

4. Briefly describe each model, highlighting its strongest and weakest characteristics. Take care that you do not introduce a lot of unnecessarily detailed matter that will detract from your main reason for selecting or rejecting each machine (tell the reader that he can find all these minor details in an appendix). Be very persuasive; try to convince the reader that your choice is correct.

Project No. 2—Fleet Ownership or Rental?

The possibility of purchasing or renting a 10-car fleet is discussed at an H.L. Winman and Associates management meeting, during which your department head volunteers to conduct a study into the economic advantages and disadvantages of fleet ownership compared with U-drive rentals. He assigns the project to you, with instructions that you prepare an informal evaluation report that he can present at the next management meeting. He asks you not only to evaluate your findings but also to recommend what action the company should take.

During your study you establish that:

1. Ten cars will be required:
    Engineering Department   4
    Field Project Engineers   3
    Architects   1
    Special Projects Group   1
    Materials Testing Laboratory   1
2. The vehicles will average 15,000 miles annually.
3. Maintenance costs (including tire replacement) for company-owned vehicles would be:
    First year of operation $320 per vehicle
    Second year of operation $410 per vehicle
    Subsequent years $410 + $40 increase per year
4. The Accounting Department would prefer to depreciate the vehicles over a five year period.
5. Insurance premiums will cost $160 per vehicle per year.
6. Licensing for each vehicle will cost $26 per year.
7. Depreciation on company-owned vehicles over two years is estimated to be 50% of purchase cost; over four years, 85% of purchase cost; over five years, 100%.
8. The lowest unit cost of vehicles that conform to your company's requirements is $3200 (including a 15% fleet discount). You obtain these figures by writing to the major car dealers in the area.
9. Similar vehicles may be leased from a U-drive agency at a unit cost of $125 per month, plus an additional cost of 2¢ per mile for every mile over 12,000 per year. These prices are based on a guaranteed two year minimum rental period.
10. Licenses, insurance, and maintenance are included in the prices

quoted for a leasing arrangement; gasoline and oil are not included.

11. The leasing agency will replace the fleet with new cars every two years.
12. The leasing agency will provide a replacement vehicle while regularly leased vehicles are being serviced.
13. The company's name cannot be painted onto leased vehicles.
14. The head of the Mechanical Department thinks it would be more efficient to hire a maintenance technician to do the servicing of company-owned vehicles. His salary would be about $6600 per year. The saving in maintenance costs would be approximately $250 per vehicle the first year, and $300 per vehicle thereafter. But maintenance would be better because it would be continuous.

Use a diagram, chart, or graph in your report to illustrate your findings.

Project No. 3—Is There a Market for Mini-Inns?

When Wayne D. Robertson, president of H.L. Winman and Associates' Canadian subsidiary, visited head office in Cleveland late in March, he discussed a possible company project with Harvey Winman. Wayne has a business associate in Toronto who would like to extend his empire of "Mini-Inns" into the United States. At present his friend's ideas are vague, but Wayne knows that with a little prompting he would most likely take the step. His hesitancy stems from lack of knowledge of the U.S. restaurant market.

Mini-Inns were started in 1954 by Mr. Paul Schmidt, now president of Mini-Inns Ltd., 563 Wadena Drive, Toronto 17, Ontario. There are now 83 restaurants in operation in Canada, 56 of them located in southern Ontario. A Mini-Inn is a small 16-seat counter-type restaurant built to a standard pattern. It is 24 feet wide by 20 feet deep and has a single door in one corner. Serving light meals at reasonable prices, and open from 6 a.m. to 2 a.m. daily, it caters mostly to a transient public looking for a quick snack. Parking facilities, either on the road or preferably on the lot itself, are therefore essential to its successful operation.

Wayne suggests to Harvey that he do some private research into business possibilities in a number of locations across the United States. Harvey decides to sponsor a private project and assigns staff members to research residential areas to identify possible Mini-Inn sites, consumer response, and business possibilities. He instructs you to research sites within a one mile radius of your home address and to prepare an evaluation report.

Harvey Winman will assess your evaluation report, together with those from other company members, to determine the overall market possibilities for Mini-Inns. If the business prospects are promising, he will prepare a feasibility study for Mr. Schmidt in which he will:

1. Describe the advantages of extending Mini-Inns into the United States.
2. Propose specific sites for setting up trial Mini-Inns.

3. Recommend that H.L. Winman and Associates undertake design, planning, and construction management of Mini-Inns across the country.

He will then attach the evaluation reports describing the most promising sites and mail them to Wayne Robertson, who will in turn give them to his Mini-Inn friend, Paul Schmidt.

For the evaluation report you will have to obtain the legal description of your site, the name of the present owner, and the price he wants either for outright purchase or long-term lease; you will have to establish the availability of public utilities and services such as gas, water, electricity, and sewers; you will have to study competition, traffic movement, and whether parking space is available on and off the street; and you will need to prepare a sketch of the area showing the Mini-Inn, parking space, and means of entry onto and exit from the restaurant site. To this information you should add your assessment of the site as a paying location for a Mini-Inn (i.e. will the cost of the land, plus construction costs and annual taxes, balance anticipated revenue and provide a reasonable profit over a 20 year period?).

You may assume that you have a firm price proposal from Mager Construction Company to build the Mini-Inn and approaches to it for $22,500. Your company's fee for the initial proposal will be 1½% of the cost of the land, and as Project Engineers during the construction stage their fee will be 6% of the contractor's price. You may also assume that a Mini-Inn requires a staff of two persons all the time, and that the average customer's bill amounts to 52¢.

Three reports are to be written for this project:

1. An informal progress report to Harvey Winman, briefly informing him of the site you have chosen and your reasons for selecting it.
2. A semiformal evaluation report for your particular Mini-Inn location.
3. A letter-form feasibility study establishing reasons why Mini-Inns should be extended into the United States. It can be addressed either to Wayne Robertson or to Mr. Schmidt.

You are also to describe your site before a meeting of H.L. Winman and Associates' engineering staff (see "Meetings Combined with Other Projects" in the Assignments section at the end of Chapter 8).

Project No. 4—Relieving a Parking Problem

You are a recent graduate of Montrose Community College, now employed by H.L. Winman and Associates.

Lorne Freeman, a high school friend, calls you one weekend and asks you to help him. He is now attending the college and has been elected to the Student Administrative Council (SAC). One of his election promises was to do something to improve car parking facilities for students, and he has to make good his promises. He has a plan, but he needs guidance in presenting it effectively.

During the weekend you study his notes (see illustration) and discuss his ideas. Briefly these are:

1. There is parking space on college property for 1500 vehicles. Members of the staff require 400 parking stalls, which leaves 1100 for students.
2. A survey of students shows that:
   a. 1426 students come by car daily.
   b. The first 1100 grab parking spots on the college grounds.
   c. The remainder park on neighboring streets.
   d. A further 118 students would bring their cars, but cannot be bothered to fight for space.
3. Bus service is adequate to and from downtown, but awkward from many of the residential areas of Montrose.
4. The college administration is continually receiving complaints from property owners (both commercial and private) that students' cars have blocked their driveways.
5. There is no more space on the college grounds that could be used for parking. The only remaining open spaces are athletic fields.
6. The Phys. Ed. Department would emphatically resist any attempt to convert its playing fields into parking pads.
7. A poll of students indicates they would accept a levy if it were necessary to rent additional space. Students consider $3 per month to be a "reasonable" charge for parking.
8. Lorne Freeman has talked to the owners of the Starlite Drive-In Theater, who would permit students to park without charge on the theater property between 7 a.m. and 5:30 p.m. But there are two conditions: (a) the SAC would have to accept financial responsibility for damage; (b) cars left after 5:30 p.m. would be towed away at the owner's expense.
9. The distance from the Drive-In entrance to the school entrance is 940 yards—about a 12 minute walk.
10. There is a narrow strip of land along the railway tracks immediately west of Cameron Cartage. It could hold two lines of 135 cars each. The property is owned by the railway, which will make it available for a token $1.00 per year.
11. Distance from the railway lot to the school entrance (around northern end of Cameron Cartage building) is 180 yards (three minutes).
12. The railway lot is very rough and will require fill. Costs:
    a. Fill $470
    b. Gravel surface $380
    c. Asphalt surface $2300
    d. Fencing $400
    (either b or c would be used, not both)

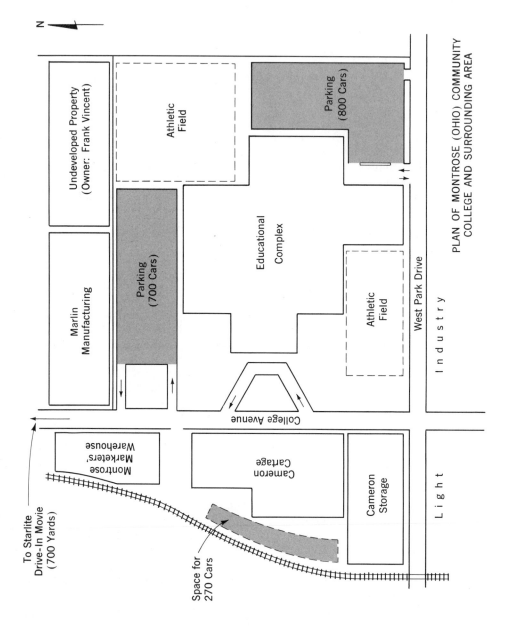

N

Undeveloped Property
(Owner: Frank Vincent)

Athletic Field

Parking
(800 Cars)

Marlin Manufacturing

Parking
(700 Cars)

Educational Complex

Athletic Field

West Park Drive

College Avenue

I n d u s t r y

Montrose Marketers' Warehouse

Cameron Cartage

Cameron Storage

L i g h t

To Starlite
Drive-In Movie
(700 Yards)

Space for
270 Cars

PLAN OF MONTROSE (OHIO) COMMUNITY
COLLEGE AND SURROUNDING AREA

13. The undeveloped property northeast of the college is owned by Frank Vincent, an elderly farmer. It is arable land that he has not farmed for several years. (Lorne thinks he bought it as an investment, hoping the college would need the land so badly the Administration would have to pay his high price.) The western half of the property would hold 460 vehicles. Walking time: one and one-half minutes.

14. Vincent will lease half the property to the SAC for $800 per year. Development costs would be:
    a. Asphalt surface $3200
    b. Gravel surface $ 600
    c. Entrance from college north
       parking lot $ 700
    (either a or b would be used, not both)

15. Approximately 12 staff parking stalls would have to be given up to build an approach road to Vincent's property. The college Administration would have to approve this.

16. Cost of rented parking should be borne by all students using college parking facilities, not just by those using the additional area.

Lorne wants to combine all this data into a letter-type feasibility study that the SAC can present to the Administration. You spend the remainder of the weekend helping him do this. Your role is to write the feasibility study; his is to prepare the drawing that will accompany it.

TECHNICAL BRIEFS AND PROPOSALS

Project No. 5—Cutting the Cost of Company Meetings

Part 1. A lunchtime conversation topic at H.L. Winman and Associates is "meetings." Everybody agrees that they last much too long, but nobody seems to know how to shorten them. The company has circulated a bulletin describing how to run and take part in meetings (similar to the ideas presented in Chap. 8), but it seems to have had no effect upon some garrulous individuals.

You half-seriously, half-humorously suggest that the company needs a timer in the conference room—something to show just how long a meeting has gone on, and, if possible, how much it has cost. Enthusiasm mounts as you make a rough sketch of the gadget:

Andy Rittman, senior field project engineer, joins the group and listens to the conversation. He suggests you prepare a brief describing your idea and send it into your department head for evaluation by management.

When you return to your desk you jot down a few headings:

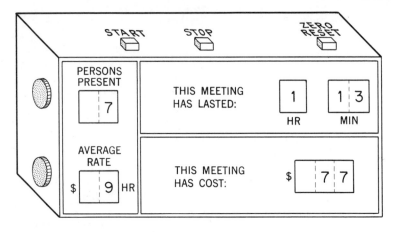

1. Reason for device
2. What it looks like
3. Approximate size: 12 in.W × 8 in.H × 4 in.D
4. For mounting on wall or setting on table
5. "Average Rate" Window could be covered by a hinged flap
6. Square push-buttons on top for starting and stopping meter, and for resetting time and total cost counters to zero when meeting is over
7. A rough estimate of the cost to build a single unit would be $300, made up of:

Manufacturing case	$ 60
Counters, timer, mechanism, etc.	$170
Assembly and wiring	$ 70

You may, if you wish, incorporate additional features to improve either the concept or the design. If you do, allow for the additional cost your improvements will incur.

Part 2. Management generally has mixed feelings about your idea, but Harvey Winman is enthusiastic. He authorizes the Mechanical Engineering Department to prepare production drawings and to build two prototypes: one for H.L. Winman and Associates, the other for Robertson Engineering Company.

While this is being done, your department head requests you to prepare a letter to Fred Stokes, Robertson Engineering Company's Chief Engineer. In it you are to:

1. Introduce the gadget to him (assume that the gadget and production drawings accompany the letter), and tell him its purpose and how it works.
2. Ask him to use it and then send your department head a report

on (a) its acceptance by persons attending meetings, and (b) its effectiveness.

3. Ask for his opinion of its suitability as a marketable product.
4. Ask him to prepare a cost estimate based on the manufacture of 1000 units.

Project No. 6—Random Lighting of Older Homes

You are to write an informal report to Jim Perchanski, describing a feasibility study you have conducted to establish whether a device can be installed in older homes to switch several lights on and off at random during the absence of the owner. The device can be real or imaginary. If real, your report should contain your own design drawings and calculations of cost. If imaginary, you can either devise your own information or use the following ideas:

1. The device uses a modified Poliwaxor RG-7 Module mounted on the switchboard of the dwelling.
2. It can control up to three lighting circuits, switching lights on and off independently.
3. The major modification to the RG-7 Module is the addition of an electric timer.
4. The timer can be installed remotely (e.g. in the kitchen or by the back door).
5. Before going out, the householder has to turn "on" all lights being controlled, and set the timer.
6. Time for installation: about two hours.
7. Cost of RG-7 Modules:

100	$ 42.70 each
1000	$ 36.25 each
10,000	$ 29.50 each

8. Cost of Timer:

100	$ 21.60 each
1000	$ 17.45 each
10,000	$ 13.10 each

9. Cost to assemble and package kits: $8.45 each.
10. Manufacturer's selling price: cost of module, timer, and assembly, plus 20%.
11. Distributor's mark-up: manufacturer's price plus approximately 30% (varies from 15% to 45%, depending on distributor).
12. An electrical contractor's installation charge averages $8.00 per hour.
13. A market survey conducted two years ago measured consumer acceptance of or resistance to household security gadgets as follows:

Price	
Below $40	excellent acceptance
$40–$70	moderate acceptance

$70–$99   fair acceptance
$100–$130   moderate resistance
Over $130   strong resistance

A chart, graph, or diagram should be used to illustrate your report.

NOTE: Refer to Project No. 7 of Chapter 6 (page 228) for background information relating to this project.

6

Formal Reports

A formal report is an important document that requires more careful preparation than the informal reports described in the previous chapters. Because it will be distributed outside the originating company, its writer must consider the impression it will convey of himself and his company. Harvey Winman recognized long ago that a well-written, aesthetically-pleasing report can do much to convince a prospective client that H.L. Winman and Associates should handle his business, whereas a poorly-written, badly-presented report can cause a client to question the company's capability. Harvey also knows that the initial impression conveyed by a report can influence a reader's readiness to plough through its heavy technical details.

The presentation aspect must convey the originating company's "image," suit the purpose of the report, and fit the subject it describes. For instance, a report by a chemical engineer evaluating the effects of diesel fumes on the interior of bus garages would most likely be typed on standard bond paper, and its cover, if it had one, would be simple and functional. At the other end of the scale, a report by a firm of consulting engineers selecting a college site for a major metropolis might be printed professionally and bound in an artistically-designed book-type folder. Regardless of the appearance of a report, its internal arrangement will be basically the same.

Formal reports are made up of several standard parts, not all of which appear in every report. Each writer uses the parts that best suit his subject and his intended method of presentation. There are six basic and seven subsidiary parts on which he can draw, as shown in the table at the top of page 154. Opinions differ throughout industry as to which is the best arrangement of these parts. The two arrangements I suggest under The Full Report on page 170, and illustrate in the mini-report in Figures 6–3 and 6–4, are those most frequently encountered. Knowledge of these parts and the two basic arrangements will help new engineers and technicians to adapt quickly to the variations in format preferred by their employers. The parts are:

BASIC PARTS	*SUBSIDIARY PARTS*
Summary	Cover
Introduction	Title Page
Discussion	Table of Contents
Conclusions	Bibliography or References
Recommendations	Distribution List
Appendix	Cover Letter
	Letter of Transmittal

THE SIX BASIC PARTS

Because the six basic parts form the central structure of every formal report, a technical writer must understand their purpose and function if he is to use them effectively.

Summary

The summary is a brief synopsis that tells the reader quickly what a report is all about. Normally it appears immediately after the title page, where it can be found easily (readers may be confused if it is positioned elsewhere). It identifies the most important features of the report, states the main conclusion, and sometimes makes a recommendation. It says this in as few words as possible, condensing the narrative of the report to a handful of succinct sentences. It also has to be written so interestingly—so enthusiastically—that it encourages readers to delve further into the document.

The summary is considered by many to be the most important part of a report—and the most difficult to write. It has to be informative, yet brief. It has to attract the reader's attention, but must be written in simple, non-technical terms. It has to be directed to the executive reader, yet it must be readily understood by almost any reader.

Generally, the first person in an organization who sees a report is a senior executive, who may have time to read only the summary. If his interest is aroused by what he reads, he will pass the report on to his technical staff to read in detail; if it fails to convince him of its value, he will put it aside to read "later." When this happens, the writer has failed in his task to attract the reader's interest, and his report will probably not be read by anyone.

Because it is so important, the summary should be written last, after the remainder of the report has been completed. Only then will the writer be fully aware of the highlights of his report, of his main conclusions, and of his recommendations. Attempting to write the summary first will prove difficult and frustrating because he will not yet have hammered out many of the finer points that frequently remain unresolved until the report has been written and revised several times. Only with the knowledge of a soundly developed report firmly fixed in his mind will he be able to fashion an effective summary.

If a summary is to be interesting, it must be informative; if it is to be informative, it must tell a story. It should have a beginning, in which it states why the project was carried out and the report written; a middle, in which it highlights the most important features of the whole report; and an end, in which it reaches a conclusion and possibly makes a recommendation. The examples below illustrate how the interest is maintained in an informative summary:

Informative Summary No. 1

A specimen of steel was tested to determine whether a job lot owned by Northern Railways could be used as structural members for a short-span bridge to be built at Peele Bay in northern Alaska. The sample proved to be G40.12 structural steel, which is a good steel for general construction but subject to brittle failure at very low temperatures.

Although the steel could be used for the bridge, we consider that there is too narrow a safety margin between the −60°F temperature at which failure can occur, and the −52°F minimum temperature occasionally recorded at Peele Bay. A safer choice would be G40.8C structural steel, which has a minimum failure temperature of −80°F.

Informative Summary No. 2

A unique midcourse objective test that was devised, answered, rated, and analyzed by students provided the data for a standard Item Analysis. The analysis revealed that two-thirds of the 100 test questions must be discarded, either because they are insufficiently difficult or because they fail to correlate with the answers obtained from students writing the test. Approximately half of these questions can be retained if they are rewritten to resolve ambiguities. A second test and Item Analysis will identify any revised questions that need to be discarded.

Other informative summaries preface the two formal reports at the end of this chapter, and the semiformal report in Figure 4–6.

Some writers prefer to write a topical summary for reports that describe history or events, or that do not draw conclusions or make recommendations. As its name implies, a topical summary simply relates the topics covered in the report without attempting to draw inferences or captivate the reader's interest. It therefore conveys much less information than an informative summary. The following is an example of a topical summary:[1]

Topical Summary

Construction of Manitoba Hydro's Kettle Generating Station was initiated in 1966, and first power from the 1,224 MW plant is scheduled for 1971. A general description of the structures and problems peculiar to construction of this large development in the subarctic climate of Northern Manitoba is presented. The river diversion program, permafrost foundation conditions, and major equipments are described. The latter include the 12 propeller turbines, among the largest yet installed, each rated at 140,000 horsepower at 98.5-foot head.

In a formal report the summary should be centered on a page by itself, and prefaced by the word SUMMARY. If it is short it may be indented equally on both sides to form a roughly square block of information. If it is very short it may also be double-spaced.

Introduction

Some report writers confuse the summary with the introduction. There is no need for this if they remember that the summary provides a *brief synopsis of the whole report*, whereas the introduction *introduces the subject* to the reader so that he may read the remainder of the report more intelligently.

The introduction begins the major narrative of the report by preparing the reader for the discussion that follows. It orients him to the purpose and scope of the report, and provides sufficient background information to place him mentally "in the picture" before he tangles with technical data. A well-written introduction contains exactly the correct amount of detail to lead the reader quickly into the major narrative.

The length of an introduction and its depth of detail depend mostly on the reader's existing knowledge of the topic. A writer who knows that his ultimate reader is technically knowledgeable will want to use technical terminology immediately, but he must also be aware that he has to cater to the executive reader who is probably only partly technical. To overcome this disparity in reading levels, skillful report writers often write the summary in layman's language and the introduction, conclusions, and recommendations in semitechnical language. They thus permit the semitechnical executive to gain a reasonably comprehensive understanding of the report without devoting time and attention to the technical details contained in the discussion.

[1]Knight, A.W. and J.G. Macpherson, *Kettle Generating Station: A General Description*, Canadian Electronic Association Conference, Montreal, P.Q., March 1969.

Most introductions contain three parts: *purpose, scope,* and *background information.* Frequently the parts overlap, and occasionally one of them may be omitted simply because there is no reason for its inclusion. The introduction normally is a straightforward narrative of one or more consecutive paragraphs; only rarely is it divided into distinct sections preceded by headings. It always starts on a new page (normally identified as p. 1 of the report) and is preceded by the report's full title. This is followed by the single word INTRODUCTION, which can be either a center heading or a side heading, as shown below.

Part of the introduction describes the *purpose* of the project: the reason why it has been carried out and the report is being written. It may indicate that a problem exists and the project has been authorized to investigate it and recommend a solution, or it may describe a new concept or method of work improvement that should be brought to the client's attention. In the first case it may be assigned to the author as a specific project that is his entire responsibility, or to a project group in which he is assigned the role of report writer. In the second case it is more likely to be author-motivated.

The *scope* defines the parameters of the report. It describes the ground that it covers and outlines the method of investigation used in the project. If there are limiting factors, it identifies them. If 18 methods for improving packaging are investigated in a project, but only 4 are discussed in the report, the scope indicates what factors (such as cost, delivery time, and availability of space) limited the selection. Sometimes the scope may include a short glossary of terms that need to be defined before the reader starts to read the discussion.

Background information comprises facts the reader must know if he is to understand fully the discussion that follows. Conditions or events that caused the project to be authorized, plus details of previous investigations and studies on the same or a closely related subject, fall under this heading. In a very technical report, or when a significant time lapse has occurred between it and previous reports, background information may include a brief theory review and references to other documents. If the theory review is lengthy, it

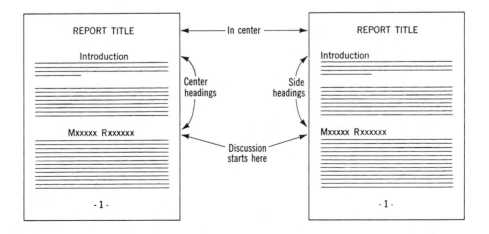

is relegated to the appendix and only a brief theory summary is inserted in the introduction; similarly, detailed references are listed at the end of the report rather than in the introduction.

The introductions shown here, plus those forming part of the two sample reports at the end of this chapter, are representative of the many ways a writer can introduce his topic.

Introduction No. 1

Northern Railways plans to build a short-span bridge one-half mile north of Lake Peele in northern Alaska, and has a job lot of steel they want to use in its construction. In letter } *Background*

NR-70/LM dated 20 March 1971, Mr. David L. Harkness, Northern Area Manager, requested H.L. Winman and Associates to test a sample of this steel to determine its properties, and to assess its suitability for use as structural members for a bridge in a very low temperature environment. } *Purpose*

The test performed was the Charpy Impact Test. Two series of tests were run, one parallel to the grain of the test specimen, and the other transverse to the grain, at 10° increments from +70°F down to −90°F. } *Scope*

Introduction No. 2

H.L. Winman and Associates was commissioned by Mr. D.C. Scorobin, President and General Manager of Auto-Marts Inc. of Dallas, Texas, to select a Cleveland site for the first of a proposed chain of drive-in groceries to be built in the northeastern United States. This area was chosen since it represents an average community in which to assess customer acceptance of this revolutionary form of service. } *Purpose*

Aside from exterior facade and foundation details, Auto-Marts are built to a standard 72 by 24 foot pattern with an order window at one end of the 72 foot wall, and a delivery window at the other. Auto-Marts carry only a limited selection of household goods, but boast 90-second service from the time of placing an order to its delivery at the other end of the building. For this reason, Auto-Marts attract the householder hurrying homeward from work, the impulse buyer, and the late night shopper, rather than the selective shopper. The store depends more on volume of customers than on the volume of goods sold to each individual. } *Background*

The chief consideration in selecting a site must
therefore be a location on the homeward-bound
side of a main trunk road into a large residential
area. The site must have quick and easy entry
onto and exit from this road, even during peak
rush hour traffic. The residential area should be   } *Scope*
occupied mainly by upper and middle class people
who own cars and have money to spend. And
there should be little competition from walk-in
grocery stores.

Discussion

Although the discussion normally is the longest part of a report, it
needs little description. It simply presents as much evidence (facts, arguments,
details, data, and test results) as the reader will need to understand the subject.
The writer must develop this evidence in an organized, logical, and orderly
manner if he wants to avoid confusing his reader; he must present it imaginatively if he wants to retain his interest.

In the discussion the writer tells the reader what he tried to do, what
he actually did, and what he found out. He can present the evidence in the
order in which it occurred (*chronological* development); he can divide it into
natural subject areas, then present it in a predetermined order for each subject
(*subject* development); or he can organize and present it as a series of ideas
that reveal imaginatively, coherently, and logically how he reasoned his way
through the project (*concept* development). Reports using the chronological or
subject method tend to be less interesting than those using the concept method,
mostly because they depend on a straightforward presentation of information.
The concept method can be very persuasive. By identifying and describing the
writer's ideas and thought processes, it helps the reader to organize his thoughts
along the same lines.

But just because I have said the concept method is the most interesting, I do not mean it is suitable for every report. Many topics are ideally suited
to a simple, straightforward presentation of facts in chronological or subject
order, and will be most effective when presented this way. As a report writer
you must decide early in the planning stages which method you intend to use,
basing your choice on which is most suitable for the evidence you have to
present. For further suggestions on selecting the most appropriate method,
refer to "Writing the Discussion" on page 176.

Every piece of information in the discussion must be relevant. All supporting information, unless its inclusion is essential for reader understanding,
should be banished to the appendix. This includes tables, graphs, illustrations,
large photographs, statistics, and test results. A reference to the appended data
must be included in the discussion. For instance, if the accuracy of a rotating
antenna is being discussed, and a series of tests at five-degree intervals has

indicated where the error exists, the statement inserted in the discussion should summarize the appended data:

> The test results attached at Appendix "C" show that aircraft on a bearing of 265°T experienced considerably weaker reception than aircraft on any other bearing. This was attributed to . . . .

A reader interested only in *results* would consider that this statement tells him enough, and continue reading the report. A reader interested in knowing *how the results were obtained,* who wants to see the overall picture, will turn to Appendix "C" to find out how the writer went about performing the tests. He probably will not do so immediately, but will return to it when he has finished the report or reached a suitable stopping place. Thus in neither case is the reader slowed down by supporting information which, though relevant, is not immediately essential to an understanding of the report.

Unless the discussion is very short, it will consist of a series of sections, each containing information on a major topic. An informative heading at the beginning of each section should indicate the section's contents. If the section is long or naturally subdivides into subtopics, subordinate headings should be inserted. Major topics normally demand center or side headings, while minor topics need only side or paragraph headings, as shown below (also see Fig. 10–4). These headings eventually form the table of contents for the whole report.

At the end of each major section the writer should insert a concluding statement that summarizes the result of his discussion within that section. From these section conclusions he can later draw the main conclusions to his whole report.

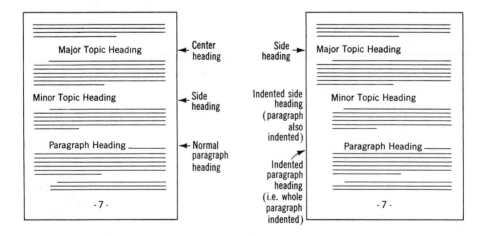

Conclusions

Conclusions state in brief terms the major inferences that can be drawn from the discussion. They must be based entirely on previously stated information, and never introduce new material or evidence to support the writer's argument. If there is more than one conclusion, the main conclusion is stated first, followed by the remaining conclusions in decreasing order of importance. This is shown in the two examples below, which present the same conclusions in both narrative form and tabular format.

Narrative Conclusion

The operational life of QK 3801 magnetrons can be increased 30% by installing modification XL–1 on our radar transmitters. This will save $4300 over the next 12 months, but will require us to increase magnetron run-in time from 4 to 8 hours.

Tabular Conclusion

Installation of modification XL-1 on our radar transmitters will result in:
1. A 30% increase in the operational life of QK 3801 magnetrons.
2. A $4300 saving over the next 12 months.
3. A change in operating procedures to increase magnetron run-in time from 4 to 8 hours.

Whether a writer uses narrative or tabular conclusions depends on his writing style. Both are acceptable, although the advantage of the tabular arrangement is that its separate conclusions are easier to identify. Perhaps the best rule is to use narrative style when there are very few conclusions, and to tabulate them when there are several.

Because conclusions are *opinions* (based on the evidence presented in the discussion), they must never suggest or recommend future action.

Recommendations

Recommendations appear in a report only when the discussion and conclusions have indicated that further work needs to be done, or that a specific solution to a problem has evolved from the writer's investigations. They must be written in strong, definite terms to convince the reader that the course of action advocated by the writer is valid. Preferably, they should be written in the first person using active verbs, as has been done here:

> *Strong:*   I recommend that we build a five-station prototype of the
>   Microvar system and test it operationally.

Compare this with the same recommendation written in the third person,
using passive verbs:

> *Weak:*   It is recommended that a five-station prototype of the
>   Microvar system be built and tested under operational
>   conditions.

The first version is more convincing because the writer has allowed himself
to be personally involved in the recommendation. The second version is weak
because there does not seem to be an author who really wants to build a
prototype.

Unfortunately, in many industries, and often in government circles,
personal involvement is severely frowned upon. (No one seems to know whether
this stems from natural reticence supposedly displayed by technical writers, or
from fear that a faulty recommendation might be traced to a specific individual!) As a result, many reports fail to contain really strong recommendations
because their writers are afraid to use the first person. They hesitate to say
"I recommend. . . ," and so use the indefinite "It is recommended. . . ." Surely
a technical man who has researched his subject well enough to advocate
definite action can present his recommendations more convincingly than this.
He doesn't need to use the personal "I" (in most reports prepared for clients
he would be expected to use the plural "we", to indicate that he represents
the company's viewpoint). For example:

> *Strong:*   We recommend building a five-station prototype of the
>   Microvar system. We also recommend that you:
>   1. Install the prototype in Railton High School.
>   2. Commission a physics teacher experienced in writing
>   programed instruction manuals to write the first pro-
>   grams.
>   3. Test the system operationally for three months.

Because recommendations must be based solidly on the evidence presented in the discussion and conclusions, they can never introduce new evidence
or new ideas.

Appendix

I said earlier that related data not necessary to an immediate understanding of the discussion should be placed further back in the report, in the
appendix. The data can vary from a complicated table of electrical test results

to a simple photograph of a blown resistor. The appendix is a suitable place for manufacturer's specifications, graphs, comparative data, drawings, sketches, excerpts from other reports or books, and correspondence. There is no limit to what can be placed in the appendix, providing it is relevant and reference to it is made in the discussion.

The term "appendix" applies to only one set of data. In practice there can be many sets, each of which is called an appendix and assigned a distinguishing letter, such as "A," "B," or "C." The sets of data are collectively referred to as either "Appendixes" or "Appendices," the former term being preferred in the United States, and the latter in Canada and Great Britain.

The importance of an appendix has no bearing on its position in the report. Whichever set of data is mentioned first in the discussion becomes Appendix "A," the next set becomes Appendix "B," and so on. Each appendix is considered a separate document complete in itself, and is page-numbered separately, with its front page labeled 1. (Not every organization follows this practice. Although most companies now number appendix pages separately, some continue to number them consecutively as an integral part of the report.) Examples of appendixes appear at the end of the two sample reports included with this chapter.

THE SEVEN SUBSIDIARY PARTS

In addition to the six basic report parts just described, there are seven additional parts that perform more routine functions. Although I refer to these as "subsidiary," they nevertheless contribute much to the effectiveness of a report. A writer must never assume that these seemingly peripheral details will look after themselves. They so affect the image he conveys of himself and the firm that he represents that he cannot afford to prepare them any less carefully than the rest of his report.

Cover

Almost every major formal report has a cover. It may be made of cardboard printed in multiple colors and bound with a dressy plastic binding, or it may be only a light cover of colored fiber material stapled on the left-hand side. In any case it has one obvious purpose: to inform the reader of the main topic of the report. Its more subtle purposes are to convey an immediate image of the type of company that has produced the report, and to reflect the type of subject that the report describes. This "matching" of subject matter and company image plays an important part in setting the right tone for a report.

The cover should contain very few words. The report title is an obvious choice, since a good title will invite a prospective reader to open the report and discover what it is about. These words must also stand out clearly on the page. Printing a title like a newspaper banner headline in two-inch high letters may attract attention, but will have a negative effect on the reader's opinion of the author's technical competence. A few words tastefully arranged and well separated from other printed matter will help to present the correct image. The only other words should be the originating company's name and perhaps the name of the author, if he is well known enough to lend weight to the technical level of the report and advance the company's image.

Choice of an informative title is specially important. It should be short, yet imply that the report has a worthwhile story to tell. Compare the vague title below with the more informative rewritten version beneath it.

Original vague title:

RADOME LEAKAGE

Revised informative title:

POROSITY OF FIBERGLAS CAUSES RADOME LEAKS

A technical officer at a radar site would glance at the first title with only a muttered, "Looks like somebody else has the same problem we've got." But the second title would encourage him to stop and read the report. He would recognize that someone may have found the cause of (and, hopefully, a remedy for) a trouble spot that has bothered him for some time.

Note how the following titles each *tell* something about the reports they precede:

REDUCING AMBIENT NOISE IN AIR TRAFFIC CONTROL CENTERS
EFFECTS OF DIESEL EXHAUST ON LATEX PAINTS
SALT EROSION OF CONCRETE PAVEMENTS

These titles may not attract every potential reader who comes across them, but they will certainly gain the attention of anyone interested in the topics they describe.

Title Page

The title page normally carries four main pieces of information: the report title (the same title that appears on the cover); the name of the person, company, or organization to whom the report is directed; the name of the company originating the report (with, sometimes, the author's name); and the date it was completed. It may also contain the authority or contract under which the report has been prepared, a report number, a security classification such as

CONFIDENTIAL or SECRET, and a copy number (important reports given only limited distribution are sometimes assigned copy numbers to control and document their issue). All this information must be tastefully arranged on the page, as has been done in the sample reports in this chapter (see pages 184 and 201).

Table of Contents

All but very short reports contain a Table of Contents (T of C). The T of C not only lists the report's contents, but also shows how the report has been arranged. Just as a prospective book buyer will scan the contents page to find out what a book contains and whether it will be of interest to him, a potential reader will scan a report's T of C to assess how the author has organized his work. He will be searching for a clue to the author's competence to arrange technical information in a logical fashion. If he can sense that the report writer knows his business both as a technical man and as a writer, then he will delve into the report.

A satisfactory arrangement for a T of C is shown on page 186. This page also contains a list of appendixes attached to the report, with each identified by its full title. In long reports, a list of illustrations with their page numbers may also be helpful. If used, it should be inserted between the T of C and the list of appendixes.

Bibliography, Footnotes, and Endnotes

Reference documents, textbooks, previous reports, and even correspondence used by the writer which he feels will be useful to the reader are listed in a bibliography inserted immediately after the recommendations. Rules for preparing a bibliography vary, depending on the type of document. For technical papers and articles, the bibliography should conform to the style adopted by the society or journal that is to publish the material. For technical reports, the bibliography may have to conform to a style stipulated by the originating company, or the style may be left to the discretion of the author. A suitable style for the more common bibliography listings is illustrated in Figure 6-1.

A report writer may sometimes wish to refer to a statement made in a previous report, or to a conclusion reached by another person engaged in similar work. He wants the reader to be able to refer to the document, but does not want to clutter up his discussion with copious cross-references. He can overcome this either by describing the document in a footnote, or by treating all the documents as endnotes in a list of references at the end of the report. In either case, the references contain all the details the reader needs to order the documents.

BIBLIOGRAPHY

(1)   Agnew, Jeremy. Thick Film Technology, Rochelle Park, N.J.,
          Hayden Book Co., 1973.

(2)   Felber, Stanley B. and Arthur Koch. "Group Communication,"
          in What Did You Say?, Englewood Cliffs, N.J., Prentice-
          Hall, Inc., 1973.

(3)   Lovatt, Frederick G. et al. Management is No Mystery,
          2nd Edn, ed. Robt B. Arpin, Trent, Ariz., Bonus Books
          Inc., 1975.

(4)   Milnark, S.W. "TACS -- A New Approach to Systems Design,"
          The Airconditioning and Refrigeration Business, Vol. 25,
          No. 2, Feb. 1968, p. 36.

(5)   Talbot, J. RATCON Noise Reduction, Report No. 6745-1,
          CAE Industries Ltd., Electronics Division (Western),
          Winnipeg, Man., Feb. 1967.

(6)   Wood, J., Head, Materials Testing Laboratory, H.L. Winman
          and Associates, Cleveland, Ohio. Letter to Wayne
          D. Robertson, President, Robertson Engineering Company,
          Toronto, Ont., 11 March 1972.

Fig. 6-1   A Typical Bibliography.

166

Comments on Bibliography Entries:

1. Reference to a book by one author contains: author's name, <u>Book Title</u> (underlined), city of publication, publisher's name, date of publication.

2. Reference to a specific chapter in a book, book by two authors: authors' names (name of primary author entered with surname first, those of additional authors entered with given name or initial[s] first), "Chapter Title" (in quotation marks), <u>Book Title</u>, city of publication, publisher's name, date of publication, page number(s).

3. A second edition of a book by more than two authors: only primary author's name and editor's name are listed; et al means "and others".

4. Magazine article: author's name, "Title of Article," <u>Magazine Name</u>, volume and issue numbers, date of issue, pertinent page number.

5. Report or technical paper: author's name, <u>Title of Report</u>, report number, name and location of company or organization publishing report, date of report.

6. Correspondence: names of originator and recipient, details of their organizations, date of letter.

NOTE: Bibliography entries are normally listed in alphabetical order.

Both footnotes and endnotes are used in modern reports. I recommend endnotes because they are much simpler to handle. Footnotes, which have to be inserted at the foot of the pages on which the references occur, can cause difficulties for the typist, since she constantly has to adjust the length of the typing area to make room for them. An endnote requires no more than a number inserted in the text, to act as a cross-reference to the item in the list of references at the end. For instance, the discussion may contain the following statement:

> The diagnostic engineer described by John Dent[5] uses a computer to assist him to prepare a diagnostic program.

The "5" raised slightly above the rest of the line refers to the fifth endnote in the list of references, which appears like this:

> 5. Dent, J. "Diagnostic Engineering," *IEEE Spectrum*, Vol. 4, No. 7 (July 1967).

You will probably have noticed that a list of references and a bibliography are very similar. The chief difference is that the entries in a list of references are presented in the order in which they are mentioned in the report, numbered consecutively from 1. In a bibliography, the entries can appear in any order (most often alphabetically by authors), and normally are not numbered.

References, whether footnotes or endnotes, have to be prepared according to a generally accepted pattern. This pattern will vary in detail among companies, although the essentials are basically the same. I suggest the beginning report writer use endnotes and follow the rules illustrated in Figure 6–1 until he enters industry, when he can adopt the method preferred by his employer.

Typical footnotes appear at the foot of the pages on which this section is printed.[2] They may be explanatory notes, as in footnote 2, or references to other documents, as in footnotes 3 through 6. The full entry[3] for a reference is used when a bibliography is not included as part of the report. Under these conditions, abbreviations of Latin terms may be used to shorten the footnotes. The two most common are *ibid.*, which means "in the same place" and is inserted only when the footnote entry *immediately before it* is identical except for the page number[4]; and *op.cit.* ("in the work cited"), which refers to a specific page of a document that has been listed in a footnote earlier in the report. In the latter case the author's name is repeated[5].

Methods for handling footnotes are subject to change and, as in the case of bibliography listings, the rules differ among various organizations and authorities. The trend today is to do away with the Latin terms altogether, or to print them in the same typeface as the rest of the narrative. If they are omitted, footnote 4 becomes

[4]Piper and Davie, p. 29.

and *op.cit.* is deleted from footnote 5.

Sometimes a report writer will utilize a combination of endnotes and footnotes when he wants to make several references to specific pages of a document that is listed in full in the bibliography. A typical footnote entry[6] would then list only the author's name (or authors' names) and the page number.

Distribution List

A report sent to several readers will often contain a distribution list that identifies all the persons who are to receive it. The position of the distribution list depends on company practice, the preference of the author, and the length of the list. Probably the most suitable position is immediately after the bibliography, although a short distribution list may appear at the foot of the

[2]In some typeset documents the footnotes are printed in a smaller typeface.
[3]Piper, H.D. and F.E. Davie, *Guide to Technical Reports* (New York: Holt, Rinehart and Winston, Inc., 1959), p. 22.
[4]*Ibid*, p. 29.
[5]Knight and Macpherson, *op.cit.*, p. 68.
[6]Shere and Love, p. 378.

T of C page, and a long list may have a page to itself and be inserted as a separate appendix.

Cover Letter

The cover letter is a brief courtesy letter that identifies the report and states briefly why it is being forwarded to the addressee. It may be either a letter or an interoffice memorandum, and is always attached to the outside front of the report (*never* bound inside). A letter is used for reports distributed outside the orginating company, an interoffice memorandum for reports confined to internal distribution. A typical cover letter precedes the elevator selection report on page 200.

Letter of Transmittal

A letter of transmittal is rather like a foreword or preface to a report. Normally written and signed by top management, it reflects company policy and sometimes management's interpretation of the report's findings. Because it repeats much of the information contained in the introduction, conclusions, and recommendations, it usually takes the place of the summary. Many companies consider it an unnecessary embellishment and seldom use it. I am describing it briefly here to prevent new report writers from confusing it with the cover letter, because its name seems to describe the latter's role.

The letter of transmittal can also serve a useful purpose when it is used with a technical proposal, which is a form of technical report in which a company describes how it can successfully tackle a proposed task at an economical price for the government or another company. In this case it is used chiefly to discuss financial and legal implications, such as specifications and statements that may be open to more than one interpretation. Because it is mainly a management document, it is not likely to be encountered immediately by a newcomer to industry.

A letter of transmittal is always bound *inside* a report, behind the cover or title page.

THE FULL REPORT

The Main Parts

My description of the full formal report covers only the ten parts that contribute directly to the report itself. The distribution list, cover letter, and letter of transmittal are omitted because they add nothing to the technical content. Other parts that sometimes may be omitted are the cover and T of C (if the report is very short), the recommendations (if the writer has none to make), and the bibliography and appendix (if there are no references or supporting data).

Normal Arrangement

(Conclusions and Recommendations *after* Discussion)

These ten main parts, complete with capsule descriptions, are listed in Figure 6–2. The order in which they are presented is their order in most formal business and technical reports, and is the format illustrated in the mini-report in Figure 6–3. Note that with this arrangement there is a logical flow of information. It starts with an *introduction* that leads into the *discussion*, from which the writer draws his *conclusions* and makes his *recommendations* (the two latter parts sometimes are referred to jointly as the terminal section). The sample formal report, "Evaluation of Sites for Montrose Residential Teachers' College" (p. 184) uses the normal arrangment of report parts.

Alternative Arrangement

(Conclusions and Recommendations *before* Discussion)

In recent years, more and more report writers have altered the organization of their reports to serve the needs of their readers. This alternative arrangement brings the conclusions and recommendations forward, positioning them immediately after the introduction so that the executive reader does not have to leaf through the report to find the terminal section. The advantages of this format are immediately evident: The busy reader has only to read the initial pages to learn the main points contained in the report, and the writer can help him along by gradually increasing the technical content of the report, catering to the semitechnical executive reader up to the end of the recommendations, and to the fully technical reader in the discussion and appendix. Although the natural flow of information that occurs in the normal arrangement is disrupted, it is replaced by three compartments, each containing progressively more technical detail:

SUMMARY	NONTECHNICAL
INTRODUCTION CONCLUSIONS RECOMMENDATIONS	PARTLY TECHNICAL
DISCUSSION APPENDIX	FULLY TECHNICAL

THE MAIN PARTS OF A FORMAL REPORT

Cover:	Outside jacket of report; contains title of report and name of originating company; its quality and use of color reflect company "image"
Title Page:	First page of report; contains title of report, name of addressee or recipient, author's name and company, date, and sometimes a report number
Summary:	An abridged version of whole report, written in nontechnical terms; *very* short and informative; normally describes salient features of report, draws a main conclusion, and makes a recommendation; always written last, after remainder of report has been written
Table of Contents:	Shows contents and arrangement of report; also includes a list of appendixes and, sometimes, a list of illustrations.
Introduction	Prepares reader for discussion to come; indicates purpose and scope of report, and provides background information so that reader can read discussion intelligently.
Discussion:	A narrative that provides all the details, evidence, and data needed by the reader to understand what the author was trying to do, what he actually did, what he found out, and what he thinks should be done next
Conclusions:	A summary of the major conclusions or milestones reached in the discussion; conclusions are only opinions, so can never advocate action
Recommendations:	If the discussion and conclusions suggest that specific action needs to be taken, the recommendations state categorically what must be done
Bibliography/ References:	A list of reference documents used by the author to conduct his project, which he considers will be useful to the reader; contains sufficient information for the reader to correctly identify and order the documents
Appendix(es):	A "storage" area at the back of the report that contains supporting data (such as charts, tables, photographs, specifications, and test results) which rightly belong in the discussion but, if included with it, would disrupt and clutter the major narrative

Fig. 6–2   Capsule Descriptions of Formal Report Parts.

This gradually increasing development of the topic in three separate stages is similar to newspaper technique. If you read any well-written news item, you will find that the first paragraph or two contain a capsule description of the whole story. The next three or four paragraphs contain a slightly more detailed description, and then the newspaper invites you to continue reading

FORMAL REPORT – ARRANGEMENT OF PAGES

PRELIMINARY PAGES

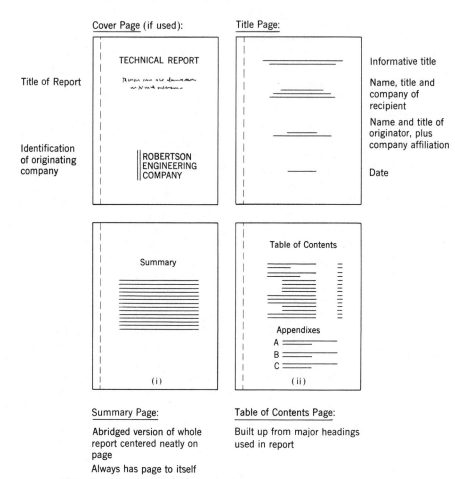

Cover Page (if used):

Title of Report

Identification of originating company

Title Page:

Informative title

Name, title and company of recipient

Name and title of originator, plus company affiliation

Date

Summary Page:

Abridged version of whole report centered neatly on page

Always has page to itself

Table of Contents Page:

Built up from major headings used in report

NOTE: Preliminary pages are identified by lower case roman numerals

Fig. 6–3   Formal Report–Normal Arrangement.

FORMAL REPORT – ARRANGEMENT OF PAGES
(Conclusions and Recommendation following Discussion)

Introduction, Discussion, Conclusions, Recommendation and Bibliography

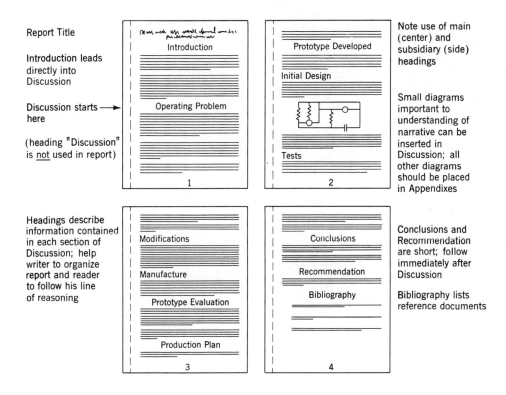

Report Title

Introduction leads directly into Discussion

Discussion starts → here

(heading "Discussion" is not used in report)

Headings describe information contained in each section of Discussion; help writer to organize report and reader to follow his line of reasoning

Note use of main (center) and subsidiary (side) headings

Small diagrams important to understanding of narrative can be inserted in Discussion; all other diagrams should be placed in Appendixes

Conclusions and Recommendation are short; follow immediately after Discussion

Bibliography lists reference documents

NOTE: Page numbers are arabic numerals either at foot of page as shown here or at top right-hand corner

Left-hand margin is wider to allow for binding edge

SUPPORTING DATA (Appendixes)

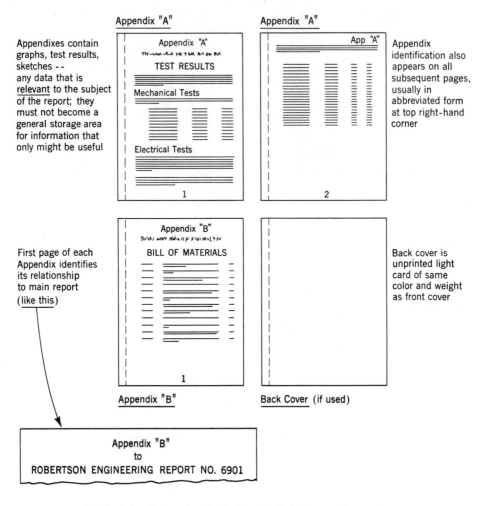

Appendixes contain graphs, test results, sketches -- any data that is <u>relevant</u> to the subject of the report; they must not become a general storage area for information that only might be useful

Appendix identification also appears on all subsequent pages, usually in abbreviated form at top right-hand corner

First page of each Appendix identifies its relationship to main report (like this)

Back cover is unprinted light card of same color and weight as front cover

NOTE: Appendixes are inserted in the order that they are referenced in the report

Each Appendix is considered a complete document in itself, thus is numbered separately using arabic numerals

Fig. 6–3 Formal Report—Normal Arrangement.

on a subsequent page. The final eight or nine paragraphs repeat the same story, but this time with more names, more peripheral information, and more details. The newspapers are catering to the busy reader who may not have time to read more than the opening synopsis, and also to the leisurely reader who wants to read all the available information.

A mini-report layout of this alternative arrangement appears in Figure 6–4. This figure contains only the central page of the layout (containing the introduction, conclusions, recommendations, and discussion). The remaining two pages, with the layout for the preliminary information and appendixes, are

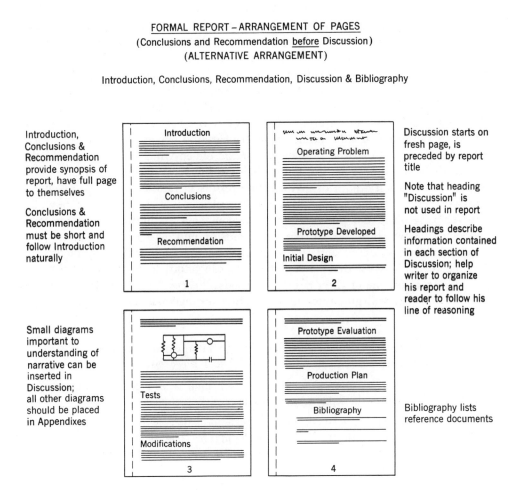

FORMAL REPORT – ARRANGEMENT OF PAGES
(Conclusions and Recommendation <u>before</u> Discussion)
(ALTERNATIVE ARRANGEMENT)

Introduction, Conclusions, Recommendation, Discussion & Bibliography

NOTE: Page numbers are arabic numerals either at foot of page as shown here or at top right-hand corner

Left-hand margin is wider to allow for binding edge

Fig. 6–4   Formal Report—Alternative Arrangement.

the same for both arrangements and can be obtained from sheets 1 and 3 of Figure 6–3. An example of the alternative arrangement of report parts can be seen in the technical report, "Selecting New Elevators for the Merrywell Building," which starts on page 201.

WRITING THE DISCUSSION

Regardless of the arrangement of report parts, the discussion remains the same. It is the "meat" of a report, the part in which the writer describes fully what he set out to do, how he went about it, what he actually did, and what he found out as the result of his efforts. He may consider the writing of these facts as a chore, and simply record them dutifully but unenthusiastically. Or he can respond to the challenge by organizing the information so interestingly that the reader feels almost as though he is reading a short story.

There is no reason why the discussion should not have a plot in much the same way as a short story. It can have a beginning, in which the writer describes the problem and outlines some background events; a middle, in which he tells how he tackled the problem; dramatic effect, which he can use to hold the reader's interest by letting him anticipate his successes and share his disappointments when a path of investigation results in failure; a climax, in which the writer permits the reader to share his pleasure as he describes how he eventually resolved the problem; and a denouement (a literary word for final outcome), in which he ties up the loose ends of information and evaluates the final results. Storytelling and dramatization can be such important factors in holding reader interest that no report writer can afford to overlook them.

By using one of three proven methods, the new report writer can build up a report that will guarantee an interesting arrangement of information. The methods are:

> Chronological development—in which the information is arranged in chronological sequence (in the order that the events occurred).
>
> Subject development—in which the information is arranged by "subjects," with the subjects grouped in a predetermined order.
>
> Concept development—in which the information is organized by "concept," with the writer developing his line of reasoning in roughly the order in which he thought the project through to a logical conclusion.

Chronological Development

A discussion that uses the chronological method of development is simple to organize and to write. It requires minimal planning. The writer simply arranges the major topics in the order that they occurred, eliminating irrelevant topics as he goes along. It is useful for very short reports, for laboratory reports showing changes in a specimen, for progress reports showing cumulative effects or describing advances made by a project group, and for reports of investigations that cover a long time and require visits to many locations to collect evidence.

But the simplicity of the chronological method is offset by some major disadvantages. Because it reports events sequentially, it tends to give equal emphasis to each event, regardless of importance. It can also be dull (unless expertly handled), with the result that readers quickly lose interest in the day-to-day record of the writer's investigative life. For example, if you read a report of an imaginary astronaut's third day in orbit, you do not want to read about *every* event in exact order. It may be chronologically true to report that he rose at 7:15, ate concentrated corn flakes and dehydrated milk for breakfast, sighted the second stage of his rocket at 9:23, carried out metabolism tests from 9:40 to 10:50, extinguished a cabin fire at 11:02, passed directly over Cape Kennedy at 11:43, and so on, until he retired for the night. But it makes deadly dull reading. Even the exciting moments of a cabin fire lose impact when they are sandwiched between routine occurrences.

Even when he is using chronological development, a writer must manipulate events if he is to hold his readers' attention. He must emphasize the most interesting items by positioning them where they will be noticed, and deemphasize less important details. Note how this has been done in the following passage, which groups the previous events in descending order of interest and importance:

> The highlight of Sam Smith's third day in orbit was a cabin fire at 11:02. Rapid action on his part brought the fire, which was caused by a short circuit behind panel C, under control in 38 seconds. His work for the day consisted mainly of metabolism tests and. . . .
> He sighted the first stage of his rocket on three separate occasions, first at 9:23, then at. . . , and passed directly over Cape Kennedy at 11:43. Meals were similar to those of the previous day.

This narrative uses *chronological* order to describe the events on a day-to-day basis, and a modified *subject* order to describe the events for each day.

Suppose that over a five year period one of H.L. Winman and Associates' technicians conducted an investigation into the effects of salt on concrete pavement. His final report would lend itself to chronological development because in the course of the investigation he would have recorded the extent of concrete erosion at specific intervals. He could describe how the erosion increased annually, perhaps in direct relation to the amount of salt used to melt snow each year, and then draw conclusions at suitable intervals. This chronological development would demonstrate the cumulative effect of the salt on the concrete.

Subject Development

If the previous investigation had been broadened to include tests on different types of concrete pavements, or if both pure salt and various mixtures of salt and sand had been used, then the emphasis would shift to an analysis of erosion on different surfaces or caused by various salt/sand mixtures, rather than a direct description of the cumulative effects of pure salt. For this type of report, the writer needs to arrange his topics in subject order.

The subject order could be based on different concentrations of salt and sand. The technician would first analyze the effects of a 100% concentration of salt, then continue with salt/sand ratios of 90/10, 80/20, 70/30, and so on, describing the results obtained with each mixture. Alternatively, he could select the different types of pavement as the subjects, arranging them in a specific order and describing the effects of different salt/sand concentrations on each surface.

A draftsman instructed to investigate blueprinting machines and recommend the most suitable model for installation in his department also would write his report using the subject method. He would tell the reader that he intends to evaluate the speed, economy of operation, and usefulness of each blueprinter, then emphasize the most suitable machine by discussing it either first or last. If he chooses to describe it first, he can state immediately that it is the best blueprinter, and say why. He can then discuss the remaining models in descending order of suitability, comparing each to the selected machine to show why it is less suitable. If he prefers to describe the best machine last, he can discuss the others in ascending order of suitability, stating why he has rejected each one before he starts describing the next. With this arrangement, he can also guide the reader's thinking so that he is ready to agree that the final model is the best.

Treatment of this topic by chronological development would be less satisfactory, because there are few events that could be reported in sequence. Possibly the draftsman could describe the blueprinters in the order that he examined them, but this rather dull approach might easily emphasize the wrong machine.

Another topic suitable for subject development would be an analysis of insulating materials conducted by a test laboratory. In his report, the technician who conducted the tests would first establish the parameters that the ideal material must meet (e.g. maximum permissible cost, insulation capability, resistance to moisture). He would then arrange the materials into groups having similar basic properties or falling within a specified price bracket. Within each group he would describe the test results for each material, drawing a conclusion as to its relative suitability for meeting the established parameters. The order in which he discusses the subjects in each group will be in ascending or descending order of material quality.

The report selecting a Residential Teachers' College for the City of Montrose (p. 184) uses the subject method of development. In this example the table of contents provides an immediate clue to the writing method used by the report writer.

Concept Development

By far the most interesting reports are those using the concept method of development. True, they need to be organized more carefully than reports using either of the previous methods, but they give the writer a tremendous opportunity to devise an imaginative arrangement of his topic—which is the best way to hold his reader's attention.

They can also be very persuasive. Because the writer organizes his report in the order that he reasoned his way through the investigation, the reader is able to understand the subject much more quickly. He will readily appreciate the difficulties encountered by the writer, and frequently draw the correct conclusions even before he reads them. This helps the reader to feel he is personally involved in the project.

The best advice I can offer on how to write a report using the concept method is to refer you to the elevator selection report that starts on page 201. The author has gradually developed his topic a step at a time, so that the reader can see the logic of his argument. There are many combinations of elevators that could fit the Merrywell Building, but we are ready to accept the author's selection because he has effectively persuaded us that his choice is the only sound one. You can apply his approach to your reports by thinking of each project as a logical but forceful procession of ideas that form a total concept. If you are personally convinced that the results of your investigation are valid, and remember to explain in your report *how* you reached the results and *why* they are valid, then you will probably be using the concept method properly.

Even though the concept method challenges the writer to fashion interesting reports, it is not always the best reporting medium. For instance, it could possibly have been used for the blueprinter report mentioned previously, but I doubt whether the depth of the topic would have warranted full analysis of the writer's ideas. When a topic is fairly clear-cut, and the writer can base his selection on direct analysis of each subject, there is no need to lead the reader through a lengthy "this is how I thought it out" discussion. Only when the subject is complex enough for the writer to help the reader by describing how he reasoned his way through the investigation is it necessary to use the concept method.

Whichever method the report writer uses, he must strike a happy medium between providing too much and too little information. His "story" must be fully developed, so the reader can follow it easily. However, it must not include any irrelevant information which may confuse and bore the reader.

The writer must make sure that his report is complete, but he must also know when to stop.

TYPICAL FORMAL REPORTS

I have selected two formal reports that are typical of the type of work expected of graduate engineers, technologists, and technicians by the time they apply for a job. These are:

Report No. 1: *Evaluation of Sites for Montrose Residential Teachers' College*
The writer of this report has placed the conclusions and recommendations after the discussion, following the normal arrangement of report parts. He has used the subject method of development.

Report No. 2: *Selecting New Elevators for the Merrywell Building*
In this report the writer has adopted the alternative arrangement of report parts, placing the introduction, conclusions, and recommendations as a single unit ahead of the discussion. He has used the concept method of development.

On the pages preceding each report I have inserted a few comments, plus some of the data that initiated each project and resulted in the report being written. Both reports have been typed single-spaced. Double-spacing is more common in industry, although some companies prefer to single-space their reports and leave two lines between main paragraphs, as in these two examples.

Formal Report No. 1
Evaluation of Sites for Montrose
Residential Teachers' College

This H.L. Winman and Associates' report was prepared originally as a student writing project in which the terms of reference closely paralleled a real situation. The original assignment instructions are printed here so that you may see how the writer used the information available to him, plus his own imagination, to write a very realistic report.

Assignment Details

You are employed by H.L. Winman and Associates, a Cleveland firm of consulting engineers, for whom you have been conducting a survey to determine the best site for a Residential Teachers' College in the City of Montrose, Ohio (population 275,000). Your task is to prepare a formal report that describes your findings and recommends the best location for the college.

H.L. Winman and Associates was engaged by the College's Board of Governors (composed mainly of local businessmen) to resolve the impasse caused by a three-way split over the choice of a site. Some members of the Board want to build the college at Bluff Heights, so that it will be a prominent feature in an urban redevelopment area; others would like to locate it at Varsity Valley beside the University of Montrose, where it will become part of a proposed city education center; the third group want to place it 22 miles out of the city, at Hartland Point, where it will develop an entity of its own. The three sites are described briefly below.

> *Bluff Heights.* Part of urban development area; triangular; very few trees; close to expressway; on a medium hill; commands a good view of city. Boundary to the south borders Heritage Industrial Park (light industry); to the east is mainline railway (Northern Railways route from Montrose to Waverly); to the northwest, Highway 79. Favored by three governors because it will rejuvenate a depressed area.
>
> *Varsity Valley.* Shallow valley between Montrose University and Varsity Heights residential area; although new, this residential area promises to become shoddy and run-down (already evidenced by decreasing property prices); valley tends to be damp and hold diesel and automotive fumes from expressway. Varsity Valley is a rectangular area bounded by Elm Creek and Varsity Heights to the south and east, and an expressway and U. of M. to the north; low vegetation and bush with scattered oaks and elms. Favored by three governors because it is close to U. of M. and so will help form an "Education Complex" for the city.
>
> *Hartland Point.* Superb location; rolling countryside; well-treed (conifers, elm, and poplar); drops gently down from northern boundary to picturesque Hartland River that bounds south and east edges of area. Boundary to the north is Highway 101, main highway from Montrose to Hartland (population 3265); to the west, Hartland Golf Course. Hartland is three-quarters of a mile east of the site. Favored by three governors because it would truly become a "residential" college; has ample space for expansion; can also provide space for building houses for staff. Disliked by others because it will be too remote from city.

Plans for the college call for immediate acquisition of 80 acres on which the main instructional complex and residence building will be erected. The availability of 80 additional acres adjacent to the first 80 is considered desirable: 40 for future college expansion and 40 for an additional residence building and a social activities center.

(*NOTE:* A rough sketch of the three sites and a table of comparative data were included with the original assignment. Since these also appear in the report, I have not repeated them here.)

Reading the Report

Try reading this report in the order in which the members of the Board of Governors would read it:

1. Scan the summary first to get a capsule comment.
   Does it make you want to read further?
2. Read the introduction, then turn to the conclusions and recommendations.
   By now you should have a good grasp of the report's main conclusions. Do you agree with them?
3. Now read the discussion.
   If you previously did not agree with the writer's conclusions, do you now? Has he used a logical and convincing method to develop his discussion?

Comments on the Report

The writer of this report has been very diplomatic. He has recognized that only three of the nine members of the Board of Governors will agree readily to his selection, and that he must persuade the remaining six that he has made a valid choice. By stating in the beginning of the summary that *all three* sites could be used successfully, he is forcing all the members to agree with him, and to read further to find out which site he has selected (and also to turn to the discussion to establish why it is better than the other two sites). Then, in the discussion, he writes an unbiased description of each site, presenting facts and figures without trying to persuade the reader that any one has greater advantages than the others (he even refrains from placing them in the normal ascending or descending order of preference). He may not fully persuade all the reluctant members, but at least he will convince them that he has reached an unprejudiced decision.

Some specific comments on the report are:

a. Words like "sociological" and "utilitarian" may make some readers of the summary reach for their dictionaries. Are they too technical? In this context I think not: the Board of Governors will expect to be faced with words that reflect the professional level of the study. They are the right words used in the right place.
b. The subject arrangement used for writing the discussion is clearly evident in the T of C. A reader turning to it from the summary can see immediately how the author has organized his report.

    c. The three parts of the introduction are easily recognizable: paragraph 1 is the *purpose*, paragraph 2 the *background information*, and paragraph 3 the *scope*.

    d. The writer should probably have discussed the sites in descending order of preference (the order used in the conclusions). He may have chosen not to do so simply to follow the order in which the board of Governors listed them.

    e. Two subtle negative phrases in the discussion of character on page 7 provide the first real hint that the author will select Varsity Valley. The first is "a *questionable* combination" at the end of the comment on Bluff Heights; the second is "activities would be *limited*" at the end of the Hartland Point comment. There is no negative comment concerning Varsity Valley.

    f. The rating system used to identify the best site at the end of the discussion seems artificial, but helps stress that there still is no element of bias in the writer's selection.

    g. The conclusions provide a convincing terminal summary that follows naturally from the introduction. The first paragraph is the main conclusion, and the remaining paragraphs are three subsidiary conclusions listed in decreasing order of importance.

H. L. Winman and Associates

PROFESSIONAL CONSULTING ENGINEERS
475 Reston Avenue-Cleveland, Ohio, 44104

FEASIBILITY STUDY

EVALUATION OF SITES

FOR

MONTROSE RESIDENTIAL TEACHERS' COLLEGE

Prepared for:

Board of Governors
City of Montrose Residential Teachers' College
Montrose, Ohio

10 October 1971

SUMMARY

Sound sociological and utilitarian reasons can be
found for recommending any one of the three sites
proposed as the permanent location of the Montrose
Residential Teachers' College. We have therefore
analyzed the suitability of each site on tangible
factors that would influence the initial construction
and eventual growth of the college. By comparing environment, accessibility, servicing, ambient noise,
soil condition, and land availability, we found that
although the margin of selection between the sites is
narrow, Varsity Valley offers the best all-round
location with maximum growth potential.

To our assessment we added one other less-tangible
factor: the effect that each site would have on
the character of the college. Again, Varsity Valley
offers the best location. A Teachers' College located
here would be able to develop a character of its own,
separate from but still within the overall educational environment already established by the University
of Montrose.

(i)

185

TABLE OF CONTENTS

EVALUATION OF SITES FOR MONTROSE RESIDENTIAL TEACHERS' COLLEGE

INTRODUCTION

This report is the result of our investigation of three possible sites for a Residential Teachers' College at Montrose, Ohio. The study was commissioned by the Board of Governors in May 1971 and was to be completed for their year end meeting on 10 December 1971.

The requirements for a proper site are dictated by the College development plan previously approved by the Board. This plan calls for an immediate land acquisition of 80 acres with an additional 80 acres to be available for future expansion.

As stipulated in the terms of reference we limited our study to three sites: Bluff Heights, Varsity Valley and Hartland Point. These sites are compared on the basis of environment, accessibility, servicing, land availability and cost, soil conditions, and noise level. The effect that each site would have on the character of the new college also is considered.

Appendix A contains a comparative analysis of the main features of each site.

BLUFF HEIGHTS

Bluff Heights was selected as a potential site because a Residential Teachers' College located here would contribute much to the redevelopment of the area.

Environment

The site is part of Urban Renewal Area No. 1, which is to begin construction 1 January 1972. The site is elevated some 300 feet above the general city area and commands an excellent view of the Wabagoon River and the city park to the east. It is a triangular site bounded on the east by the railway to Waverly, on the north-west by Highway 79 and on the south by Heritage Industrial Park.

- 1 -

187

In spite of being downwind from the industrial park there is no air pollution problem as the industries are of a "light" nature. That is, they are warehouses, assembly plants, printing plants etc, of a non-smoke or odor-producing variety.

Accessibility

Access to the Bluff Heights site is excellent. Highway 79 Expressway provides a direct route from the Montrose business area, a distance of 3½ miles, and there is a main arterial road in the industrial park that could be extended directly north into the site.

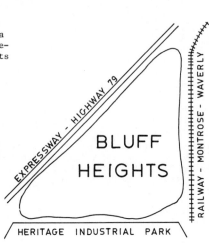

Servicing

All utilities including sewer, water, power, gas and telephone services are installed and available for immediate connection to the College buildings.

Land

Only 120 acres of land are available at this site, with no chance for future expansion beyond this size. Although this would provide the 80 acres required for immediate needs it could supply only half the land necessary for future expansion. The cost of the site would be $775 per acre for a total of $93,000.

Soils

The site is dry and well drained. A series of test holes revealed dry sandy clay which should present no particular foundation problems. Bedrock is at an average depth of 24 feet. Poured in place concrete piles could be used but the sandy soil may present some difficulties in boring. Driven timber, steel or precast concrete piles are recommended.

Noise

The average weekday noise level of 46 decibels at this site is rather high for a residential/educational community. Most of the noise is caused by the high speed truck traffic on the expressway and by the numerous trains. Traffic volume on both these routes is expected to increase by at least 20% over the next five years, so that the noise level also is likely to increase.

- 2 -

We took a series of noise readings over a seven-day period to determine the variations at night, during the day, and at weekends. These readings are shown in Appendix B for this and the other two sites.

VARSITY VALLEY

Varsity Valley was selected as a possible site for the Residential Teachers' College because of its proximity to the University of Montrose. It was considered that the two seats of higher learning would together provide the start of a comprehensive education complex for the city.

Environment

This is a rectangular site in a shallow valley in the predominantly residential area of Varsity Heights. Montrose University is directly north of the site but separated from it by the Circular Expressway. Elm Creek curves around the east and south sides of the site and runs between it and the Varsity Heights residential area. Although this residential area is only five years old, present decreasing property values may indicate future deterioration into a second class district. The presence of a high class education facility could avert this deterioration.

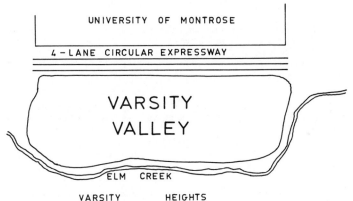

Because of its proximity to the University full advantage could be taken of all the University facilities, both technological and social. A full rapport could be established between faculties of the two establishments as well as the student bodies.

The valley tends to be damp and as a result holds the exhaust fumes from the expressway. This should be relieved to a large extent when the present

- 3 -

drainage program in the valley is completed. The green belt of shrubs and trees already planted along the expressway will trap much of the expressway fumes and noise after they have acquired a two or three year growth.

Accessibility

Access to the Varsity Valley site is almost as good as that to the Bluff Heights site. It is six miles from the city center to Varsity Valley, with good street connections from all areas of the city to the expressway. The University bus line runs to Montrose via the expressway from 6:00 a.m. to 2:00 a.m., and could easily loop through the college site.

Access to the University could be established by building an underpass or overpass across the expressway.

Servicing

The area is presently 40% serviced, with hydro, water and sanitary sewers now available. Gas lines, telephone lines and storm sewers can be readily extended from either the University or Varsity Heights and would cause no delays in construction.

Land

There are 150 acres of land available for present and future requirements -- 10 acres less than the desired ultimate area. This need not be a problem since at this site some of the proposed residences and recreation facilities could be eliminated by combining with and making use of the University facilities.

The total cost of the site would be $102,000 at an average price of $680 per acre. At present there is little demand for land in Varsity Valley and the price is low. Prices are likely to rise if a large portion of this land is acquired for the College.

Soils

A recently constructed ditch system now provides surface drainage into Elm Creek. This should relieve the present pooling conditions caused by the relatively impermeable dense clayey silt type soil which covers the area. Bedrock is quite close, being at an average depth of 12 feet. Foundation construction would therefore be quite cheap.

Noise

This site has a lower weekday noise level than Bluff Heights, being in the medium-low range of 40 decibels. The major cause of noise is vehicular

- 4 -

traffic on the expressway which, although traffic volume is expected to increase annually, will be screened by the now-growing shelter belt. The
present noise level is causing no problems at the University on the north
side of the expressway.

HARTLAND POINT

Although Hartland Point lies some distance east of Montrose, it was selected
as a potential site for the Teachers' College because it was considered that
its remoteness would contribute much to the residential nature of the training establishment.

Environment

The Hartland Point site is a country setting approximately three quarters of
a mile west of Hartland (population 3265). It is a rectangular site adjacent to and south of highway 101 which connects Hartland with Montrose.
The Hartland River flows along the south and east sides of the site, while
the Hartland Golf Course, which is a private members-only club, acts as the
west boundary. The site slopes from the highway to the river in a series of
gentle undulations. It is well treed with a variety of conifers, elms and
poplar.

HIGHWAY 101   MONTROSE - HARTLAND

HARTLAND
GOLF
COURSE

HARTLAND

POINT

HARTLAND   RIVER

The surrounding area is
completely agricultural,
although some fringe
commercial development is
now spreading west from
Hartland along highway
101 and could reach the
site area within five
years.

Accessibility

Road access to the site is good via highway 101 from either Hartland or
Montrose. However, the latter's city center is 22 miles away and travel
time especially in the winter would be a problem.

Bus service is now poor with only the H-B Bus Lines running twice daily past
the site. Undoubtedly this service would be improved if the College is
situated here, otherwise it will become almost isolated for persons without
their own means of transportation. Isolation could also raise overall

- 5 -

191

development costs by causing too high a proportion of the student body to seek on-campus accommodation.

Servicing

At present no services are available at the site, except for a combined power-telephone line along highway 101. All other services such as gas, water and sanitary sewers would have to be extended from Hartland. This would entail considerable expense which likely would have to be borne by the College. Even then water pressure would be insufficient and some type of booster and storage facilities would be required at the site.

Surface drainage could be run directly into the Hartland River.

Land

The proposed site comprises approximately 200 acres, which at the present price of $350 per acre would cost $70,000. This would be more than adequate for present and anticipated future needs. If further expansion is necessary, it would have to be north of highway 101.

Soils

The soil, which is of a generally sandy clay underlain with glacial till, provides excellent drainage. Below the till is bedrock at an average depth of 32 feet. Foundation requirements would vary for each building, depending on its loading and excavation depth, with a separate foundation study being necessary in each case.

Noise

Noise at the site is virtually nil, varying from 17 to 30 decibels. The chief cause of noise is traffic on highway 101, which occurs mainly between 8:00 a.m. and 6:00 p.m.

The site is, however, directly under the aircraft approach line for the east-west runway of the Montrose airport. With the expected increase in jet aircraft (particularly the Jumbo jets) the noise level could double in five to ten years.

COLLEGE CHARACTER

After investigating each site on the basis of its physical features and the facilities offered, we felt there was one more point that must be considered: This is the college character, or the personality that the college would develop in each location. This character will depend to a great extent on the college site and its immediate surroundings:

The Bluff Heights site will develop no real character of its own but will be a focal point for the urban rejuvenation of the area. It will be an educational complex in an industrial area; a questionable combination.

A college at the Varsity Valley site would fit in with the existing University complex to form a general educational center. It would become another one of the many colleges which now make up the University of Montrose.

An entirely separate college would develop at the Hartland Point site, completely unrelated to any particular city or University. Because of its relative remoteness it would be a residential college in every sense of the word. Outside activities would be limited.

SITE COMPARISONS

Because location of the college at any of the sites offers definite advantages either to the college or to the surrounding area, we decided to run a comparative analysis based on the factual conditions we have already discussed. We did this by rating the factual condition for each site as first, second or third, and awarding 1, 2 or 3 points according to our selection. We totaled the points for each site, then added one other important but less tangible factor -- college character -- to our analysis. In chart form the ratings are:

Factual Condition	Bluff Heights	Varsity Valley	Hartland Point
Environment	3	2	1
Accessibility	1	2	3
Servicing	1	2	3
Land	3	2	1
Soil	2	1	3
Noise	3	2	1
Total	13	11	12
Character	3	1	2
Total points	16	12	14

- 7 -

The site with the lowest total points score represents the best choice. This is Varsity Valley, which is third choice in none of the comparison points. In several cases where it is second choice, strong arguments can be made to bring it close to first choice.

CONCLUSIONS

Selection of the best location for the Montrose Residential Teachers' College is difficult because sound reasons can be found for building it at all three sites. Even when the physical advantages of each location are taken into account the margin of selection between the sites remains quite narrow.

At Varsity Valley the college would form part of a comprehensive education complex for the City of Montrose. Since this site offers the best combination of physical features, and also promises to contribute most to the development of college character, it would be our first choice.

At Hartland Point the college would become a truly residential educational institution in a pastoral setting. For some this would represent an ideal location, while for others it would mean isolation. Because its physical features and ability to contribute to college character are slightly less suitable than those at Varsity Valley, we would make this site our second choice.

At Bluff Heights the college would help to rejuvenate a depressed area of the city. This, we feel, would be the wrong reason for building the college at this site, because the relative isolation of an educational institution in an industrial setting would inhibit development of college character. Although the site offers excellent physical features, it has insufficient land to permit adequate future expansion. We therefore would rate this site as our third choice.

RECOMMENDATIONS

We recommend that Varsity Valley be selected as the permanent site of the Montrose Residential Teachers' College. We further recommend that the entire 150 acres be purchased immediately in readiness for future expansion, to avoid future speculative increases in land prices.

- 8 -

APPENDIX "A"
to

H. L. WINMAN AND ASSOCIATES REPORT

Evaluation of Sites for Montrose Residential
Teachers' College

ANALYSIS OF FEATURES

	LOCATION		
	Bluff Heights	Varsity Valley	Hartland Point
Environment	Generally industrial	Generally residential	Countryside
Land Available	120 acres	150 acres	200 acres
Cost per acre	$775	$680	$350
Distance from Montrose center	3½ miles	6 miles	22 miles
Accessibility	Excellent; good roads and bus service	Good-Very good; good roads and average bus service	Fair; good roads but poor bus service
Utilities & Services	Installed	40% installed	None; need to be extended from town of Hartland (3/4 mile)
Soil Conditions	Dry sandy clay; bedrock 24 ft (ave)	Dense clayey silt; bedrock 12 ft (ave)	Sandy clay underlain with glacial till; bedrock 32 ft (ave)
Noise level (average)	High (46 decibels)	Med-Low (40 decibels)	Very low (less than 30 decibels)

APPENDIX "B"

to

H. L. WINMAN AND ASSOCIATES REPORT

Evaluation of Sites for Montrose Residential Teachers' College

NOISE LEVEL READINGS RECORDED AT EACH SITE

Readings were noted over a 5-minute period each hour and then were averaged for entry in the table below.

	BLUFF HEIGHTS		VARSITY VALLEY		HARTLAND POINT	
Time	Weekdays	Weekend	Weekdays	Weekend	Weekdays	Weekend
0100	38 dB	36 dB	31 dB	30 dB	23 dB	21 dB
0200	36	35	30	30	22	19
0300	37	36	30	29	19	17
0400	34	33	29	30	18	18
0500	35	37	30	29	19	17
0600	39	36	36	31	23	21
0700	48	37	41	34	28	24
0800	53	39	46	34	34	26
0900	54	41	48	35	35	27
1000	52	47	47	36	34	26
1100	53	48	46	37	33	28
1200	53	49	48	36	35	29
1300	54	51	48	38	33	32
1400	53	51	46	39	34	31
1500	52	52	46	36	34	30
1600	54	48	49	37	36	29
1700	55	49	50	39	36	31
1800	47	46	46	36	34	31
1900	46	42	42	34	33	30
2000	43	40	38	35	28	27
2100	41	38	36	33	27	27
2200	39	38	34	32	26	23
2300	40	38	34	32	26	22
2359	39	37	32	31	24	22
Average:	45.6 dB	41.8 dB	40.1 dB	33.9 dB	28.9 dB	25.3 dB

Formal Report No. 2
Selecting New Elevators
for the Merrywell Building

The writer of this report has used the concept method of development to guide the reader through the maze of possible elevator combinations to the one choice that offers the most advantages to his client. It is a persuasive report, and quite convincing.

Before reading the report, read the client's letter authorizing H.L. Winman and Associates to initiate an engineering investigation (see p. 198). By comparing it with the report, you can assess how thoroughly the engineer responsible for the project has answered the client's requests. Also read the details describing the Merrywell Building and some of the data used by the engineer.

Data Used by H.L. Winman and Associates

1. The Merrywell Building is owned by Merrywell Enterprises Inc., which occupies the top two floors. It is a nine story commercial and light manufacturing building located at 617 Carswell Avenue, a busy thoroughfare in the Montrose, Ohio, downtown business district. It has a frontage of 180 feet, is 130 feet deep, and is 61 years old. The present owners purchased it in 1954 and made some moderate renovations to bring it up-to-date. These included installation of up and down escalators between the ground floor and the second floor, which is occupied by a popular cafeteria patronized by both the building's occupants and the staffs of offices occupying neighboring buildings. Two rather slow manually operated passenger elevators and an even slower freight elevator were left intact, although it was recognized that new elevators eventually would have to be installed.

2. The existing two passenger elevators are slow and cumbersome. They can carry only 11 passengers each, plus the operator, and their speed is only 10 seconds per floor. The freight elevator is 10 feet wide by 8 feet deep, but because of its age government inspectors have limited its load capacity to 2000 lb.

3. Merrywell Enterprises Inc. has set aside a budget appropriation of $250,000 for the purchase and installation of new elevators. They would like the cost to be less than budgeted, but would accept a slight increase above this figure if the right combination of elevators can be found to satisfy the needs of all the companies occupying the building.

4. Top management of Design Consultants Inc. and Vulcan Oil and Fuel Corporation would like an executive elevator for the sole use of executives of each major company in the building. Although such an elevator would be costly and have limited use, David P.

MERRYWELL ENTERPRISES INC.

617 Carswell Avenue

Montrose, Ohio, 45287

File: WDR/71/007

April 27, 1971.

Mr. Martin Dawes, P. Eng.
Head - Civil Engineering,
H.L. Winman and Associates,
Professional Consulting Engineers,
475 Reston Avenue,
Cleveland, Ohio, 44104

Dear Mr. Dawes:

The elevators in the Merrywell Building are showing their age. Recently we
have experienced frequent breakdowns and, even when the elevators are
operating properly, it has become increasingly evident that they do not provide adequate service at the start of work, at noon, and at the end of the
working day. I have therefore decided to install a complete range of new
elevators, with work starting in mid-August.

Before proceeding further I would like you to conduct an engineering investigation for me. Specifically, I want you to evaluate the structural condition
of my building, assess the elevator requirements of the building's occupants,
investigate the types of elevators available, and recommend the best type or
combination of elevators that can be purchased and installed within a proposed budget of $250,000.

Please use this letter as your authority to proceed with the investigation.
I would appreciate receiving your report by the end of June.

Yours sincerely,

David P. Merrywell,
President,
Merrywell Enterprises Inc.

DPM/ib

Merrywell indicates that he might accept the idea, but hesitates because of the problems it would introduce.

Comments on Report

A cover letter travels with this report. It simply refers to the client's letter of request and offers further consulting services.

The summary is short and direct because it is written primarily for one man: the President of Merrywell Enterprises Inc. It encourages him to read the report immediately, and to accept its recommendations, by offering the chance to save $25,000. (Barry Kingsley, the astute writer of this report, believes he can best appeal to an executive's heart by offering to place coins in his pocket!)

Although the background information contained in the first two paragraphs of the introduction seems to repeat details the client already knows, Barry recognizes it is necessary for the benefit of other readers who may not be fully aware of the situation in the Merrywell Building. He then defines the purpose and scope of the investigation by stating the client's terms of reference in paragraph 3 of the introduction. (Note that he has copied them almost verbatim from Mr. Merrywell's letter.)

The conclusions present Barry's answers to Mr. Merrywell's four requests. Their order is different from that in paragraph 3 of the introduction because it is normal to present the main conclusion first (in this case, the best combination of elevators that can be purchased within the stipulated budget), and to follow it with subsidiary conclusions in descending order of importance.

By using the concept method of development for the discussion, Barry permits the reader to follow his line of reasoning. He first evaluates the condition of the building to convince the reader that it is worthwhile to install new elevators. He then establishes the needs of the owner and tenants, and at the foot of page 4 summarizes them as a conclusion to the section. Then in the third section he establishes that it is reasonable to purchase the elevators from only one manufacturer. Having conditioned the reader to accept the factors that will eventually limit his selection to only one combination of elevators, he examines other combinations and demonstrates why each is unsuitable. Gradually he eliminates each combination until he leads the reader to the one that not only fits almost all the requirements tabulated earlier, but also offers a substantial saving in cost.

Conceivably, Barry could have stated in the beginning of the discussion that only one combination would be suitable, and then proved it in the rest of the report. This would have been a very direct method, but might not have been nearly as effective. It would not have had dramatic effect because the result would have been known from the beginning, and it would not be nearly as persuasive. A reader likes to be challenged, and appreciates a strong, convincing line of reasoning. He is less likely to resist a report's recommendations if the writer does not try to force them down his throat.

H. L. Winman and Associates

PROFESSIONAL CONSULTING ENGINEERS
475 Reston Avenue-Cleveland, Ohio, 44104

17 June 1971

Merrywell Enterprises Inc.
617 Carswell Avenue,
Montrose, Ohio, 45287

Attention: Mr. David P. Merrywell, President

Dear Sir:

We enclose our report No. 71-23 "Selecting Elevators
for the Merrywell Building", which has been prepared in response to your letter WDR/71/007 dated 27 April 1971.

If you would like us to submit a design for the en-
larged elevator shaft or to manage the installation project on
your behalf, we shall be glad to be of service.

Yours sincerely,

Martin Dawes, P. Eng.
Head, Civil Engineering

BVK:wp

encl.

H. L. Winman and Associates

TECHNICAL REPORT

SELECTING NEW ELEVATORS

FOR

THE MERRYWELL BUILDING

Prepared for:

Mr. David P. Merrywell, President

Merrywell Enterprises Inc.
Montrose, Ohio

Prepared by:

Barry V. Kingsley, P. Eng.

Senior Structural Engineer
H. L. Winman and Associates

Report No. 71-23

17 June 1971

SUMMARY

The elevators in the 63-year old Merrywell Building are to
be replaced. The new elevators must not only improve the
present unsatisfactory elevator service, but must do so within
a purchase and installation budget of $250,000.

Of the many types and combinations of elevators considered,
the most satisfactory proved to be four 8 ft by 7 ft deluxe
passenger elevators manufactured by the YoYo Elevator
Company, one of which will double as a freight elevator during
non-peak traffic times. This combination will provide the
fast, efficient service requested by the building's tenants
for a total price of $225,000, which will be 10% less than the
budgeted price.

(i)

TABLE OF CONTENTS

APPENDIXES

SELECTING NEW ELEVATORS FOR THE MERRYWELL BUILDING

Introduction

When in 1954 Merrywell Enterprises Inc. purchased the 46-year old Wescon
property in Montrose, Ohio, they renamed it "The Merrywell Building" and reno-
vated the entire exterior and part of the interior. The building's two
manually-operated passenger elevators and a freight elevator were left intact,
although it was recognized that eventually they would have to be replaced.

Recently the elevators have been showing their age. There have been frequent
breakdowns and passengers have become increasingly dissatisfied with the inadequate service provided at peak traffic hours.

In a letter dated 27 April 71 to H. L. Winman and Associates, the President
of Merrywell Enterprises Inc. stated his company's intention to purchase new
elevators. He authorized us to evaluate the structural condition of the
building, to assess the elevator requirements of the building's occupants, to
investigate the types of elevators available, and to recommend the best type
or combination of elevators that can be purchased and installed within the
proposed budget of $250,000.

Conclusions

The best combination of elevators that can be installed in the Merrywell
Building will be four deluxe 8 ft by 7 ft passenger models, one of which will
serve as a dual-purpose passenger/freight elevator. This selection will provide the fast, efficient service desired by the building's tenants, and will
be able to contend with any foreseeable increase in traffic. Its price at
$225,000 will be 10% below the proposed budget.

Installation of special elevators requested by some tenants, such as a full-
size freight elevator and a small but speedy executive elevator, would be
feasible but costly. A freight elevator would restrict passenger-carrying
capability, while an executive elevator would elevate the total price to at
least 20% above the proposed budget.

The quality and basic prices of elevators built by the major manufacturers
are similar. The YoYo Elevator Company has the most attractive quantity
price structure and provides the best maintenance service.

The building is structurally sound, although it will require some minor modifications before the new elevators can be installed.

- 1 -

Recommendations

We recommend that four Model C deluxe 8 ft by 7 ft passenger elevators manufactured by the YoYo Elevator Company be installed in the Merrywell Building.
We further recommend that one of these elevators be programed to provide
express passenger service to the top four floors during peak traffic periods,
and to serve as a freight elevator at other times.

SELECTING ELEVATORS FOR THE MERRYWELL BUILDING

Evaluating Building Condition

We have evaluated the condition of the Merrywell Building and find that it is
structurally sound. The underpinning done in 1948 by the previous owner was
completely successful and there still are no cracks or signs of further
settling. Some additional shoring will be required at the head of the elevator
shaft immediately above the 9th floor, but this will be routine work that the
elevator manufacturer would expect to do in a 63-year-old building.

The existing elevator shaft is only 24 feet wide by 8 feet deep, which is unlikely to be large enough for the new elevators. We have therefore investigated relocating the elevators to a different part of the building, or enlarging the existing shaft. Relocation, though possible, would entail major
structural alterations and would be very expensive. Enlarging the elevator
shaft could be done economically by removing a staircase that runs up the
center of the building immediately east of the shaft. This staircase is used
very little and its removal would not conflict with fire regulations since
there are also fire staircases inside the east and west walls of the building.
Removal of the staircase will enlarge the elevator shaft to 35 feet wide by
8 feet deep, which will provide just sufficient space for the new elevators.

Establishing Tenants' Needs

To establish the elevator requirements of the building's tenants we asked a
senior executive of each company to answer the questionnaire attached at
Appendix "A". When we had correlated the answers to all the questionnaires
we identified five significant factors that would have to be considered
before selecting the new elevators. (There were also several minor suggestions
that we did not include in our analysis, either because they were impractical
or because they would have been too costly to incorporate.) The five major
factors were:

> Every tenant stated that the new elevators must eliminate the lengthy
> waits that now occur. We carried out a survey at peak travel times
> and established that passengers waited for their elevators for as
> much as 70 seconds. Since passengers start becoming impatient after
> 32 seconds, we estimated that at least three, and probably four,
> faster passenger elevators would have to be installed to contend with
> peak-hour traffic.

- 3 -

Although all tenants occasionally carry light freight up to their offices, only Rad-Art Graphics and Design Consultants Inc. considered that a freight elevator was essential. However, both agreed that a separate freight elevator would not be necessary if one of the new passenger elevators was large enough to carry their displays. They initially quoted 9 feet as the minimum width they would require, but later conceded that with minor modifications they could reduce the length of their displays to 7 feet 6 inches. All tenants agreed that if a passenger elevator doubled as a freight elevator they would restrict freight movements to non-peak travel times.

The three companies occupying the top four floors of the building requested that one elevator be classified as an express elevator serving only the ground floor and floors 6, 7, 8 and 9. Because these companies represent more than 50% of the building's occupants, we consider that their request is justified.

Three companies expressed a preference for deluxe elevators. Rothesay Mutual Insurance Company, Design Consultants Inc, and Rad-Art Graphics all stated that they had to create an impression of business solidarity in the eyes of their customers, and they felt that deluxe elevators would help to convey this image.

The management of Rothesay Mutual Insurance Company and Vulcan Oil and Fuel Corporation requested that a small key-operated executive elevator be included in our selection for the sole use of top executives of the major companies in the building. We asked other companies to express their views but received only marginal interest. The consensus seemed to be that an executive elevator would have only limited utilization and the privilege of using it could easily be abused. However, we retained the idea for further evaluation, even though we recognized that an executive elevator would prove costly in terms of passenger usage[2].

We decided that the first two of these factors are requirements that must be implemented, while the latter three are preferences that should be incorporated if at all feasible. The controlling influence would be the budget allocation of $250,000 stipulated by the landlord, Merrywell Enterprises Inc. In decreasing order of importance, the requirements are:

1. Passenger waiting time must be no longer than 32 seconds.

2. At least one elevator must be able to accept freight up to 7 ft 6 in. long.

3. An express elevator should serve the top four floors.

4. The elevators should be deluxe models.

5. A small private elevator should be provided for company executives.

- 4 -

Researching Elevator Manufacturers

We asked the three major United States elevator manufacturers to furnish specifications of the types of elevators they would recommend for the Merry-well Building, plus price quotations and details of their maintenance policies. The catalogs at Appendix "B" show that except in appearance each company's elevators are basically the same and model for model carry very similar price tags. Significant differences are apparent only in discount policies and maintenance capabilities.

The YoYo Elevator Company of Chicago, Ill. offers a 10% discount to purchasers contracting for a multiple installation in which all elevators are identical. We queried the other two manufacturers, Jackson Elevators Inc. of Detroit, Michigan, and Matson Building Equipment Manufacturers of New York, N.Y. but neither would agree to incorporate a discount into their contract. Although such a discount agreement might limit the range of elevators from which we could make a selection, we consider that the incentive of a $20,000 to $25,000 reduction in price should not be overlooked.

The YoYo Elevator Company also is the only manufacturer with a service office in Montrose and thus can provide immediate response to calls for emergency maintenance. Both the other manufacturers contract with a local representative to carry out routine maintenance and state that they will fly in a maintenance team within 24 hours if major problems occur. We checked their reputations with the plant superintendents of local buildings using their elevators, and in both cases found that almost invariably delays of three or four days occurred before a maintenance team showed up.

Because of their advantageous pricing policy and superior maintenance service, we consider the new elevators should be selected from the range offered by the YoYo Elevator Company.

Selecting a Suitable Combination of Elevators

There are five basic elevators manufactured by the YoYo Elevator Company that are suitable for installation in the Merrywell Building. These comprise three sizes of passenger elevators that can be supplied in both standard and deluxe versions, a freight elevator, and an executive elevator (see Table I for a summary of their specifications and Appendix "C" for a more detailed description). From these five basic elevator types we derived numerous combinations of elevators that could be installed in the 35 ft by 8 ft space, and calculated the total cost of the most suitable combinations. Our aim was to select a range of elevators that would meet as many as possible of the tenants' requests within the proposed budget of $250,000.

By calculation, we determined that any three of the passenger elevators described above together could provide an adequate service for the present population of the Merrywell Building, but would be able to handle only a 4%

- 5 -

```
                              TABLE I


    Summary of Elevators manufactured by the YoYo Elevator Company
    that are suitable for installation in the Merrywell Building


    MODEL                  SIZE (w x d)     CAPACITY        SPEED
                              (ft)            (lb)        (sec/flr)


      A    Passenger       6 x 6            2000             5

      B    Passenger       7 x 7            2600             6

      C    Passenger       8 x 7            3000             7

      E    Executive       5 x 5            1500             2
           Passenger

      F    Freight         10 x 7.5         5000            15


    Note:  A more detailed description of these elevators is
           attached at Appendix "C".
```

increase in passenger traffic before again becoming overloaded. Since we
estimate that future changes in tenants could possibly increase the population by as much as 23%, we consider that four passenger elevators should be
installed. It would be a sounder proposition, both economically and
structurally, to anticipate this increase and install the fourth elevator
now, than to install it at a later date.

The installation of four passenger elevators would, however, impose a limit
on the size of freight elevator that can be installed. Even four of the
smallest (model A) passenger elevators would leave only 9 feet for a freight
elevator, thus automatically eliminating the 10-ft wide 5000-lb capacity
model F freight elevator. The only alternative would be to install one of
the largest standard passenger elevators (model C) as a freight elevator,
and accept its load restriction of 3000 lb and maximum interior width of
8 feet. (We have already established that such limitations would be acceptable to the building's tenants.) The price of this combination would,
however, exceed the budget by more than $43,000, as shown in Table II.

To bring the cost down we considered making the model C elevator a dual-
purpose unit, serving as a passenger elevator during peak traffic hours, and
as a freight elevator at other times. With this arrangement only three
regular passenger elevators would have to be purchased, and they could be
selected from any of the three models available. Table II shows space utilization and the price of the three possible combinations, all of which
would cost less than the proposed budget.

- 6 -

209

TABLE II

COST AND SPACE UTILIZATION
FOR FEASIBLE ELEVATOR COMBINATIONS

Combination:	1 "C" Freight PLUS 4 "A" Passenger Units	1 "C" Passenger/Freight Unit PLUS 3 Passenger Units:		
		3 "A"	3 "B"	3 "C"
Standard Passenger Units	$293,200	$235,200	$240,000	$244,800
Deluxe Passenger Units	$298,400	$239,100	$243,900	$248,700
Total Space Occupied	34.5 ft	28 ft	31 ft	34 ft

The combination proposing three model C passenger elevators and a model C passenger/freight elevator fills the available space most efficiently and has the greatest passenger capacity. It also has a significant advantage in that it qualifies for the 10% discount offered by the manufacturer if all elevators installed are exactly the same model and type. By purchasing four identical model C elevators the total purchase price would be:

	4 Model C Standard Elevators	4 Model C Deluxe Elevators
List price	$244,800	$250,000
Discount	24,480	25,000
Purchase price	$220,320	$225,000

With this combination the need to provide an express elevator serving the ground floor and floors 6, 7, 8 and 9 is easily satisfied. Probably the best elevator to use would be the "freight" elevator, since it could be programed to stop only at the preselected floors during peak travel hours, and could revert to its normal role as a freight elevator during off-peak hours (thick protective padding "curtains" could be used to protect the interior paneling when freight is being carried). A second passenger elevator could be similarly programed if peak-hour passenger loads dictate more than one express elevator is needed. We do not consider that an express elevator is necessary during non-peak travel times.

The combination of four deluxe model C elevators would therefore satisfy all but one of the tenants' requirements established at the beginning of this report. It reduces waiting time to less than 32 seconds; it provides a freight elevator able to handle articles up to 8 ft long; it provides an express elevator serving the top four floors; and it offers prestige. Not only does it meet the budget appropriation of $250,000 established by Merrywell Enterprises Inc, but it also qualifies for a discount of $25,000 which will bring the total purchase price down to $225,000.

This arrangement does not, however, satisfy the request for a small private elevator for company executives. The only combination in Table II that leaves enough room in the elevator shaft to install an executive elevator consists of 3 model A passenger units and 1 model C passenger/freight unit. With the addition of a model E executive elevator, the occupied space would increase to 33.5 feet, and the cost would rise to $306,500 or $310,400, depending on whether standard or deluxe passenger elevators are installed. Although this combination would provide superior service for all of the building's occupants, we consider that its extra cost of at least $56,500 more than the budget, and $81,500 more than the cost of four model C deluxe passenger elevators, is too expensive to warrant the limited extra convenience it offers.

REFERENCES

1. Johnson, Byron. "Don't Keep Passengers Waiting" in Elevators in the Industrial Complex. New York: Antrim Book Company, 1971, p. 75.

2. Krauston, K. K. "The Executive's Private Elevator" in Business and Industry, Vol. 27, No. 5, May 1970, p. 43.

APPENDIX "A"

to

H. L. WINMAN AND ASSOCIATES REPORT NO. 71-23

QUESTIONNAIRE

To assist us in selecting a new range of elevators for the Merrywell Building you are asked to complete the following questionnaire:

Company Name:..................................... Floor:...................

1.   Do you consider the present elevator service satisfactory? ....

      ....

2.   What is the average length of time your employees have to wait for an

      elevator ....

3.   Do you need a freight elevator? (If so please state dimensions and

      weight of largest loads carried). ....

4.   What percentage of your employees' travel is between your floor and

      the ground floor?......%; between your floor and other floors?.......%

5.   Would a small private elevator be of use to your company's executives?

      .... If so, how many members of your staff would use it?........

6.   If you have additional comments or suggestions, please insert them

      here:

APPENDIX "B"

to

H. L. WINMAN AND ASSOCIATES REPORT NO. 71-23

ELEVATOR MANUFACTURERS' CATALOGS

This appendix contains catalogs supplied by the three major North American elevator manufacturers:

1.   YoYo Elevator Company, Chicago   Page   2

2.   Jackson Elevators Inc., Detroit, Mich.   Page 13

3.   Matson Building Equipment Manufacturers, New York   Page 27

Price quotations and maintenance policies appear on the final page of each catalog.

(Note to reader: because of their bulk, the catalogs have been omitted from this report)

- 1 -

H. L. WINMAN AND ASSOCIATES REPORT NO. 71-23

DESCRIPTION OF SUITABLE ELEVATORS

The new elevators will be selected from the following range offered by the YoYo Elevator Manufacturing Company. Standard elevators have brown metal paneling, vinyl-tiled floors, and plain trim on doors and exterior; deluxe units have deep-pile carpeting, mahogany paneling, fancy trim inside and out, and an emergency telephone. All passenger elevators are fully automatic.

TYPE A

Size 6 ft wide x 6 ft deep; capacity 2000 lb; speed 5 seconds per floor. Price: standard unit $58,000; deluxe unit $59,300.

TYPE B

Size 7 ft wide x 7 ft deep; capacity 2600 lb; speed 6 seconds per floor. Price: standard unit $59,600; deluxe unit $60,900.

TYPE C

Size 8 ft wide x 7 ft deep; capacity 3000 lb; speed 7 seconds per floor. Price: standard unit $61,200; deluxe unit $62,500.

TYPE E (Executive)

Deluxe unit 5 ft wide x 5 ft deep; capacity 1500 lb; very high speed of 2 seconds per floor (on non-stop runs of 6 floors or more). Price $71,300. Two versions available:

        Type EB - push-button operated
        Type EK - key-operated

TYPE F (Freight)

Rugged unit 10 ft wide x 7 ft 6 in. deep; manually operated; 5000 lb; slow speed (15 seconds per floor) Price: $65,000

Elevator sizes quoted are exterior dimensions. An additional clearance of 6 in. is required between each elevator, and 3 in. between each end elevator and the wall. Prices include purchase, installation, and one year's free maintenance.

ASSIGNMENTS

The nine formal report writing assignments on the following pages require varying levels of technical knowledge. Generally, the technical level increases with each assignment. All the data required to write a report is supplied in projects 1 to 3. You will need to spend time researching information, either in a technical library or from resources in your local area, for projects 4 to 9.

Project No. 1—Moving a Bus Stop

H.L. Winman and Associates has been carrying out a transportation study for Montrose (Ohio) Community College, and will shortly be issuing a report recommending changes in traffic patterns and improvements in transit services. These, it is hoped, will ease the traffic bottlenecks that occur in the mornings and evenings at major intersections near the college. Project coordinator is Ian Bailey.

In response to a telephone call from Darryl Kane, the college's Technical Services Administrator, Ian Bailey assigns you to a separate project at the college. He tells you that Mr. Kane will give you all the details, and that your time is to be charged against Montrose Community College Project No. M71-C.

When you visit Darryl Kane the following morning he tells you that he has received a series of noisy telephone calls from an indignant cafe owner, Georges V. Picolino, who operates a restaurant at 227 Wallace Avenue, just south of Main Street. Yesterday, Mr. Picolino visited the College.

"Indignant" hardly described Mr. Picolino adequately. He was bristling with anger. He had been to the Transit Office; he had been to the Montrose Streets and Traffic Department; he had even called at the home of his City Councillor; and to all he woefully related his tale. They all did the same thing: told him they cannot help him and that he should go and see someone else. Angry at being shuffled from person to person, none of whom showed any real interest in his problem, Mr. Picolino finally turned to the college, because it is college students who are the source of his annoyance.

With some difficulty, Mr. Kane calmed him down and elicited a collection of details from which he pieced together the cause of Mr. Picolino's indignation. Apparently, college students traveling from east and west Main Street disembark from their buses at Wallace Avenue and wait for the college bus at a special stop on Wallace Avenue immediately in front of Picolino's Pantry. On cold or wet mornings (which at this time of year means every morning), the students crowd into the Pantry to keep warm while they wait. Georges Picolino's complaint is: (1) they don't buy anything (other than an occasional cup of coffee); and (2) they discourage other customers from coming in. The cafe becomes so crowded that passers-by think it is full and turn elsewhere to seek their breakfast. This is the time of day when Mr. Picolino used to do a lot of business, but since the bus stop was put in five months ago his early morning business has dropped to virtually nil.

Mr. Kane soothed Mr. Picolino's ruffled surface by promising that the problem would be resolved shortly, and then placed the telephone call to Ian Bailey that resulted in your visiting the college this morning. He asks you to evaluate Mr. Picolino's complaint and, if it is valid, to investigate ways to resolve it. He asks for a report of your findings that he can present to city and college officials.

.   .   .

In your investigation you take the following approach, asking yourself pertinent questions as you go along:

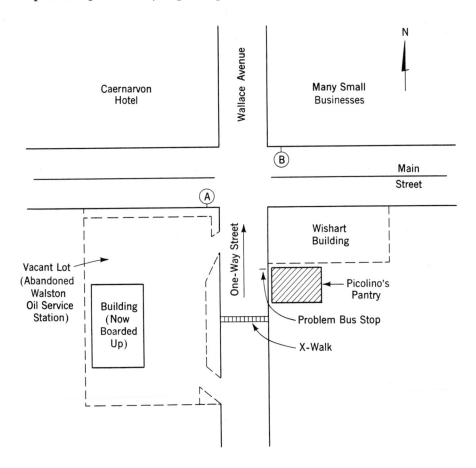

1. Is Mr. Picolino's complaint warranted?
   Yes, it seems to be. You check on two consecutive mornings:

Time	No. of Students in Cafe on Wednesday	No. of Students in Cafe on Thursday
0700	0	0
0710	5	0
0720	8	7
0730	13	10
0740	15	18
0750	27	36
0800	42	39
0810	23	31
0820	6	5
0830	1	0

   a. Mr. Picolino informs you that these are quieter mornings than usual. You check up and find that this week 35% of the college's student population are writing special tests in the afternoon and are unlikely to be attending classes in the morning.
   b. There are seats for 36 customers in Picolino's Pantry.
2. How frequent are the buses?
   A bus loads passengers at the bus stop at 0715, 0732, 0745, 0755, 0805, 0815, and 0828.
3. What is the bus route in this area?
   The bus travels north along Wallace Avenue (a one-way street), then turns left to travel west along Main Street.
4. Could the bus stop be moved?
   There are four possible directions:
   a. South: Yes. But it would have to be at least 120 yards south; there is a crosswalk immediately south of the present stop.
   b. North: No. No room between the existing stop and Main Street, and Wallace Avenue north of Main Street is not on the bus route.
   c. East : No. Not on bus route.
   d. West : Yes. By Caernarvon Hotel.
5. What factors affect the location of the bus stop?
   a. Connecting bus routes drop passengers at points A and B on Main Street (refer to diagram).
   b. Montrose Rapid Transit System insists that the connecting stop must be within 100 yards of the intersection of Main Street and Wallace Avenue.
   c. There is a cab rank outside the Caernarvon Hotel, but Montrose Streets and Traffic Department is adamant: It will not authorize a bus stop in that block.
   d. The bus stop could be moved directly across the street, to the west side of Wallace Avenue, where there are no stopping re-

strictions. But this poses two problems: Bus doors would open onto the traffic side of the bus rather than onto the sidewalk, and it is still too close to Picolino's Pantry.

6. You note that the lot at the southwest corner of the Main and Wallace intersection is vacant. Previously a Walston Oil service station, it is now abandoned and has been for ten months; the building is boarded up. You investigate ownership. The land was purchased by the City of Montrose six months ago. The city plans to build a car park on the lot, but not for five years, when it plans to acquire adjoining lots.

7. An idea begins to form: Why not build a bus shelter on the vacant lot and route the bus through the lot? (It could use the old entry and exit ramps built for the service station.)

8. You discuss your idea with the Traffic Engineer at Montrose Streets and Traffic Department (D.V. Botting), who says his Department would not object providing the shelter conforms to City of Montrose, Transit Division, Building Construction Specification BCS 232. You also check with Ken Whiteside of the Montrose Rapid Transit System, who says his department will go along with it. (He also says Transit will pay the cost of building a shelter on the lot, providing the cost is no greater than $1200.) Then you check with the City of Montrose (Planning Division) and get approval to use the land for five years. The city will also authorize expenditure of $300 a year for general upkeep of the lot. You ask all three parties to state their agreement with your plan in writing, and you receive letters from them indicating their willingness to participate.

9. You write to three contractors asking them to give you a price quotation, telling them that the shelter must be able to hold at least 40 people, that the price quotation must include an adequate electric heating system (minimum temperature +50°F), seats, and lighting, and that the work must be completed within three weeks of receipt of an order. You also inform them that the work must conform to City of Montrose Bylaw 61A.

10. The lowest price quotation you receive is submitted by Donovan Construction Company, 2821 Girton Boulevard, Montrose, whose price of $1945 is still $745 above the maximum stipulated by Montrose Transit. (Other quotations were $2367 and $2488.) You visit Gavin M. Donovan, owner-manager of Donovan Construction Company, to see if he can bring his price down to within Montrose Transit's limit.

"Impossible" he says. "No way! Nobody could do it."

Then, as an afterthought, he adds: "But tell you what: We could do as good a job by renovating the service station."

He digs out some files from his desk. "See here," he says, "we ran an estimate on renovating the existing building and installing seats. Electric heat and light are already in it. The price would be $965—and I'll throw in some landscaping."

Since this resolves the problem, you return to your office and prepare a report for Darryl Kane. You decide to prepare a formal investigation report, since he will want to submit copies to all parties involved, and to use it as the document for implementing conversion of the Walston Oil building and moving the bus stop.

Project No. 2—Evaluating Employment Possibilities

H.L. Winman and Associates has been engaged by your local Department of Education to study employment patterns for technical graduates of junior colleges and technological institutes over the past six years. The intent is to assess which fields offer the best employment opportunities for graduates. The project is authorized by Dennis M. Carstairs, Coordinator of Special Services in the Department of Education.

At H.L. Winman and Associates the project has been directed by Vic Braun, head of the Special Projects Department. Several months have been spent conducting a survey of a random selection of graduates (some difficulty was experienced in tracing graduates). Results of the survey have been tabulated, and Vic Braun has assigned you to analyze them and prepare a report of your findings. Factors you should consider are:

> Employment patterns and trends.
> Probable future trends for particular fields.
> Fields that offer the best and worst employment opportunities.
> Which groups of employers offer the best opportunities from the
> point of view of:
>> 1. Company growth.
>> 2. Salary.
>> 3. Stability.
> The number of graduates likely to be hired in the next two years.
> Whether colleges are churning out more graduates than the employment
> market can bear.

The survey divided employers of graduates into the following general categories:

> 1. Engineering and Research Organizations
> 2. Project Management Organizations
> 3. Manufacturing Organizations
> 4. Education
> 5. Government
> 6. Service Organizations
> 7. Miscellaneous Technical Employers

SURVEY OF TECHNICAL GRADUATES BY EMPLOYMENT CATEGORIES

Employment Category	Total No. Currently Employed In Category	Number Currently Employed in Each Category (according to year of graduation) Plus Current Average Monthly Salary — Number of Years Since Graduation					
		6 years	5 years	4 years	3 years	2 years	1 year
T	1666	140	195	241	323	368	399
1	274	31 ($725)	42 ($700)	45 ($675)	48 ($640)	57 ($600)	51 ($575)
2	181	2 ($675)	5 ($650)	12 ($630)	32 ($600)	58 ($570)	72 ($540)
3	115	7 ($655)	21 ($630)	30 ($615)	25 ($575)	18 ($545)	14 ($520)
4	174	23 ($650)	28 ($640)	26 ($630)	30 ($610)	32 ($585)	35 ($565)
5	109	18 ($625)	11 ($605)	15 ($595)	28 ($570)	16 ($550)	21 ($520)
6	219	36 ($715)	38 ($690)	32 ($665)	47 ($645)	35 ($600)	31 ($570)
7	205	15 ($660)	29 ($635)	34 ($610)	51 ($585)	37 ($570)	39 ($550)
0	291	8 ($525)	20 ($530)	39 ($525)	41 ($510)	86 ($485)	97 ($455)
U	98	0	1	8	21	29	39

In the survey of each year's graduates in the table on page 220, the following coding was used:

T—Total number of graduates
1 to 7—Fields in which employed
0—Graduates employed in fields other than that for which they were trained
U—Graduates currently unemployed
Two figures appear against each coding in the table:
  The number of graduates *currently* employed in that category
  The average monthly salary *currently* earned by these graduates

To ensure accurate comparison between years, the number of graduates surveyed for each year was maintained as a constant percentage of the total technical graduates for that year.

A factor not evident in the table applies to employment category 6—Service Organizations. A breakdown of this category shows that in one group of employers (Computer Companies), the statistics contradict the results for the other employee groups:

Graduates Currently Employed by Service Organizations—Category 6

No. Years Since Graduation	Category 6 Total	Computer Companies	Other Companies
6	36	4	32
5	38	7	31
4	32	15	17
3	47	29	18
2	35	26	9
1	31	28	3

Analyze the results of the survey and then write a formal report of your findings for Mr. Carstairs. Prepare charts and graphs to illustrate your report.

Project No. 3—Stormwater Disposal, Cayman Flats

The Fairview Development Company wants to develop an area of virgin land known as Cayman Flats on the southern perimeter of Montrose, Ohio. The City of Montrose displays interest in the proposal, and asks for a formal presentation of the company's development plans.

The land is flat, and Fairview Development Company soon realizes that it has a stormwater drainage problem to overcome before it can complete its presentation. It calls in H.L. Winman and Associates to resolve the problem (see letter). The project is assigned to you.

CHAPTER 6

FAIRVIEW DEVELOPMENT COMPANY
212 Bligh Street
Montrose, Ohio
45287

Harvey L. Winman, President
H.L. Winman and Associates
475 Reston Avenue
Cleveland, Ohio, 44104

Dear Mr. Winman:
   We are preparing a feasibility study for the City of Montrose, in
which we are proposing to develop the Cayman Flats area to the
south of the city as a new residential district. This low, flat land offers
drainage problems because of its distance from the Wabagoon River.
The storm sewers of the intervening developed areas cannot be used
since they have insufficient capacity to handle the additional stormwater run-off that Cayman Flats will generate.
   We are asking you to conduct an engineering investigation into
this stormwater disposal problem, and to recommend an economical
method that we can include in our presentation to the City of
Montrose.

                                        Yours sincerely,

                                        Frederick C. Magnusson,
                                        President
FCM/jms

.   .   .

   You start your investigation by examining Cayman Flats. It is generally
flat, low-lying, and frequently waterlogged. The maximum variation in height
is 9 feet; its area is 1594 acres. To the east is a railway line (Northern Railways) into Montrose center, to the north a residential area, and to the west
and south lie arable land 70% cultivated (see map).
   You calculate the maximum stormwater run-off for the land in its present
condition. (Stormwater is rainwater that must be drained from the land
quickly to prevent flooding of low-lying areas and basements. Maximum
stormwater run-off is the largest amount of water likely to occur; it is calculated on past records of maximum precipitation accumulated from the
heaviest rainstorms.) Using the Rational Formula, you calculate that a 21 foot
diameter culvert would be required to handle the heaviest peaks.
   Since such a large culvert would offer construction problems, use a lot of
property, represent an eyesore, and be a source of danger to small children,
you look for other methods. The most obvious is to construct storm sewers
throughout Cayman Flats before development starts. Because the run-off would

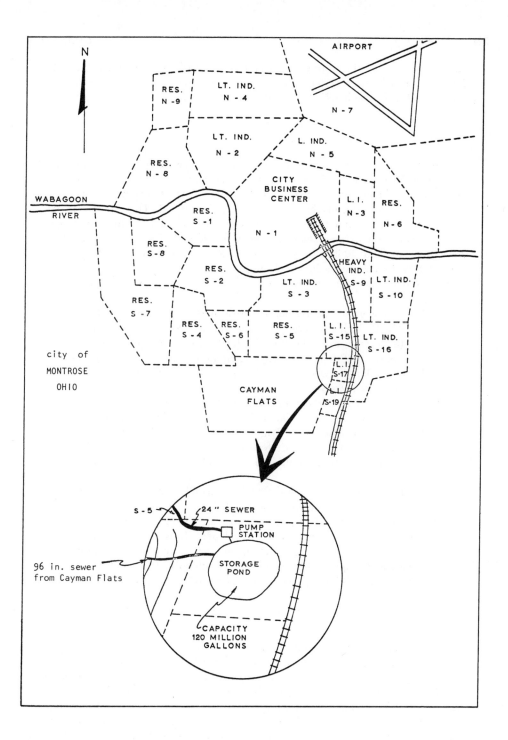

N

AIRPORT

RES.
N-9

LT. IND.
N-4

N-7

LT. IND.
N-2

L. IND.
N-5

RES.
N-8

CITY
BUSINESS
CENTER

L.I.
N-3

RES.
N-6

WABAGOON

RIVER

RES.
S-1

N-1

RES.
S-8

RES.
S-2

HEAVY
IND.
S-9

LT. IND.
S-10

RES.
S-7

LT. IND.
S-3

city of
MONTROSE
OHIO

RES.
S-4

RES.
S-6

RES.
S-5

L.I.
S-15

LT. IND.
S-16

L.I
S-17

CAYMAN
FLATS

S-19

S-5

24" SEWER

PUMP
STATION

96 in. sewer
from Cayman Flats

STORAGE
POND

CAPACITY
120 MILLION
GALLONS

then be channeled, the stormwater would be "staged" (that is, reach the out-fall, or disposal sewer system, as a series of peaks). You estimate that the largest peak could be handled by a 96 inch storm sewer.

.   .   .

Your next step is to examine the territory between Cayman Flats and the Wabagoon River. The whole area has been developed; to build a sewer of the size required directly to the river would be phenomenally expensive, and you would likely have difficulty in getting municipal approval. The distance is too great to build it around the southwest perimeter of the city. But there is a 40 foot wide belt of open land on the west side of the railway line. It is owned by Northern Railways, who agree to lease it to the city for $1200 per year, and will guarantee a 20 year lease.

Some other factors you discover are:

1. Storm sewers of the residential areas north of Cayman Flats are of only 18 inch diameter in zones S-4 and S-6, and 24 inch diameter in zone S-5. Those in S-4 feed into the 30 inch sewers in S-2; those in S-6 run directly down to the river, following the boundary line between S-2 and S-3; those from S-5 feed into the 36 inch sewers in S-3.

2. The municipalities view with alarm your inquiries to find an easy route to the river for a 96 inch storm sewer through zones S-3, S-5, and S-6. They fear (rightly) that you could not excavate deeply enough to avoid interfering with the existing services. (Too deep an excavation would foil the natural flow into Wabagoon River.)

3. Cayman Flats slopes very slightly downhill from the southwest corner to the northeast corner. Total drop is 9 feet 3 inches.

4. Although zones S-15, S-17, and S-19 are all classed as "light industry," S-17 has never been developed. It is a small zone of approximately 95 acres, lower in elevation than the surrounding zones, and hence rather swampy. Undoubtedly the amount of fill needed to build it up has hindered its development.

.   .   .

At this point an idea occurs to you. If zone S-17 is already acting as a collection point for some of the stormwater from the surrounding zones (principally Cayman Flats), why not excavate it even further and use it as a quick run-off storage pond? If it were big enough you could hold all the stormwater run-off from Cayman Flats, and then pump it at a controlled rate into a smaller diameter outfall sewer from S-17, along the railway line, to the river. It would mean building a pump station (it would have to be automatic), and fencing the storage pond, but it would be feasible. You make some quick calculations:

Size of storage pond required: 120,000,000 gallons
Size of outfall sewer to river: 24 inch diameter

To evaluate the ideas so far you work up some cost estimates:
a. Cost of installing storm sewer system throughout Cayman Flats $487,000
b. Cost of 96 inch diameter storm sewer along railway line from Cayman Flats to Wabagoon River $672,000
c. Cost of 24 inch diameter storm sewer over same route $427,000
d. Cost of purchasing zone S-17, excavating a 120 million gallon storage pond and fencing it, and building an automatic pumphouse and pump system $321,500
e. Pumphouse annual maintenance and operating costs $ 3,500

At this stage you feel your problem has been resolved and you have a reasonable proposal to offer Fairview Development Company. Your report is almost finished before another obvious and logical idea occurs to you: If you are going to *control* the flow of water into the outfall storm sewer, why not control it even further and pump the stored stormwater into the *existing* 24 inch storm sewers of zone S-5? You would have to wait until the stormwaters from S-5 had been fed into the river, but this would be no problem because you have planned an oversize storage pond that could hold the water from the heaviest storms recorded during the past 30 years. All you would have to do is obtain municipal approval and build a 24 inch sewer from the pumphouse beside the pond to one of the storm sewers in Zone S-5 (see sketch).

You approach the City of Montrose with your unusual idea. The City Engineer hesitates briefly (for three weeks), and then gives approval in principal. You then calculate one more cost:

f. Cost of building a 24 inch storm sewer from the pumphouse in zone S-17 to southeast tip of zone S-5, and connecting it to the storm sewer in zone S-5 $ 82,000

When you are preparing your report for Fairview Development Company, you should bear in mind how they intend to use it. They will probably attach it to their proposal to the City of Montrose as evidence that they have researched and resolved the stormwater disposal problem. Hence you must write with the knowledge that although you are addressing it to Fairview Development Company, the ultimate readers are likely to be the City Councillors of the City of Montrose.

Project No. 4—Installing Dial-a-Wash in the United States

You are to assume that when Mr. Ralph Rosenkratz, President of Dial-a-Wash Inc., 3300 Mountain Drive, Topeka, Kansas, visited H.L. Winman and Associates earlier this month, he discussed his company's plans to build a series of semiautomatic dial-operated car wash systems in major centers in the United States, with the first to be installed in your city. He was looking not only for a suitable site but also for a firm of consulting engineers to research other sites and manage the construction of Dial-a-Wash systems.

Dial-a-Wash systems have been very successful in Great Britain. They are built to a standard design, each installation having five bays. For 50 cents, a customer washes his own car with a high pressure hose and brush, using a special type of detergent patented by the Dial-a-Wash company. The remainder of the operation is automatic, the customer dialing the pre-rinse, after-rinse, dry and wax operations from inside a telephone-type booth.

Location of the Dial-a-Wash systems is important. They must be easily accessible from a major trunk road, have space for entry and exit ramps, and be away from competition. Each bay is 14 feet wide by 25 feet long, and the whole unit measures 90 feet wide by 25 feet deep.

Mr. Rosenkratz asked H.L. Winman and Associates to assess sites in your area and to recommend the one most suitable for development. He requested a proposal that justifies the choice of site, indicates its availability for either purchase or lease, and quotes a firm price for building the Dial-a-Wash unit.

Martin Dawes assigns the project to you, with instructions to do a thorough research job. He stresses that your proposal must convince Mr. Rosenkratz that he should appoint H.L. Winman and Associates as project engineers, since there will be many more Dial-a-Wash units to follow.

This assignment will require you to identify several sites and to assess their suitability with regard to traffic movement, competition, parking, means of exit and entry, availability, zoning, and price (either for purchase or lease). You should also establish whether the cost of the land, plus annual taxes, will balance anticipated revenue to make the Dial-a-Wash a paying proposition at each location.

You may assume that you have a firm price proposal from Pierce Construction Company to build the Dial-a-Wash and approaches to it for $27,450, and that the wash equipment will cost $16,500. Your fee for the initial proposal will be 1.75% of the cost of the land, and as Project Engineers during the construction stage, 6.5% of the contractor's price.

Project No. 5—Conducting a College Travel Survey

You are to assume that the Streets and Traffic Department of the municipality in which your college is located has engaged H.L. Winman and Associates to carry out a mode of travel and route study of all students and staff at the college. Additionally, the company is to identify when traffic peaks occur at a major intersection (or major intersections) near the college. You and several other H.L. Winman and Associates' employees have been assigned to this project.

The Streets and Traffic Department wants a 100% survey of all students and staff at the college. If a 100% survey is not feasible, then a smaller sample may be taken with the results scaled up to represent a 100% survey.

Information required by the Streets and Traffic Department:

A. For each person surveyed:
   1. General location of residence
   2. Mode of travel (car, bus, bicycle, walk)
   3. If mode of travel is by car, whether a driver or passenger
   4. Normal time of arrival at the college
   5. Normal time of departure from the college
B. For each intersection surveyed:
   1. Traffic counts during normal arrival and departure hours of college students and staff
   2. Counts of traffic traveling in all major directions at the intersection
   3. Separation of traffic counts into:
      3.1. Traffic bound to or from the college
      3.2. All other traffic
   4. Times and locations of traffic tie-ups (with, if possible, the cause)

This project requires the class to:

a. Organize the overall task.
b. Assign project groups to plan, organize, and administer difficult phases of the task.
c. Coordinate the work done by each project group.
d. Adopt a task force approach for the two major surveys.
e. Sort, sift, and document all the information collected.
f. Draw conclusions.

At the end of the project, each participant is to write a formal report on a specific aspect of the assignment, the results obtained from the surveys pertaining to this aspect, and the conclusions that can be drawn from them. (This report should contain illustrations to supplement the discussion.) He is also to present his findings in a briefing to representatives of the Streets and Traffic Department.

> *NOTE:* This is a comprehensive project that can be used to investigate more than has been outlined here. If carried out in the depth suggested, it will duplicate many of the steps taken in a real-life project. Suggestions for coordinating the project are contained in the comments on "Meetings" in the Assignments section at the end of Chapter 8.

Project No. 6—A Supervisory Training Program for RamSort Corporation

You are to assume that H.L. Winman and Associates has been engaged by the RamSort Corporation of 2123 Vince Street, Montrose, Ohio, to investigate supervisory training methods and to recommend a management training plan that it can adopt. When you are handed the assignment, you are told that the RamSort Corporation is very similar in size and structure to the .... company (*a company in your home city to be suggested by your instructor*), and that you should interview a management member of this company to assess the efficacy of its supervisory training methods.

You will later pool the information you obtain with that obtained by other participants in this project, who will be visiting other companies. (For details see the comments on "Meetings" in the Assignments section of Chap. 8.) You will then prepare a report for your client, in which you will identify the methods used by the company you visit, compare these methods with those of other companies, and recommend what methods you think the RamSort Corporation should adopt.

When interviewing the company assigned to you, you should search for answers to the following questions:

1. What is the company's overall philosophy toward upgrading the administrative and supervisory skills of its technical staff?
2. How are engineers, technologists, and technicians selected to undertake supervisory positions?
3. What methods are used to develop the supervisory capabilities of the company's technical staff?
4. How are existing supervisory staff encouraged to groom themselves for upgrading into higher management positions?
5. What kind of supervisor rating scheme or system does the company employ?
6. How do the company's supervisory training methods compare with those of other companies?
7. Are any supervisory training schemes sponsored by the company?
8. Does the company utilize management training schemes sponsored by outside sources (i.e. University, Community College, Technical Institute, Management Institute)?
9. Does the company consider that its supervisory development and training methods are adequate? Do you?

In making your comparisons, you should relate the answers obtained from the company you interview to the average answers obtained from all the companies surveyed.

Project No. 7—A Random Lighting Device for New Homes

This project requires that you devise an electrical device as well as prepare a report. You are to assume that you are a member of the Electrical Engineering Department of H.L. Winman and Associates. Jim Perchanski, your Department Head, has had a request from the Happy Valley Home Development Company to design a device for installation in new homes. By automatically switching lights on and off, this device will guard against theft while the owners are absent. Your task is to design a suitable gadget, test it in a partly-built home, and prepare a formal report for submission to the client, whose letter of request appears below.

HAPPY VALLEY HOME DEVELOPMENT COMPANY
"Builders of Quality-Engineered Homes"
1223 Grayside Drive
Boston, Mass.

H.L. Winman and Associates
475 Reston Avenue
Cleveland, Ohio, 44104
Attention: Mr. J. Perchanski, Head, Electrical Engineering
Dear Sir:

Design for New Electrical Device

As part of our continuing program to build the best quality houses in the Eastern United States, we are always looking for new features to incorporate into our range of Happy Valley Homes. A feature that we would like to include in next year's plans would be a built-in automatic system for controlling lights during the absence of the owner. We feel that we would have a distinct selling advantage if in our new homes we offered a device that could control up to three lights in different parts of an empty house, switching them on and off at random intervals during an evening as a deterrent to thieves.

We are therefore requesting you to propose a simple, inexpensive device that could be built in as part of the electrical system. We estimate that if your device is suitable we would make an initial purchase of 250 units from a supplier you can recommend. Will you therefore prepare a rough cost estimate per unit based on this quantity?

Since this new device will be most advantageous to us if next year's Happy Valley Homes are the only homes so equipped, please ensure that development of the device is treated as "Confidential."

Yours truly,

Harold G. Kowalchuk,
Manager, Home Marketing

HGK/lb

You are to research information to find suitable components, and then devise a system that contains:

A switching device
A 24 hour timing device
A remote control switch

The system you design should be:

a. Basically simple
b. Reliable
c. Inexpensive
d. Easy to install
e. Easy to operate

In your report to Mr. Kowalchuk you should:

1. Describe the system and how it works.
2. Emphasize the advantages of your system over other possible systems.
3. Include a sketch and a circuit design.
4. Inform him that it has been "use-tested" (you may assume that this has been done in a house under construction, and that the prototype system has been left *in situ* if he wishes to inspect it).
5. Include a cost analysis. Although Mr. Kowalchuk has asked for a price for only 250 units, I suggest you also include prices for 500 and 1000 units if there is a significant advantage in a quantity purchase.
6. Recommend that the units be manufactured by Robertson Engineering Company. (You may assume that H.L. Winman and Associates will act as a sales outlet for marketing in the U.S.)

If some of the information you need for your report is not available, you may quote imaginary data.

.   .   .

Jim Perchanski suggests that you should also consider designing a device for installation in existing homes, which H.L. Winman and Associates could develop as a product to be manufactured and marketed by Robertson Engineering Company, the organization's Canadian subsidiary. He realizes that it may not be a practicable suggestion, but he would like you to conduct a feasibility study and submit an informal report and cost estimate to him. (No mention of this device must appear in the formal report for Happy Valley Home Development Company.) Instructions for writing this informal technical proposal are contained in Project No. 6 of Chapter 5.

Project No. 8—A Warehouse for Roper Corporation

This project will require you to research industrial parks in your city to identify a suitable site for a warehouse. When you have found it you will have to determine:

1. Whether the site is zoned for construction of a warehouse.
2. The legal description of the site.
3. The title to the land, and the name of its present owner.
4. Whether caveats exist describing rights of way, liens, and so on, on the site.
5. Price of the land ($ per acre or $ per foot frontage, plus total price).
6. Availability of utilities/services, such as gas, water, electricity, and sewer.
7. Means of egress onto and exit from the site, and whether the city has placed restrictions on the building of entrances and exits.

This project starts with the visit by Mr. Harold J. Speers, General Manager of the Roper Corporation, 227 Waskesiu Avenue, Superior, Wisconsin, to H.L. Winman and Associates to discuss a design plan for a warehouse to be built in your city. His requirements are:

A warehouse with an initial 10,000 square feet of warehouse space, 1000 square feet of office space, and parking facilities for 10 cars, to be completed in 10 months.

Total acquisition of 40,000 square feet of land, to accommodate expansion needs over the next 20 years.

A proposal from H.L. Winman and Associates suggesting a suitable structure having an exterior finish in character with the surrounding area, accompanied by a site plan.

A firm price for erection of the warehouse. This must indicate cost of land, cost of warehouse, H.L. Winman and Associates' fee, plus an estimate of taxes that will be levied on the area when it is developed.

Mr. Speers says he would prefer his warehouse to be located in an established industrial park, that he expects all shipments will be brought in by semitrailer, and that the warehouse will be used to store electronic parts and components.

To tackle this project you will first have to identify a suitable site, then design a warehouse to suit it and the environment (or you may be able to adapt an existing design to suit your client's needs). In addition to a site plan showing the warehouse and projected additions, your client will also need a narrative description of the site and its present condition (i.e. does it presently have buildings or fill on it? will it need fill?); a photograph might be useful here.

Your report will have to be thorough and complete. Remember that you want your client to accept your proposal and to appoint H.L. Winman and Associates as Project Engineers. Also remember that Roper Corporation is expanding rapidly, and there will probably be many more warehouses to follow.

You may assume that you have a firm price proposal from Stout Construction Company to build the warehouse and approaches for a total of $110,000.

Your fee for the initial proposal will be 1.5% of the cost of the land, and as project engineers during the construction stage 6% of the contractor's price.

> *NOTE:*   This project can be combined with speaking practice, as outlined in the Assignments section of Chapter 8.

Project No. 9—Inspecting Aluminum Castings

You are employed in the Engineering Department of H.L. Winman and Associates. Peter Bell, Department Head of Mechanical Engineering, calls you into his office and asks you to research a technical project for Basil Wright, the Casting Superintendent of Clovelly Foundries, 2120 Dundas Road, Cleveland, Ohio. He outlines the problem to you:

1. In recent months the management of Clovelly Foundries has become increasingly concerned about the number of aluminum castings that have proved defective. The quantity has not been overly large, but the extent of customer dissatisfaction has resulted in several orders being lost to other light metal foundries.
2. Local machine shops contract with Clovelly Foundries to supply a specific number of aluminum castings, only to find that some are defective. The castings appear satisfactory, but frequently a machinist will hit porosity within the casting. This means that the job must be rejected, often after considerable machining has been done. The cost of the original casting can be recovered from the foundry; the machinist's time is not recoverable. Consequently, the machine shops experience an unnecessary increase in job costs, and are dissatisfied with the foundry's workmanship.
3. Management of Clovelly Foundries has authorized Basil Wright to purchase an instrument that will examine castings internally before they leave the foundry. Basil has identified two types of equipment that can do the work: an X-ray tube and a gamma camera. He knows little about these instruments and has asked Peter Bell to recommend one of them.

Peter Bell instructs you to conduct a small research project and prepare a formal report of your findings for submission to Clovelly Foundries. He reminds you that although Basil Wright will want a decision, he also will want to read some technical information supporting your recommendation. Important features from Basil Wright's viewpoint will be method of operation, safety, and cost.

Assume that during your investigation you jot down the following notes:

X-ray tube: 140 kilovolts, manufactured by G-X Corporation, Atlanta, Ga., costs $2980.

Gamma camera: for 16 curies of Iridium 192, built by Nucleonics Instruments Inc., Paramus, N.J., costs $1300.

X-ray tube and gamma camera are both portable and have the same weight.

Gamma camera: Ir-192 gives a fixed and invariable radio activity.

X-ray tube needs to be plugged into 120 volt outlet, while gamma camera is completely self-contained.

X-ray tube gives a better quality cardiograph and is more sensitive to material flaws, especially in light metals.

X-ray tube will burn out in time, like any vacuum tube. Cost of replacing the tube is $1200 (perhaps every two years).

X-ray tube is somewhat fragile. Gamma camera is almost indestructible.

Gamma camera must be supplied periodically with 16 curies of Ir-192, costing $300. Half-life of Ir-192 is 73 days, i.e. after 73 days the exposure time for taking a radiograph is doubled, after 146 days is quadrupled, after 219 days is eightfold, and so on.

Assume that:

a. If an X-ray tube is bought it will be used 4 hours in every 8 working hours.
b. Castings normally are no longer than 30 inches square and 5 inches thick.
c. Clovelly Foundries has no experience in using or handling equipment that emanates dangerous rays.

7

Other Technical Documents

In our technological era, with its increasingly complex range of equipment and processes, there is a growing need for manufacturers to write clear technical manuals to accompany their products. Most equipment is issued with a book of Operating Instructions that contains: (1) a brief description of the equipment; (2) instructions on how to use it; and (3) a list of likely replacement parts. Also available, but usually only to qualified repair specialists, is a set of Maintenance Instructions with detailed maintenance and repair procedures plus a comprehensive parts list describing every item in the equipment. Both publications perform the same task for a different type of reader. Operating Instructions assume that the reader has only slight technical knowledge, whereas Maintenance Instructions assume that he is fully competent.

This chapter describes how to write a technical description and instruction, and prepare a parts list. It assumes that you will be knowledgeable in the subject area if you are assigned such a task, and hence may find it difficult to write in simple terms on the topic. It also offers suggestions on how to convert your knowledge of a process, equipment, or a new technique into an interesting magazine article or technical paper.

TECHNICAL DESCRIPTION

The suggestions that follow apply to any type of technical description. Whether you are describing a piece of heavy construction machinery or a delicate instrument, a site plan for a major building or an intersection of two streets, a photograph of faulty machining or a schematic of an electronic circuit, a chart depicting output, a new batching process, or a revised assembly method, you must organize your description so that it follows a distinguishable, coherent pattern.

Spatial Arrangement

Suppose Harvey Winman asks you to write a description of a new business complex (known as "Tower Twenty-One") that he can give to visitors to the construction site. The bottom five floors are ready for occupancy, while the remainder are in varying stages of completion, ranging from nearly ready at the sixth floor to the final concrete-pouring stages on the twenty-first. A logical approach would be to describe the floors in sequence from the basement up, or, alternatively, from the penthouse down. The reader would immediately comprehend your approach because it would be natural for him to think of a vertical building as a series of floors arranged in sequence from 1 to 21. But if you started by describing the bottom five floors, jumped to the top, then back to floors 10, 11, and 12, then up to 17 and 18, and so on, you would confuse him. He would find your description difficult to follow because it would be *incoherent*.

Coherence means arranging information in such an order that the logic is evident to the reader. There can be many patterns. A spatial arrangement (*spatial* means "arranged in space") depends on the shape of the subject:

> *Vertical Subjects.* As indicated by the "Tower Twenty-One" example, tall, narrow subjects lend themselves to a vertical description, each item being described in order from top to bottom, or bottom to top. Vertically arranged control panels and electronic equipment racks fall into this category.
>
> *Horizontal Subjects.* Long, relatively flat subjects demand a horizontal description, with each part being described in order from left to right, or right to left. The description of a Heathkit EA–3 amplifier in Figure 7–1 is a typical example.
>
> *Circular Subjects.* Round subjects, such as a clock, a revolution counter, a pilot's altitude indicator, or a circular slide rule, suit a circular arrangement, with the description of markings starting at a specific point (e.g. "12" on a clock) and continuing in a clockwise direction until the entire circle has been described. In effect, this is the same as the horizontal description, if you consider that the circle can be broken at one point and unwound in a straight line to the right. Alternatively, if the subject has a series of items that are arranged more or less concentrically (such as the scales of a circular slide rule), they can be described in sequence starting at the center and working outwards, or from the outermost item inwards.

Regardless of the shape of the subject and the arrangement of its parts, there is almost always some pattern that can describe it. If it is not clearly

The Heathkit EA-3 high fidelity amplifier can be shelf-mounted, when it is housed in the black-and-gold vinylclad steel cover shown in the photograph, or rack-mounted with only the control panel visible. It accepts inputs from both magnetic and crystal phonograph cartridges, and from either AM or FM tuners, the input mode being selected by a single switch. Output volume can be adjusted from complete cut to a full 14 watts (rms); the volume control does not, however, incorporate the usual on-off switch. Separate controls are provided for base and treble adjustments, the treble control having an "AC OFF" position just beyond the cut level.

Fig. 7–1 Technical Description—Heathkit EA–3 Amplifier (courtesy the Heath Company, Benton Harbor, Mich.). Photo: Anthony Simmonds. Photo and instructions used by permission of the Heath Company.

left-to-right, top-to-bottom, or center-to-circumference, you may have to search for a logical arrangement. The timer control box in Figure 7–2 would pose such a problem. Probably it would be best to start with the control knob at the center and work outwards, even though the positions of the electrical outlets and the slide switch are diametrically opposite. A top left to bottom right basic arrangement might also be practical, combined with a concentric description of the items forming the central cluster. Note how this description guides the reader to the correct part of the control box by unobtrusively introducing gentle directions ("on top," "in the center," "around," and "at the bottom right"):

> *Spatial* Two electrical outlets on top of the control box connect
> *Description* the timer to the exposure lamp and the heater in the
> developing tray. The group of controls in the center con-
> tains a control knob with a pointer for setting exposure
> time, a time setting ring with a raised flange that acts
> as a stop for the pointer, and a knurled clamp screw that
> when tightened holds the time setting ring firmly in posi-
> tion. Around these controls is a dial that indicates ex-
> posure time (in seconds) and "activate," or developing,
> time. At the bottom right is a three-position slide switch
> that starts the timer for either the expose or activate
> cycles.

Fig. 7–2 Timer Control Box. Photo: Anthony Simmonds.

Sequential Arrangement

Sometimes it is more suitable to describe a subject sequentially, introducing items in a natural order that does not depend on physical layout. This method is particularly suitable for describing processes, techniques, or equipment that has to be operated. The two most common methods are operating order and cause to effect.

Operating Order. The items are introduced in the order in which they are employed when the equipment is in use. This makes a very natural description which prepares the reader for the operating instructions that follow. Its one major disadvantage is that the writer has to find a way to mention static items that the operator does not use. In the case of the timer control box, a sequential description would differ significantly from the spatial description mentioned earlier:

Sequential Description (Operating Order) The raised flange on the time setting ring rotates freely around the control knob at the center of the control box. When locked in position by the knurled clamp screw, it presets the desired exposure time and acts as a stop for the control knob and pointer. The scale around the central cluster indicates exposure time (in seconds) and "activate," or developing, time. Operation is controlled by a three-position slide switch at the bottom right, which starts the timer for either the expose or activate cycles. Two electrical outlets on top of the control box connect the timer to the exposure lamp and the heater in the developing tray.

Cause to Effect. This method is used mainly to describe processes or operations which lead to a direct result. Each step in the process is described in sequence. It can be a simple description of the action that takes place when I depress a key on my typewriter to cause an imprint on the paper, or it can be a complex description of the effect of soil disturbance on permafrost in the discontinuous zones of Alaska and Canada.

Sequential Description (Cause to Effect) When exploration crews in search of oil and mineral deposits first cleared long narrow strips of undergrowth for their seismic lines, they little knew the damage they were causing. Beneath its thin protective covering of vegetation, the soil had remained frozen for aeons at a temperature only a degree or two below freezing. Then, with the stubby spruce gone and the insulating layer of moss torn up and thrown aside, the ice in the soil began to melt. Small pools of water formed which spread into shallow ponds that in succeeding summers became a narrow stretch of muskeg. The short summer months of the subarctic offer little time for vegetation to grow, so that now, 30 years later, pools of muskeg are still visible between the thin cover of poplar and coarse grasses that have slowly replaced the spruce and moss. The seismic lines have become a permanent scar upon the landscape.

Which method should you use? In some cases the physical shape of the subject, or arrangement of its parts, will dictate the most suitable method. In other cases you may have a choice. When you do, try to find the method most natural for the specific situation, always trying to see the description through the eyes of a reader unfamiliar with the equipment or process. You can take a first step in the right direction by inserting a photograph or a drawing beside your description.

TECHNICAL INSTRUCTION

When Harvey Winman's No. 1 project engineer, Andy Rittman, wants a job done, he issues instructions in clear, concise terms: "Take your crew over to the east end of the bridge and lay down control points 3, 4, and 7," he may say to the survey crew chief. If he fails to make himself clear, the chief has only to walk back across the bridge to ask questions. But Fred Stokes, Chief Engineer at Robertson Engineering Company, seldom gives spoken instructions to his electrical crews. Most of the time they work at remote sites and follow printed instructions, with no opportunity to walk across a project site to clarify an ambiguous order.

A technical instruction tells somebody to do something. It may be a simple one-sentence statement that defines what has to be done, but leaves the time and the method to the reader. Or it may be a step-by-step procedure that describes exactly what has to be done, and tells when and how. It is the latter type of technical instruction that I will describe here.

Before attempting to write an instruction you must first define your reader, or at least establish his level of technical knowledge and familiarity with your subject. Only then can you decide the depth of detail you must provide. If he is familiar with a piece of equipment, you may assume that the simple statement "Open the cover plate" will not pose a problem. But if the equipment is new to him, you may have to broaden the statement to help him first identify and open the cover plate:

> Find the hinged cover plate at the bottom rear of the cabinet. Open it by inserting a Robertson No. 2 screwdriver into the narrow slot just above the hinge, and then rotating the screwdriver half a turn counterclockwise.

Knowing the technical competence of the reader is particularly important when writing complex instructions, such as describing the installation of a new electronic assembly into existing communications equipment. Without this knowledge the writer is faced with the continual problem of whether to tell his reader how to perform simple tasks. A mammoth document describing every possible step will probably infuriate a competent technician; yet omission of simple steps will run the risk of destroying the confidence of a lesser trained man.

Give Your Reader Confidence

A well-written technical instruction automatically instills confidence in its readers. They feel that they have the ability to do the work even though it may be highly complex and quite new to them. Consider these examples:

A. *Vague*   Before the trap is set, it is a good idea to place a small piece of cheese on the bait pan. If it is too small it may fall off and if it is too big it might not fit under the serrated edge, so make sure you get the right size.

B. *Clear and*   Cut a 17 inch length of 10-gauge wire and strip 1 inch *Concise*   of insulation from each end. Solder one end of the wire to terminal 7 and the other end to pin 49.

Excerpt A is much too ambiguous. It only suggests what should be done, it hints where it should instruct (almost inviting the reader to nip his finger), and in 31 explanatory words it fails to define the size of a "small" piece of cheese. Excerpt B is assertive and keeps strictly to the point. Such clear and authoritative writing immediately convinces its readers of the accuracy and validity of the steps they have to perform.

The best way to be authoritative is to write in the imperative mood. This means commencing each step with an active verb, so that your instructions are commands:

Ignite the mixture. . . .	*Connect* the green wire. . . .
Mount the transit on its tripod. . . .	*Excavate* 3 feet down. . . .
Apply the voltage to. . . .	*Measure* the current at. . . .
Cut a 2 inch wide strip of. . . .	*Count* the number of blips. . .

The imperative mood in excerpt B keeps the instruction taut and definite. Notice how the active verbs (*cut, strip,* and *solder*) make the reader feel that he has no alternative but to follow the instructions. Excerpt A would have been equally effective (and much shorter) if it had been written in the imperative mood, and if the vague verb "place" had been replaced by an image-conveying verb-adverb combination such as"wedge firmly":

> Before setting the trap, wedge a 3/8 inch cube of cheese firmly under the serrated edge of the bait pan.

The difference between an instruction written in the imperative mood and one that is not is evident from the following statements:

> Disengage the gear, then start the engine.
>   (*Definite; uses active verbs*)
> The gear should be disengaged before starting the engine.
>   (*Indefinite; uses passive verbs*)

The first statement is strong because it tells the reader to do something. The second is weak because it neither instructs him to do anything nor insists that it is really necessary that anything be done ("should" implies that it is only *preferable* that the gear be disengaged before the engine is started).

Although an active verb will most often be the first word in a sentence, sometimes it may be preceded by an introductory or conditioning clause:

> Before connecting the meter to the power source, *set* all the switches to "zero."

The imperative mood is maintained here because the active verb starts the statement's primary clause (in this case the clause that describes the action to be taken).

If you want to check whether a sentence you have written is in the imperative mood, ask yourself whether it *tells* the reader to *do* something. If it does, then you have written an *instruction*.

Avoid Ambiguity

There is no room for ambiguity in technical instructions. You have to assume that the person following your instructions cannot ask questions, and therefore never write anything that could be interpreted more than one way. This statement is wide open to misinterpretation:

> Align the trace so that it is inclined approximately 30° to the horizontal.

Each technician will align the trace with a different degree of accuracy, depending on his interpretation of "approximately." How accurate does "approximately" require him to be? Within 5°? Within 2°? Within ½°? It may be that even 10° either side of 30° is acceptable, but he does not know this and is left with a feeling of doubt. Worse still, his confidence in the technical validity of the whole instruction has been undermined. Vague references like this must be replaced by clearly stated tolerances:

> Align the trace so that it is inclined 30° (±5°) to the horizontal.

More subtle, but equally open to misinterpretation, is this statement:

> Adjust the captsan handle until the rotating head is close to the base.

Here the offending word is "close," which needs to be replaced by a specific distance:

> Adjust the captsan handle until the distance between the rotating head and the base is 25 mm.

Similarly, such vague references as "*relatively* high," "*near* the top," and "*an adequate* supply" must be replaced by clearly stated measurements, tolerances, and quantities.

Specifying Tolerances. Many instructions call for the reader to take a series of readings and then record his results in special places provided either on the instruction sheet itself or on a separate data sheet. This is particularly common in field testing and troubleshooting of electronic equipment, where a typical entry might read like this:

Connect the voltmeter to test points
8 and 17, then note the reading   5.5V _____V

The figure 5.5V is the measurement that the technician should obtain, and
the short line is provided for him to enter the voltage he actually records.
Since it is very difficult to obtain an exact voltage, a slight variation from the
specified voltage is permissible. This can be stated as a percentage:

5.5V (±6%) _____V

Now the technician has tolerances on either side of the specified voltage to guide him. But he may still have a minor problem. If his voltmeter
indicates, shall we say, 5.23V, he cannot easily determine whether the reading
falls within the specified range. He must make a quick calculation, which will
be open to error. But if you do the calculation for him, you will speed up the
operation and also prevent errors for all the technicians who have to perform
the test. The permissible tolerance would then be stated like this:

5.5V (±0.33V) _____V

Alternatively, you can help the technician assess quickly whether his
reading of 5.23V is within the specified limits by offering him a voltage range:

5.17 to 5.83V _____V

A range like this is not suitable, however, when the technician is instructed to
adjust controls until he obtains a reading as close as possible to the specified
voltage. In that case the tolerance must be stated as a plus or minus voltage,
as in the middle example above, so that he will have a specific reading to aim for.

Indefinite Words. Weak words such as "should," "could," "would,"
"might," and "may" so weaken the authority of an instruction that they reduce
the reader's confidence in the writer. Though their meaning is clear, they sound
indefinite:

Set the meter to the +300V range. The needle should indicate 120
volts (±2 volts).

In this example, "should" implies that it is not really essential that the needle
indicate 120 volts (although it would be nice if it did). No doubt the writer
meant that it *must* read the specified voltage, but he has failed to say so.
Neither has he told the reader to note the reading. He has forgotten the cardinal rule of instruction writing: *Tell* the reader to *do* something. To be authoritative, the instruction needs very few changes:

Set the meter to the +300V range, then check that its needle indicates 120 volts (±2 volts).

Notice how the steps in the sample instructions in Figure 7–3 are clear,
concise, and definite. You need not be a specialist in the subjects to recognize
that they would be easy to follow; you feel that the writers knew exactly what
they were doing, and if you were to abide by their instructions you would not
run into difficulty.

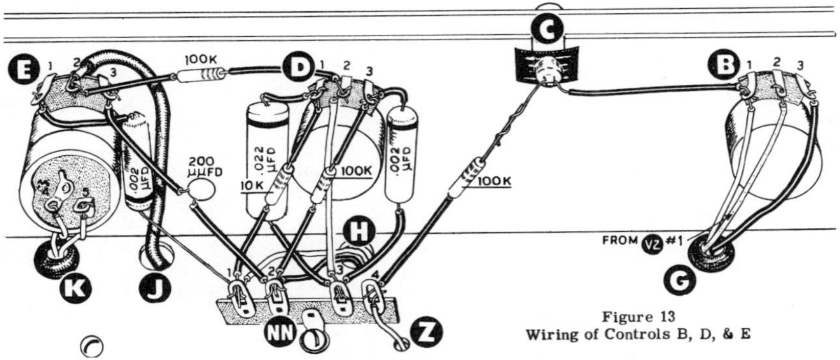

As the chain continues to stretch through further usage, the take-up bearings will eventually move back to the point where they are almost in contact with the back plate on the track. Before this happens the conveyor will shut down and the light on the Control Panel marked "Adjust Conveyor Take-up" will light up. At this point the main conveyor must be shortened, as follows:

(a) Open the take-up pit.

(b) Advance the conveyor chain to a point where a pusher dog is almost at the top of the take-up sprocket, at the position shown in Figure 5.

(c) Disconnect and lock out the main disconnect switch on the front of the Control Panel.

(d) Disconnect and lock out the conveyor disconnect switch on the wall beside the drive box.

(e) Loosen the adjusting locknuts (2-12) and back off the take-up rod adjusting nuts (2-7).

(f) Close the take-up air supply valve on the air panel.

(g) Open the take-up cylinder bleeder valve and exhaust all of the air from the take-up cylinder.

(h) Move the take-up sprocket ahead a few inches with a pry bar so that the chain ahead of the sprocket is slack.

The instruction at the left is directed at non-technical adults (they were told earlier that NS means 'do not solder' and S-1 means 'solder one lead'); it is reprinted courtesy of the Heath Company, Benton Harbor, Mich. The instruction on the right will be used by technicians maintaining mechanical equipment; it is reprinted courtesy of Washtronics, Winnipeg, Man.

Figure 13
Wiring of Controls B, D, & E

Identify the bare wire (capacitor lead) coming through hole G. Place sleeving over this wire and connect to B3 (S-1).

There are two additional wires coming through hole G. Identify the one coming from V2 pin 1 and cut to a length sufficient to reach B2 (S-1).

Now cut the third or remaining lead coming through hole G to a length sufficient to reach B1 (NS).

Select one of the pilot light (C) leads and connect to B1 (S-2) (use sleeving). Leave some slack in this lead to provide for rotation of control B.

Identify the hookup wire coming through hole Z. Cut to a length sufficient to reach NN2 (NS).

Cut the bare wire (capacitor lead) coming through hole H to a length sufficient to reach NN4 (NS). Place sleeving over this wire and connect to NN2 (NS).

Connect the hookup wire coming through hole H to NN1 (NS).

Select either lead in the twisted pair coming through grommet K and connect to E4 (S-1). Connect the other lead to E5 (S-1).

Connect a .002 µfd capacitor from E1 (S-1) (use sleeving) to NN1 (NS). Place this capacitor against the front apron as shown in Figure 13.

Connect the inside conductor of the shielded cable coming through hole J to E2 (NS).

Connect a .002 µfd capacitor from D3 (NS) to NN3 (NS). Place the body of this capacitor against the front apron. Use sleeving on both leads.

Fig. 7–3 Excerpts from Typical Instruction Manuals.

Write Bite-size Steps

A technician working on complex equipment in cramped conditions needs easy to follow instructions. You can help him by writing short paragraphs each containing only one main step. If a step is complicated and its paragraph grows unwieldly, divide it into a major step and a series of substeps. Instruction writing lends itself to subparagraphing like this:

3. List the documentary evidence in block J of Form 658. Check that blocks A to G have been completed correctly, then sign the form and distribute copies as follows:
   3.1 Attach the documentary evidence to Copies 1 and 2, and mail them to the Chief Recording Clerk, Room 217, Civic Center, Montrose, Ohio.
   3.2 Mail Copy 3 to the Computer Data Center, using one of the special 9 × 5 in. preaddressed envelopes provided.
   3.3 File Copy 4 in the "Hold—Pending Receipt" file.
   3.4 When Copy 2 is returned by the Chief Recording Clerk, attach it to Copy 4 and file them both in the "Action Complete" file.

To avoid ambiguity, use a simple paragraph numbering system. Unless your instruction is going to be very long, a straightforward system that starts at 1 cannot be bettered. Subparagraphs can be assigned decimal subparagraph numbers, such as 7.1, 7.2, and 7.3. This system also builds in a means for easy cross-referencing between steps; for example:

18. Reassemble the unit, reversing the procedure of steps 6 to 9.

Identify Switches and Controls Clearly

Opinions differ as to the best method for identifying switches, controls, and operating positions marked on equipment. Some authorities recommend using capital letters throughout, while others prefer to enclose some of the words within quotation marks. I suggest you follow these two rules:

1. Identify switches, controls, and selector positions exactly as they appear on the equipment. The loudspeaker selector switch shown in the diagram would thus be described as the SPKR switch. This helps the reader to locate the correct control on a panel containing many dials and switches. As most equipment labeling is in capital letters, much of the time you will have to use capital letters.
2. Differentiate between switches, controls, handles, and valves, and the positions to which they can be set, by enclosing the position settings within quotation marks. The loudspeaker selector switch in the diagram would be described as being set to the "BOTH" position. If you wanted to tell the reader to select only the front

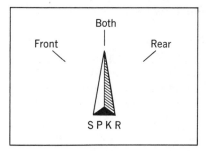

loudspeaker, you would instruct him to do so like this:

Set the SPKR switch to the "FRONT" position
or
Set SPKR switch to "FRONT".

Insert Fail-Safe Precautions

Precautionary comments are inserted in instructions to warn the reader of dangerous conditions or damage that can occur if he does not exercise care. There are three types:

Warnings To alert the reader of an element of personal danger (such as the presence of unprotected high voltage terminals).

Cautions To tell him when care is needed to prevent equipment damage.

Notes To make comments (e.g. "on some older models the valve is at the rear of the unit").

Draw attention to a precautionary comment by placing it in the middle of the text, indenting it on both sides, and preceding it with the single word WARNING, CAUTION, or NOTE. Draw a box around the words WARNING and CAUTION to give them extra prominence:

WARNING

Disconnect the power source before removing the cover plate.

A precautionary comment must always precede the step to which it refers. This will prevent an absorbed reader who concentrates on only one step at a time from acting before he reads the warning. Never assume that mechanical devices, such as indentation and the box drawn around the precautionary word, are enough to catch his attention.

Warnings and cautions must be used sparingly. A single warning will catch a reader's attention. Too many of them will cause him to treat them all as comments, rather than as important protective devices.

Insist on an Operational Check

The final test for any technical instruction is the reader's ease in following it. Since you cannot always peer over his shoulder to correct mistakes, you should find out whether he is likely to run into difficulty *before* you send it out. To obtain an objective check, give the draft instruction to a technician of a competence roughly equivalent to those who eventually will be using it, and observe how well he performs the task.

As you watch (and you must watch with a zipped lip!), note every time he hesitates or has difficulty. When he has finished, ask if he felt any parts needed clarification. Then rewrite ambiguities, and recheck your instruction with another technician. Repeat this procedure until you are confident that it can be followed easily.

PARTS LIST

A parts list helps a reader identify and describe the items he needs when ordering replacement parts. If he selects the correct components from the parts list, and describes them clearly in his request, he is likely to receive what he wants without delay. But if the parts list is so ambiguous that he misinterprets it, or if he writes vaguely for a "replacement knob for my Clarion Voltmeter," he will probably receive the wrong part or have to wait while the manufacturer asks him for clarification.

Since neither the manufacturer (who wants to keep on good terms with purchasers of his equipment) nor the user (who wants to get his equipment working again as quickly as possible) wishes to cloud the ordering procedure, both need to communicate their information clearly. There is a method to listing and ordering parts which, if adopted by both parties, will prevent supply errors from occurring.

Manufacturer's Responsibility

To help equipment users identify the parts they require, manufacturers endeavor to list parts in a logical order that will be self-evident. Common methods are to group the items according to physical location, as in the list of front panel components for the Heathkit EA-3 amplifier (Fig. 7–4), or to group them by class, e.g. all resistors, all hardware, all switches and controls, and so on. Within the groups, the items are described in three ways:

1. *By item number.* Each entry may be assigned an item number, starting at 1 for the first item in the entire list and continuing in sequential order to the end of the list. This procedure is followed

when the equipment is complex; it will help the user if an illustration is keyed to each item in the parts list (see Fig. 7–4). In large systems each major assembly may be assigned a whole number, and its parts numbered as subsections. For example, part 27 of assembly 14 would be identified as item 14–27.

2. *By part number.* Major manufacturers have such a large range of parts for their equipment, with many items being used on several different products, that they frequently set up their own numbering system to identify every item. Quoting this number is the first major step toward ensuring delivery of the correct part.

3. *By description.* Each part is described in noun-adjective order. This means identifying the noun that best describes the item and stating it first, then adding descriptive adjectives that will pin down its specific characteristics. The amber pilot lamp identified as item 26 in Figure 7–4 would be described in this order:

Lamp, pilot, amber

noun   adjectives

Similarly, the description for a cadmium plated hexagonal head No. 7 wood screw, 1¼ inches long, would be:

Screw, hex hd, cad pl, No. 7, 1¼ in.

noun   adjectives

The order in which the descriptive adjectives appear is not too important (although normally the words appear first, followed by specific dimensions and tolerances). However, the order should be consistent throughout a parts list, just as the abbreviations should be consistent. If, for instance, the first resistor in the parts list is described as "Resistor, 22 kohm, ½ watt," then the descriptions of all other resistors should state the resistive value before the wattage rating. If a dimension is quoted in decimal form, with the unit of measurement abbreviated in a particular manner (e.g. 2.75 in.), then all dimensions should adopt the same form; occasional use of a fraction and symbolic marks (e.g. 4¼″) would be inconsistent. This is no more than a technical application of parallelism, discussed in chapter 10.

The "Qty" column in the parts list shows the number of like items present in the equipment. If a unit of measurement is necessary, such as "pounds" for a quantity of nails, "feet" or "inches" for a length of wire, or "cubic centimeters" for volume of liquid, the unit is abbreviated and entered immediately after the quantity:

3 lb	30 cm³
28 ft	7 doz
14 in.	8 gal

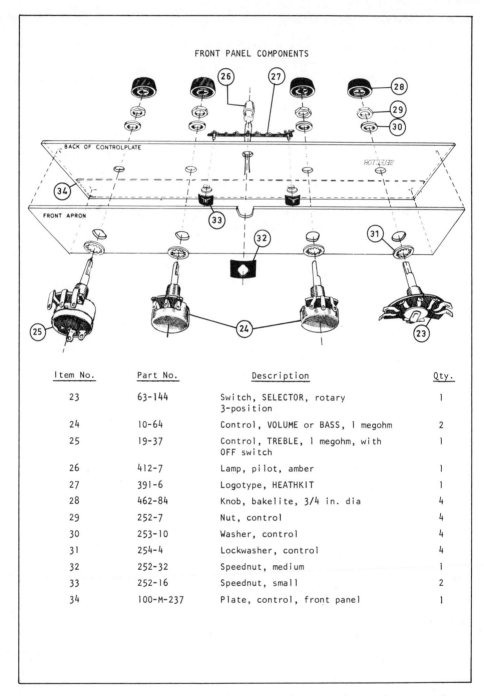

FRONT PANEL COMPONENTS

BACK OF CONTROLPLATE

FRONT APRON

Item No.	Part No.	Description	Qty.
23	63-144	Switch, SELECTOR, rotary 3-position	1
24	10-64	Control, VOLUME or BASS, 1 megohm	2
25	19-37	Control, TREBLE, 1 megohm, with OFF switch	1
26	412-7	Lamp, pilot, amber	1
27	391-6	Logotype, HEATHKIT	1
28	462-84	Knob, bakelite, 3/4 in. dia	4
29	252-7	Nut, control	4
30	253-10	Washer, control	4
31	254-4	Lockwasher, control	4
32	252-32	Speednut, medium	1
33	252-16	Speednut, small	2
34	100-M-237	Plate, control, front panel	1

Fig. 7–4 Excerpt from a Typical Parts List (front panel controls for Heathkit EA–3 amplifier; courtesy the Heath Company, Benton Harbor, Mich.).

Sometimes "each" or "only" is entered after the number in the quantity column for such simple items as capacitors, lock washers, and nuts (e.g. 7 each, 2 only). This is unnecessary, because the simple, unadorned entry is perfectly clear.

User's Responsibility

A manufacturer who issues a thoroughly detailed parts list has gone as far as he can in helping users of his products order replacements. The person needing the spare parts must do an equally thorough job when describing his needs. If the request is made on a company purchase order, spaces usually are available for copying the appropriate entries from the parts list. If done by letter, the request must be extremely explicit:

Dear Sir:
  Please ship the following replacements for our Heathkit EA-3 amplifier:

Part No.	Description	Qty Reqd
19–37	Control, TREBLE, 1 megohm, with OFF switch	1
3E–10	Resistor, wirewound, 110 ohm, 5 watt, 10%	3

The need to specify clearly exactly what parts one requires is especially important when sending in a request from the field. Too often a brief note on a scrap of paper like this is attached to a progress report:

The recipient of this missive has to decipher it before he can transcribe it onto a proper form. Unless he is technically knowledgeable he will have difficulty in executing Jack Ogilvie's request, and may easily ship the wrong type of materials, or too little or too much. A properly written request (even if it is on a scrap of paper) will always speed up delivery of the spare parts:

> Please send me the following items by air express:
>
> 13 rolls   Tarpaper, 50 yd by 36 in.
>
> 1 doz   Sealing compound, bitumastic, type ML-3, (1-lb tubes for calking gun)

TECHNICAL PAPERS AND ARTICLES

Why Write for Publication?

The likelihood that one day they might be asked to write a technical paper for publication, or even want to do so, may seem so remote to most technical students that they would be justified in skipping this section. Yet this is something they should think about, for getting one's name into print is one of the fastest ways to obtain recognition. Suddenly you become an expert in your field and are of more value to your employer, who is happy because the company's name appears in print beneath yours. You become of more value to prospective employers, who rate authors of technical papers more highly than equally qualified men who have not published. You now have positive proof of your competence, and sometimes a few extra dollars from the publisher. (The remuneration for writing technical magazine articles is only moderate, but combined with the other advantages of being in print it is adequate recompense for the time and effort expended.)

Mickey Wendell has an interesting topic to write about: As a senior technician in H.L. Winman and Associates' Materials Testing Laboratory, he has been testing concretes with various additives to find a grout that can be installed in frozen soil during the Alaskan winter. One mixture that he analyzed but discarded contained a new product known as Aluminum KL. As a by-product of his tests he has discovered that mixing Aluminum KL with cement in the right proportions results in a concrete with very high salt resistance. He reasons that such concrete could prove invaluable to builders of concrete pavements in snow-affected areas of the United States and Canada, where salt mixtures are applied in winter to melt the snow.

Mickey ought to have been thinking about publishing this particular aspect of his findings, and by now should have jotted down a few headings as a preliminary outline. But, like most of us, he "hasn't gotten around to it yet." In order for a technical paper to reach an editor's desk, Mickey will need someone like John Wood, his department head, to prod him into action, or sometimes an astute editor with his ear to the ground suggests that he prepare an outline.

Let's assume for the moment that Mickey is sufficiently motivated to try to publish his findings, and consider the four steps he must take before his ideas appear in print.

Step 1—Solicit Company Approval

Most companies encourage their employees to write for publication, and some even offer incentives such as cash awards to those who do get into print. But they expect prospective authors to ask for permission before they submit their manuscripts.

To obtain permission, Mickey must write a brief memorandum outlining his ideas to John Wood, his department head. He should ask for approval to submit a paper, explain what he wants to write about and why he thinks the information should be published, and outline where he intends to send it, as in Figure 7–5. Mr. Wood will discuss the matter at management level, then signify the company's approval or denial in writing. It is important for Mickey to have written consent to publish his findings. It sometimes takes months for an article to appear in print, and if memories are short or there have been personnel changes, it will help him to prove that he has obtained company approval.

Step 2—Consider the Market

Mickey must decide very early where he will try to place his paper. If he prefers to present his findings as a technical paper before a society meeting, as suggested by John Wood (see his comment in Fig. 7–5), he will be writing for a limited audience with specialized interests. If he decides to publish in the journal of a technical society, he will be writing for a larger audience, but still within a limited field. If he plans to publish in a technical magazine, he will be appealing to a wide readership with a broad range of technical

H. L. Winman and Associates Ltd.

INTER - OFFICE MEMORANDUM

From: .... M. Wendell .... Date: .... 24 June 1971 ....

To: .... J. Wood .... Subject: .. Approval for Proposed ..

.... Technical Article ....

May I have company approval to write an article on concrete additives for
publication in a technical journal? Specifically, I want to describe our experiments with Aluminum KL, and the salt-corrosion resistance it imparted to
the concrete samples we tested for the Alaska transmission tower project. I
believe that our findings will be of interest to many municipal engineers in
the northern United States, who for years have been trying to combat pavement
erosion caused by the application of salt during snow removal.

I was thinking of submitting the article to the editor of "Municipal
Engineering", but I'm open to suggestions if you can think of a more suitable
magazine.

MW:kr

mw

*Approval granted. Let me see an
outline and the first draft before
you submit them. Suggest you also
consider preparing a technical paper
for presentation at the Combined
Conference on Concrete to be held
in Chicago next March.*

John Wood
6 July 1971

Fig. 7–5   Soliciting Company Approval.

knowledge. His approach must therefore differ, depending on the type of publication and level of reader.

A guiding factor may be Mickey's writing capability. A paper to be published by a technical society requires high quality writing. The editor of a society journal normally does not do much prepublication editing, other than making minor changes to suit the format and style of the society's publications. A technical article will be edited—sometimes quite fiercely—by a professional editor who knows the exact style his readers expect. He prefers his authors to approximate that style, and expects them to organize their work well and to write coherently; but he is always ready to prune or graft, and sometimes even completely rewrite portions of a manuscript. Hence, the pressure on the authors of magazine articles is not so great.

A secondary consideration may be the state of Mickey's wallet: if it is thin and he needs a new set of tires for his car, he may choose to write a magazine article. Normally, there is no pay, other than recognition by one's peers, for writers of technical papers. Perhaps the most important factor is for Mickey to be able to identify a potential audience for his information. Readers of society journals and technical magazines may be the same people, but they expect different information coverage in a technical paper than they do in a magazine article.

Technical Paper. Readers of society journals are looking for facts. They neither expect nor want explanations of basic theories, and they can accept a strongly technical vocabulary. A technical paper can be very specific. It can describe a minute aspect of a large project without seeming incomplete, or it can outline in bold terms the findings of a major experiment. No topic is too large or too small, too specialized or too complete, to be published as a technical paper.

Technical Article. Most readers of magazine articles are looking for information that will keep them up-to-date on new developments. Some will have definite interest in a specific topic, and would welcome a lot of technical details. Others will be looking mainly for general information, with no more than just the highlights of a new idea. Magazine authors must therefore appeal to a maximum number of readers. Their articles should be of general interest; their style can be brief and informal; their vocabulary must be understandable; and they should sketch in background details for readers whose technical knowledge is only marginal.

Step 3—Write an Abstract and Outline

Most editors prefer to read either a summary of a proposed paper or an abstract and outline before the author submits the complete manuscript. They may want to suggest a change in emphasis to suit editorial policy, or even decline to print an interesting paper because someone else is working on a similar topic.

This type of summary is much longer than the summary at the head of a technical report, whereas the abstract usually is quite short. The summary contains a condensed version of the full paper in about 500 to 1000 words. An abstract contains only very brief highlights and the main conclusion (rather like the summary of a report), since it is supported by a comprehensive topic outline.

Some authors write the complete first draft of the paper before attempting to write a summary or abstract, then leave the revising and final polishing until after the paper has been accepted by an editor. Others prepare a fairly comprehensive outline, often using the freewheeling approach suggested in Chapter 1, and leave the writing until after acceptance. Both methods leave room for the author to incorporate changes before the final manuscript is written.

Since the abstract and outline have to "sell" an editor on the newsworthiness of his topic, Mickey Wendell must make sure that the material he submits is complete and informative. In addition to a summary, he must indicate clearly:

1. Why the topic will be of interest to readers.
2. How deeply the topic will be covered.
3. How the article or paper will be organized.
4. How long it will be (in words).
5. His capability to write and present it.

Mickey can cover the first four items in a single paragraph. The fifth he will have to prove in two ways. He can prove his technical capability by mentioning his involvement in the topic and experience in similar projects. His ability to write well he can demonstrate by submitting a clear, well-written summary or abstract.

If Mickey decides to prepare his paper for presentation before the Combined Conference on Concrete, as John Wood suggested, he will have to prepare an abstract and summary in response to a "call for papers" letter sent out by the society, and submit the material to the chairman of the papers selection committee. (The type of initial material to be submitted normally is stipulated in the "call for papers.") In order for his paper to be accepted, he has to convince the committee that the subject is original, topical, and interesting, and that he has the technical capability to prepare it. If he has presented papers previously, he can use this to demonstrate his experience in oral reporting; if he has not, he will have to prove his capability during the conference. This aspect of oral reporting is covered in Chapter 8.

Step 4—Write the Article or Paper

A good technical paper is written in an interesting narrative style that combines storytelling with factual reporting. Articles published in general interest magazines tend to be written like feature newspaper stories, while technical papers more nearly resemble formal reports. If the article deals with a factual or established topic, the writing is likely to be crisp, definite, and authoritative. If it deals with development of a new idea or concept, the narrative will generally be more persuasive, since the writer is trying to convince the reader of the logic of his argument.

The parts of an article or paper are very similar to those of a report:

Summary   A synopsis that tells very briefly what the article is about. It should summarize the three major sections that follow. Like the summary of a report, it should catch and hold the reader's interest.

Introduction   Circumstances that led up to the event, discovery, or concept that is to be described. It should contain all the facts the reader will need if he is to understand the discussion that follows.

Discussion   How the author went about the project, what he found out, and what inferences he drew from his findings. The topic can be described chronologically (for a series of events that led up to a result), by subject (for descriptive analyses of experiments, processes, equipments, or methods), or by concept (for the development of an idea from concept to fruition). The methods are very similar to those used for writing the discussion of the formal report (see p. 176).

Conclusion   A summing-up in which the writer draws conclusions from and discusses the implications of his major findings. Although he will not normally make recommendations, he may suggest what he feels needs to be done in the future, or outline work that he or others have already started if there is a subsequent stage to the project.

Illustrations are a useful way to convey ideas quickly, to draw attention to an article, and to break up heavy blocks of type. They should be used whenever it is logical, the criteria being that they be instantly clear and that they usefully *supplement* the narrative. They should never be inserted simply to save writing time; neither should they convey exactly the same message as the written words. For examples of effective illustrating, turn to any major publication in your technical field and study how its authors have used charts, graphs, sketches, and photographs as part of the story. For further suggestions on how to prepare illustrative material, see Chapter 9.

Since an article or paper is to be read by many persons, considerably more revision time is required than for formal in-plant reports. Mickey Wendell will need to give himself time between major revisions to put the article aside so that he can return to it with a fresh mind. He will also be wise to call on at least one independent reviewer to read the paper when it is nearly ready for submission. He should respect his reviewer's comments, because they are likely to be similar to those of his eventual readers.

Finally, a comment on the editor's role is in order. Mickey should not be surprised or disturbed if the editor who handles his manuscript makes some changes. These changes will affect only the arrangement of the information, and will seldom alter the technical content (if they do, Mickey will have good reason for raising his voice). The editor knows his readers well. If he feels the material is too lengthy, too detailed, or wrongly emphasized, he will make changes to bring it up to the expected standard. Mickey may feel that the alterations have ruined his carefully chosen phrases, but readers will not even be aware that changes have been made. They will simply recognize a well-written paper, for which Mickey rather than the editor will reap the compliments.

ASSIGNMENTS
DESCRIPTION WRITING

Project No. 1

Write a description of the blueprinting process, based on the information provided below. Assume that your readers will be drafting students who have not yet used a blueprinter.

This is how a technician told me he made some blueprints:

> The blueprinter was relatively simple to operate. Since the power was already on and the machine was operating, the steps involved in making my 20 blueprints were easy to follow. In the first place, for each print I took a sheet of blueprint paper and placed it on the feed table with its face side up. (The face side is the yellow side.) The original tracing, right way up, was then placed onto the blueprint paper. Before pushing the two sheets of paper into the machine, I had to set the speed selector to the correct speed, which is 15 for medium speed blueprint paper. (Mine was medium speed.) The two sheets were then pushed toward the machine until they were grasped between the belts and the exposure roller. When the two sheets emerged I had to separate them and fold back the blueprint paper and feed it up and over a developing roller. The original tracing came right out and I removed it in readiness for making another blueprint. This printing procedure was repeated until I had made all the copies I needed. At the top of the machine there was a guide bar that directed the prints to one of two delivery trays. The front delivery tray is at the front of the machine immediately in front of the operator; the back

delivery tray is used when you want to stack a succession of prints without removing them. Because I had 20 prints to make I selected "rear delivery." When I had finished printing I had to remember to turn the speed selector back to 5.

The blueprinter viewed from the side looks like this:

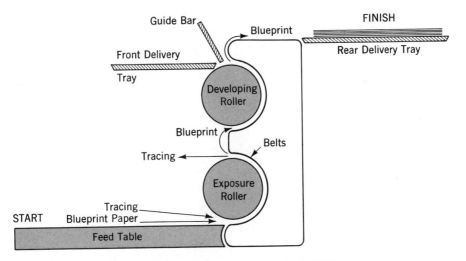

BLUEPRINTER (VIEWED FROM SIDE)

The speed selector looks like this:

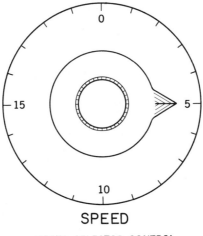

SPEED

SPEED SELECTOR CONTROL

NOTES:

1. The exposure roller is made of glass and has a high intensity light inside it.

2. The light burns off the yellow coating on the blueprint paper (except where the image on the tracing prevents light from reaching it).

3. The developing roller forces the paper through ammonia fumes (produced by heating liquid ammonia).

4. The blueprinter has a chimney to exhaust the fumes (because they are toxic).

5. Fumes "develop" the blueprint (i.e. the yellow areas turn blue).

6. The number set on the speed selector is the number of feet of paper the blueprinter produces in a minute.

7. The type of paper described here is more correctly known as "whiteprint" or "blueline" paper. In true blueprint paper the image appears white against a dark blue background.

Project No. 2

Mr. Wayne D. Robertson, president of Robertson Engineering Company, receives the following letter from his friend, Dave Kostyn, who owns a wholesale grocery company:

Dear Wayne:

Can you help me? I have a dozen high school students in the Junior Achievement group I'm working with and they have planned a project to sell packages of "resistors" to radio amateurs and hobbyists around the city. I cannot help them very much because at the moment I have no idea what a resistor is or what it looks like! All I know is that a bunch of resistors will be coming to us loose and the youngsters will have to divide them into groups of 25, package them, and arrange to sell them.

What I need right now is a paragraph or two from you describing what a resistor looks like and what it's made of (plus anything else you think I should know).

Can you do that for me? I certainly would appreciate it!

Regards,

Dave

Mr. Robertson passes the letter to the Chief Engineer, who passes it down to your supervisor, who in turn passes it on to you. Since there is no one to whom you can pass the letter, you write the description. You may assume that Dave Kostyn will pin your description to the wall where it can be read by all the Junior Achievers.

Project No. 3

In Project No. 7 of Chapter 4, you were instructed to take photographs of the general topography of the Lac le Roulet area. The Polaroid camera was lost prior to your arrival so you could not do this.

In place of the photographs, write a 200 to 250 word description of the terrain and topography of Lac le Roulet. Base your description on the map and accompanying legend on pages 110 and 111.

Project No. 4

Write a description of the layout of Radio Station DMON, Montrose, Ohio (see the sketch on p. 118). You are requested to do this by Paul Nowicki, Mechanical Engineering Department Coordinator for H.L. Winman and Associates, who wants it for members of his staff who will be visiting the radio station to investigate vibration problems.

Project No. 5

Write a 150 word description of the location and layout of Montrose Community College for the editor of the Montrose Herald. He wants to build this information into an article about the college and the programs it offers. A map of the college appears on page 148.

Project No. 6

Describe the meeting timer illustrated on page 150. Assume that the description is to be included with a press release to trade magazines.

Project No. 7

Mr. D.V. Botting, Traffic Engineer for Montrose, Ohio, Streets and Traffic Department, is aware that you have been doing an investigation into a bus stop problem at the corner of Main Street and Wallace Avenue. He writes you a note asking for recent photographs showing the properties at all four corners of the intersection. Since no photographs are available, you describe the intersection and the four properties in about 200 to 250 words. Assume that the Walston Oil Service Station building has been converted into a bus shelter, and that a bus stop is now located on the lot.

For more information turn to Project No. 1 of Chapter 6 (p.215–19).

Project No. 8

Describe the general area of Cayman Flats in relation to major features of the City of Montrose (there is a map of the city on p. 223). Assume that your readers are construction companies who will be submitting bids on:

1. Installing a storm sewer system throughout Cayman Flats.
2. Excavating zone S–17 and building a pump station on it.

3. Installing a storm sewer between the pump station and the existing storm sewers of zone S–5.

Project No. 9

In Project No. 1 of Chapter 9 you are asked to prepare illustrations depicting quantities of water sold by the City of Montrose, Ohio, to various groups of consumers during the previous year. When this has been done, prepare an analysis of water consumption for the City Engineer, who will mail it with one of the illustrations to all water consumers. In your analysis identify when each segment of the community draws most and least water, and suggest reasons for these highs and lows.

Project No. 10

You are to research information on a topic allied to your technology, and then prepare it for both written and oral presentation. The topic may be a new manufacturing material, method, or process. The written and spoken presentations must:

1. Introduce the topic.
2. State why it is worth evaluating.
3. Describe the material, method, or process.
4. Discuss its uniqueness and usefulness.
5. Show how it can be applied in your particular field.

To obtain data for your topic, you will have to research current literature and probably talk to industrial users, manufacturers, and suppliers. Typical examples of topics are: a new oil that can be used at very low temperatures; a method for supporting the deck of a bridge during concrete-pouring by building up a base on compacted fill; a new paint for use on concrete surfaces; and a new materials-handling system.

You may assume that both your readers and your audience are technicians to whom the topic will be entirely new.

INSTRUCTION WRITING

Project No. 11

Write an instruction to all installation supervisors at sites M1 through M18 telling them to test all split-bolt connectors (SBC's) on site (see Project No. 3 of Chap. 4).

NOTES:

1 Background to this project can be found on page 99.
2. The instruction is to be written as an interoffice memorandum.
3. Don Gibbon, Electrical Engineering Coordinator at H.L. Winman and Associates, will sign it.
4. Tell the site installation supervisors that they are to report their findings to Don Gibbon.
5. Tell them to test the SBC's using procedure PR27–7.
6. All SBC's on site are to be tested. (It would be best to test those in stock first, then use tested ones to replace those in use.)
7. Inform them that faulty SBC's are to be sent to the contractor with a note that they are to be held for analysis under project HW44.
8. Their tests are to be completed within 7 days.

Project No. 12

Write an instruction for using the meeting timer installed in H.L. Winman and Associates' conference room. Assume that the timer is wall-mounted. (See Project No. 5 of Chap. 5 for details and an illustration of the timer.)

Project No. 13

Write an instruction for users of a blueprint machine that is accessible to you. Assume that the machine will be running, so that users do not have to switch it on or off. Also assume that the instructions will be pinned to the wall beside the machine.

If a blueprint machine is not available to you, base your instruction on the description and illustration included with Project No. 1 of this chapter.

Include a warning of the dangers of inhaling ammonia fumes.

TECHNICAL ARTICLE

Project No. 14

Part 1. Assume that the meeting timer outlined and illustrated in Project No. 5 of Chapter 5 is to be manufactured by Robertson Enginering Company, and that you have company approval to write an article about it for publication in a magazine. Select a publication that you feel would be interested in printing your article. Describe why you have selected this particular magazine.

Part 2. Prepare an outline of your proposed article.

Part 3. Write a 500 word summary of the article, plus a letter introducing it to the editor of the magazine.

8

Technically–Speak!

I want to cover two facets of public speaking in this chapter, both concerned with the oral presentation of technical information. The first is the oral report, sometimes called the technical briefing, delivered to a client or one's colleagues. The second is the technical paper presented before a meeting of scientific or engineering-oriented persons. Both depend on public speaking techniques for their effectiveness, but neither calls for vast experience or knowledge in this field. Also included in this chapter are some suggestions on how to contribute properly to meetings that you attend.

THE TECHNICAL BRIEFING

Your department head approaches your desk, a letter in his hand, and says:

> Mr. Winman's had a letter from the RAFAC Corporation. They're sending in some representatives next Tuesday. I'd like you to give them a rundown on the project you're working on.

Every day visitors are being shown around industrial organizations, and every day engineers and technicians are being called upon to stand up and say a few words about their work. On paper, this sounds straightforward enough, but to those who have to make the oral presentation it can be a traumatic experience. Much of their nervousness can be reduced (it can seldom be entirely eliminated, as any experienced speaker will tell you) if they are given some hints on public speaking. The best training, of course, is practical experience, which can be gained only by standing up and doing the job.

Establish the Circumstances

As soon as you have been informed that you are going to deliver an oral report, you should establish some of the circumstances surrounding the visit—factors that will have a bearing on your approach. Jot down what you need to know, then ask your department head questions like these:

Who are the visitors?
You will probably be introduced to them, but at the critical moment before you speak you don't want to be concentrating on names. If you have heard them before they will be easy to remember.

How much will they know already?
You don't want to bore your visitors by repeating unnecessary details. Find out if they will come to you with no knowledge of your project, or whether management will have given them some preliminary information before you speak.

How long do you want me to talk?
Find out if management wants you to describe the project in detail, or simply touch on the highlights. It could be that the stop in your area is only a 30 second pause on a plant tour, or it may be a specific visit to study your project. The intent will directly influence your subject coverage.

Where is the briefing to take place?
Are you to address the visitors in the board room? Or will they be coming down to the project area? Availability of equipment may dictate how you tackle your subject and whether you need to make drawings to illustrate your talk.

Only when these factors have been firmly established can you start making notes. Jot down the topics you intend to discuss, and arrange them in an interesting, logical order. Think of an unenlightened person sitting in front of the equipment, and try to look at it in the same way that he will. Don't let your familiarity with the project blind you to characteristics that are unimportant to you but would be interesting to him. Use these characteristics to attract his attention and draw him into the topic.

Find a Pattern

The best technical briefings follow an identifiable pattern, just as written formal reports do. You can establish a pattern for your briefing by answering three questions that you would be likely to ask if you were a visitor to another plant.

What Are You Trying to Do? The answers to this question perform the same role as the Introduction of the formal report. They offer the listener some background information that will help him understand the technical details that follow. This information may comprise:

> How your company became involved in the project (with, perhaps, a
>   comment on your own involvement, to add a personal touch).
> Exactly what you are attempting to do (in more formal terms, your
>   objectives).
> The extent or depth of the project (i.e. the scope).

What Have You Done So Far? This would be the Discussion section of the formal report. The answers to this question should cover:

> How you set about tackling the project.
> What you have accomplished to date (work done, objectives achieved,
>   results obtained, and so on).
> Preliminary conclusions you have reached as a result of the work done
>   (if the work is complete, these will be the final conclusions).

What Remains to Be Done? (or *What Do You Plan to Do Next?*) This question is relevant only if the project is still in progress, in which case it is equivalent to the Future Plans section of a written progress report. Answers to this question would cover:

> The scope of future work.
> Results you hope to achieve.
> A time schedule for reaching specific targets and final completion.

If the project is complete, this question is not relevant and is replaced by an alternative question: "What Are the Results of Your Project?" The answers would then be combined with the final answer to the previous question, and would be equivalent to the Conclusions section of a written report.

Now we have a pattern for the main part of your briefing. But you still have to give it a beginning and an ending. The beginning should be a quick synopsis of the project in easy-to-understand terms—the equivalent of a report Summary. The ending can be a quick summing-up (a Terminal Summary) that leads into an opportunity for your listeners to ask questions. The complete pattern is shown as a flow diagram in Figure 8–1.

Prepare to Speak

Make brief speaking notes on prompt cards no smaller than 6 × 4 inches. If it is not convenient to hold the cards, place them on a makeshift speaker's stand, or even mount them on the back of a piece of equipment

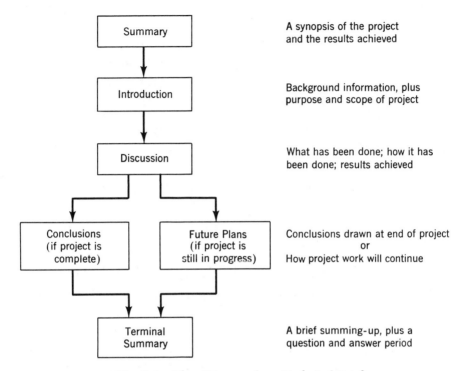

Fig. 8–1   Flow Diagram for a Technical Briefing.

where you can read them easily without straining. Write in large, bold letters that you can see at a glance, using a series of brief headings to develop the information in sufficient detail. A specimen card is shown in Figure 8–2. Prompt cards like this are scaled-down versions of the speaking notes I recommend for technical paper presentation. For comparison, a typical page of notes is illustrated in Figure 8–3 (p. 269).

Don't overlook the practical aspects of the briefing. If you have equipment to demonstrate consider its layout in relation to a logically organized description. Try to arrange the briefing so that you will move progressively from one side of the display area to the other, instead of jumping back and forth in a disorganized way. If the display is large and easy to see, let it remain unobtrusively at the back of the area. If it is small, consider moving it forward and talking from beside or behind it.

If visual aids will help you to give a clearer, more readily understood briefing, then prepare as many as you will need. They may range from a series of steps listed as headings on a flip chart to a working model that demonstrates a complex process. In most instances strive for simplicity; let the visual aid support your commentary, rather than make the commentary explain an overly complex aid. Some suggestions for preparing graphs, charts, and diagrams are contained in Chapter 9.

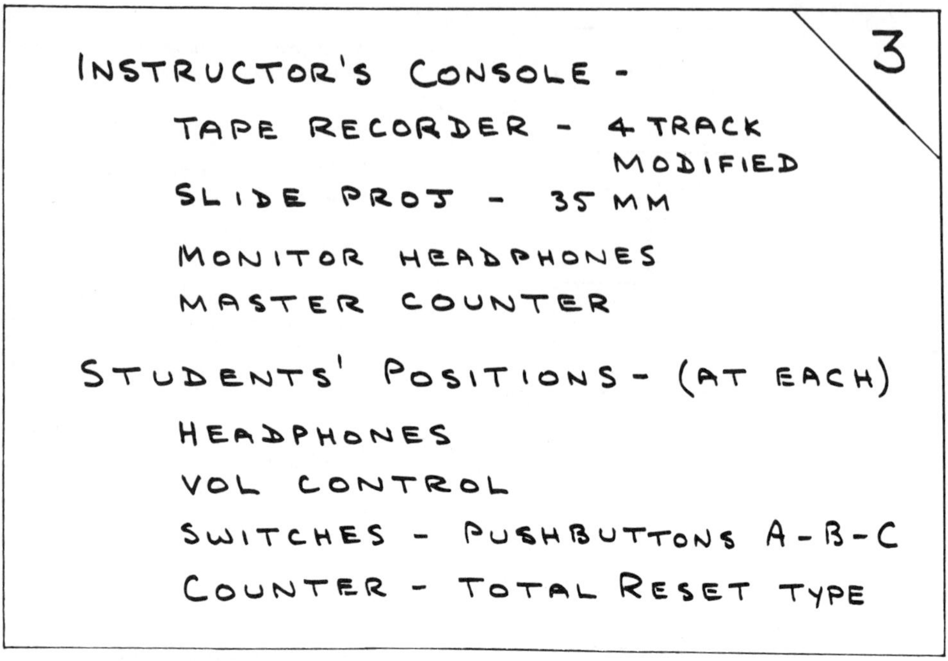

INSTRUCTOR'S CONSOLE -

    TAPE RECORDER - 4 TRACK

                      MODIFIED

    SLIDE PROJ - 35 MM

    MONITOR HEADPHONES

    MASTER COUNTER

STUDENTS' POSITIONS - (AT EACH)

    HEADPHONES

    VOL CONTROL

    SWITCHES - PUSHBUTTONS A-B-C

    COUNTER - TOTAL RESET TYPE

3

Fig. 8–2   Prompt Card for an Oral Report. (This is one of several 6 × 4 in. prompt cards made by Ron Brophy for a talk he presented on the APL System described in Figure 5–3.)

Take a leaf from the experienced technical speaker's notebook and practice your briefing. Run through it several times, working entirely from your prompt cards, until you can do so without undue hesitation or stumbling over awkward words. If the cards are too hard to follow, or contain too much detail, amend them. Then ask a fellow employee to sit through your demonstration and give critical comments.

The time and effort you invest in preparing for a briefing will depend on your confidence as a speaker and your familiarity with the topic. The more confident you are, the less time you will need. No one will expect you to give a professional briefing, but everyone—visitors and management alike—will appreciate a carefully prepared talk presented in an interesting manner.

THE TECHNICAL PAPER

In Chapter 7 I discussed the steps Mickey Wendell would have to take to publish a magazine article or a technical paper. (He is a senior technician in H.L. Winman and Associates' Materials Testing Laboratory, and he has discovered that an additive called Aluminum KL mixed with cement in the right proportions produces a concrete with high salt resistance.) In this chapter I am assuming that the papers committee of the Combined Conference on Concrete liked Mickey's abstract and summary, and the chairman of the committee has notified him that his paper has been selected for presentation at the forthcoming Chicago conference. Mickey has four months to prepare for it.

Mickey must recognize immediately that presenting a paper before a society meeting is much more demanding than delivering the same information at a technical briefing to company visitors. The occasion is more formal, the audience is much larger, and the speaker is working in unfamiliar surroundings. Many experienced engineers and scientists duck their responsibility to the audience when faced with such a situation and simply read their papers verbatim. This can result in a dull, monotonous delivery that would turn even a superior technical paper into a dreary, uninteresting recital. If Mickey is to avoid this trap, he must start preparing early.

Preparation

The key to an effective oral presentation is to have good speaker's notes, and to practice from them. This takes time. Mickey must write the publication version of his paper soon after he hears that it has been accepted (not leave it to the very last week, as so often happens), because he will need it to prepare his speaking notes.

The spoken version of a technical paper does not have to cover every point encompassed by the written version. In the 20 to 30 minutes allotted to speakers at many society meetings, there is time to present only the highlights—to trigger interest in the listeners so that they will want to read the published version. Mickey has to consider how he is to stimulate and hold this interest.

Selecting Topic Headings. His speaking notes will consist mainly of topic headings extracted from the written version of his paper. He should jot these headings onto a sheet of paper, and then study them with four questions in mind:

1. Which points will prove of most interest to the audience?
2. Which are the most important points?
3. How many can I discuss in the limited time available?
4. In what order should I present them?

When Mickey was writing his paper, he was preparing information for a reader. Now he is preparing the same information for a listener, and the rules that guided him before may not apply. The logical and orderly arrangement of material prepared for publication is not necessarily that which an audience will find either interesting or easy to digest.

It is reasonable to assume that the audience at a society meeting is technically knowledgeable, has some background information in the subject area, and is interested in the topic. In Mickey's case most of the audience will be civil engineers and technologists, with a sprinkling of sales, construction, and management people. He must keep this in mind as he examines his list of headings, identifies which points he intends to talk about, and arranges them in the order he feels will most suit his listeners.

Preparing Speaking Notes. There are many ways in which Mickey can prepare his speaking notes. He can type them onto prompt cards similar to those illustrated in Figure 8–2, print them in bold letters on 8½ × 11 sheets, or enter them in a notebook. But he should never take the shortcut of simply entering the headings in the margin of the typed copy of his paper. The temptation to read the paper may become too great if he is very nervous, and once he starts reading he will find it difficult to return to extemporaneous speech.

I prefer to use a notebook because its pages are bound together. It is reassuring to know that if I inadvertently drop my notes, I only have to turn to the correct page to continue speaking. I suggest that inexperienced speakers do the same: It could be a catastrophe for a beginner to find his carefully prepared prompt cards scattered around his feet! The vehicle Mickey chooses for his notes should be the one that offers him the most support and helps him extract all the information at a glance, so he can maintain a smooth delivery.

The notebook that I recommend should be about 9 × 7 inches when closed (slightly wider than this textbook), lie flat when opened, and have wide-spaced horizontal lines. Its left-hand page should carry speaking notes, and its right-hand page demonstration notes (see Fig. 8–3). The left-hand page is divided into four columns, the first for "time elapsed" and the remaining three each containing progressively more information. The right-hand page is a storage area for notes indicating when demonstrations are to be carried out and slides or diagrams are to be presented. It also carries comments and excerpts to be read to the audience.

In effect, Mickey Wendell would be wise to have two notebooks: one for his initial speaking notes and for speaking practice, the other, which he will prepare just before the presentation, for his final speaking notes. He should take his list of topics, abbreviate them as much as possible, and enter them in the "Topics" column of the initial notebook. He should then expand each topic into a series of brief general headings, and enter these in the "Headings" column. Finally, he should support these two columns of cryptic notes with details (information he may need to refresh his memory) which he should enter in the "Details" column. On the right-hand page he should enter comments on the visual aids he plans to use, and key them into the "Details" column at the proper points.

The amount of information he provides in these columns will depend on the complexity of the subject, Mickey's familiarity with it, and his previous

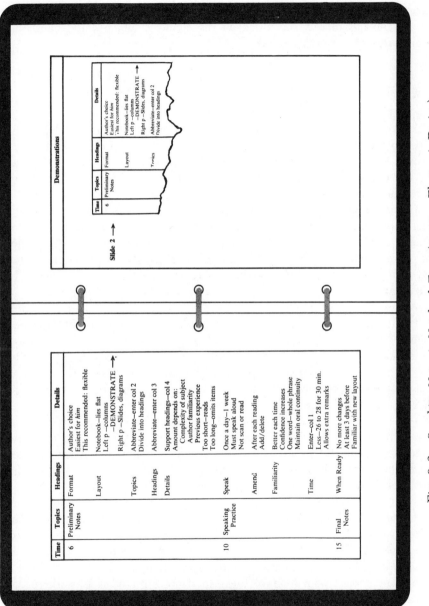

Fig. 8-3   Speaking Notes in Notebook Form (courtesy Electronic Design).

speaking experience. An experienced speaker familiar with his subject needs less information than an inexperienced speaker such as Mickey. As a general rule, the notes should not be so lengthy that he cannot extract pertinent points at a glance, for then there will again be a tendency to read. Neither should they be so brief that he has to rely too much on his memory, which could cause him to stumble haltingly through his talk.

Practice

Mickey's next task is to practice speaking from the notes at least once a day for several days. It will not be enough simply to scan or read the notes, and assume that he is thus becoming familiar with them. If he is to assess whether the notes are adequate, he must actually speak from them as though presenting the paper to an audience.

Using Preliminary Notes. After each reading he should modify the notes, including more information or deleting extraneous matter as needed. As he grows familiar with the notes his confidence will increase, and certain sentences and apt phrases will spring readily to mind at the sight of a single word or short heading. Thus he will soon find that he can easily maintain oral continuity between headings.

As he rehearses his paper, Mickey should time himself and enter time marks at suitable intervals in the left-hand column. He should aim to speak for less time than the Papers Committee allows; the ideal is to allow 26 to 28 minutes speaking time for a paper scheduled for a maximum of 30 minutes. Thus, if he wishes to make some previously unanticipated remarks when he presents his paper (possibly referring to statements in papers presented prior to his), he will have the time.

If he is using visual aids—and he should whenever possible, to insert variety into his presentation—he should practice using them, first on their own and then as part of the whole paper. This will give him a chance to check whether he has keyed them in at the right places, and whether the entries are sufficiently clear to permit him to adjust from speech to visual aid and back again without losing continuity.

Using Final Notes. When he is satisfied that his preliminary notes are satisfactory and will not need any major changes, he can prepare his final speaking notes. These should be typed on a typewriter with a large typeface, or hand-lettered (in ink) in clearly legible capital letters.

Mickey should plan to have these final notes ready at least three days before leaving for the conference. To a certain extent the headings in the first set of notes have helped to trigger familiar phrases and sentences. Now he has to familiarize himself with new pages and a new arrangement of information. During these last practice sessions, he should attempt a full dress rehearsal by presenting the paper before some of his colleagues, among whom there may be someone qualified to comment on his platform techniques. If this is not possible, he should at least try speaking the paper alone, standing at a rostrum or desk to simulate actual conditions. This "dry run" will also give him the opportunity to make a positive check of his speaking time.

Presentation

Overcoming Nervousness. There are very few speakers who are not at least a little nervous when the time comes to present their paper. Some nervous tension is perfectly normal, and can actually help Mickey give a better performance. He will find that once he starts speaking this nervousness largely disappears. A lot will depend on his speaking notes. If he has done his job well, knows that they are reliable, and is confident in using them, he will find the familiar phrases and sentences forming easily. Then he will begin to relax, and speak with even greater confidence.

Now is the moment for Mickey to start watching his listeners. Such signs as brief nods in agreement when he makes a pertinent remark, or presents a difficult point in a simple way, will tell him if they are following his talk. But if they tend to loll back in their chairs, or yawn and shift about a lot, he should consider whether his delivery is perhaps a little monotonous, and try to "liven up" his way of speaking.

Improving Platform Manner. Mickey was lucky; he was able to learn some elementary platform techniques from Ron Brophy, a member of the Winman electrical engineering staff who has been doing some research into educational training methods (his technical brief on audio-visual programed learning appears at the end of Chap. 5). Ron was present at Mickey's "dry run," and spent some time afterwards telling him how he could improve his presentation. I am repeating some of Ron's suggestions here, because they apply to any speaking situation:

* Speak at a moderate rate—120 to 140 words per minute is recommended.
* Speak up. If possible, try speaking without the aid of a microphone, since this gives you much greater freedom of movement and tonal flexibility. If a microphone has to be used, try to maintain a moderately constant speaking level.
* Pause occasionally to study your speaker's notes. Never be afraid to stop speaking for a few moments while consolidating your position and establishing that every major topic has been covered. This also gives you an opportunity to check elapsed time.
* Look at your audience. Try to speak to individuals in turn, rather than the group as a whole, picking them out in different parts of the room so that every listener will feel he is being addressed personally.

* Use humor, but only if it fits naturally into the paper and you are adept at speaking humorously. Make sure that your audience laughs *with* you and not *at* you.
* Avoid distracting habits that tend to divert audience attention. Pacing back and forth or balancing precariously on the edge of the platform (the audience will be far more interested in seeing whether you fall off than in following your paper) are examples. Nervous afflictions, such as jingling keys or coins in your pocket (put them in a back pocket, out of reach), playing with objects on the speaker's table (remove them before you start speaking), or cracking your knuckles, should also be avoided.

Ron Brophy also pointed out to Mickey that although adequate pre-platform preparation and knowledge of platform techniques can give an author confidence, they are not sufficient in themselves to break down the initial barrier between a speaker and his audience. Successful speakers and instructors develop a well-rounded personality which they use continuously and unconsciously to establish a sound speaker-audience relationship. For the author of a technical paper who faces his audience once and then only briefly, the most important personality attribute is enthusiasm.

Enthusiasm is contagious. If a speaker likes his subject, really enjoys describing it, his enthusiasm will be demonstrated in his presentation by the vigorous manner in which he tackles his material. If he is also businesslike and cheerful he will quickly reach his audience, who will respond to his approach by listening attentively.

In summary, a technical man who wants to make a good presentation before an audience should:

1. Prepare his material thoroughly.
2. Practice speaking.
3. Learn a few platform techniques.

If he is willing to devote the time and has the interest to try these methods, he will discover that he is speaking *to* his listeners, and not *at* them, as he would have done had he simply read his original paper. Even more important, he will win the support and respect of his audience, who will applaud his attempts to speak extemporaneously.

TAKING PART IN MEETINGS

We all have occasion to attend meetings. In industry you may be asked to sit on a committee set up for a multitude of reasons, from resolving technical problems that are tying up production to organizing the company's annual picnic. The effectiveness of such meetings—in terms of results—is controlled entirely by those taking part. Meetings attended by persons *aware of their role* as participants can move quickly and achieve good results; those attended by individuals who seize the opportunity to air personal complaints can be deadly dull and cripple action. Unfortunately, cumbersome, long-winded meetings are much more common than short, efficient ones.

Meetings can be either structured or unstructured, depending on their purpose. A structured meeting follows a predetermined pattern: Its chairman prepares an agenda that defines the purpose and objectives of the meeting and the topics to be covered. The meeting then proceeds logically to each point. An unstructured meeting uses a conceptual approach to derive new ideas. Only its purpose is defined, since its participants are expected to introduce suggestions and comments which may generate new concepts (this approach is sometimes known as "brainstorming"). It is the structured type of meeting that you are most likely to encounter in industry, and which I discuss here.

The Participant's Role

You can contribute most to a meeting by arriving prepared, stating clearly your facts, ideas, and opinions when called on, and keeping quiet the remainder of the time. If you observe these three basic rules you will do much to speed up affairs. Let's examine them more closely.

Come Prepared. If a meeting is scheduled to start at 3 p.m., do not wait until 2:30 to gather the information you need. Arriving with a sheaf of papers in hand and shuffling through them for the first 25 minutes creates a disturbance and makes you miss much of what is being said. Start gathering information as soon as you know what is required of you, sort it out to identify the items you need, then jot down topic headings and specific data you will have to quote. Take into the meeting only those papers you will need.

Be Brief. In your opening remarks summarize what you have to say, then follow with facts and details. Present only those items your listeners need to know. If you have a lot of statistical data to offer, print sufficient copies for each member and distribute them at the beginning of your dissertation. Be ready to answer questions and to analyze your facts in greater depth, but in doing so be sure to keep to the main topic. Finally, address your remarks to the chairman.

Keep Quiet. There are many parts of a meeting when your role is to be no more than an interested observer. At these times you should keep quiet unless you have a relevant question, an additional piece of evidence, or an educated opinion. Avoid the annoying habit of always having something to add to the information others are presenting (recognize what others already know: You are not an authority on everything!). At the same time, do not withhold information if it would be a genuine contribution. Be ready to present an opinion when the chairman indicates that a topic should be discussed, but only if you have thought it out and are sure of its validity. Recognize, too, that a discussion should be a one-to-one conversation between you and the chairman, or sometimes between you and the topic specialist. It should never become a free-for-all with each person arguing a point with his neighbor.

The Chairman's Role

Good chairmen are difficult to find. A good chairman controls the direction of a meeting with a firm hand, yet leaves ample room for the participants to feel that they are making the major contribution. He must be a good organizer, an effective administrator, and a diplomat (to smooth ruffled feathers if opinions differ too widely). Much of the success of a meeting will result from his preparation before the meeting starts, and his ability to maintain control as it proceeds.

Prepare an Agenda. Some time before the meeting starts, the chairman should prepare an agenda of topics to be discussed and circulate it to all committee members. If certain members have specific contributions to make, he should indicate by name who will be presenting the information. A typical agenda might follow the pattern in Figure 8–4, which also reminds committee members of the time and date of the meeting.

Run the Meeting. The chairman's first responsibility to his committee members is to start the meeting on time. A chairman who is known to be punctual will engender punctual attendance, whereas one who is slow in getting meetings started will encourage latecomers. His second responsibility is to keep the meeting as short as possible without seeming to "railroad" decisions. His third responsibility is to maintain adequate control.

The meeting should be run roughly according to the rules of parliamentary procedure. (Since most in-plant meetings are relatively informal, full parliamentary procedure would be too pedantic and cumbersome.) The chairman should introduce each topic on the agenda in turn, invite the person specializing in the topic to present his report, then open the topic for discussion. The discussion offers him the greatest challenge, for he must permit a good debate to generate among the members, yet be able to steer a member who wants to digress or reminisce back to the main topic. He must be able to sense when a discussion on a subject has gone on too long, and be ready to break in and ask for a decision. Similarly, he must know when strong opinions are likely to block resolution of a knotty problem, and assign a person or subcommittee to investigate further.

The best way to learn to be a good chairman is to watch others undertake the role. Study those who seem to get a lot of business done without

H. L. Winman and Associates

INTER - OFFICE MEMORANDUM

From: R. Davis   Date: 5 November 1971

To: A. Rittman   Subject: Notice of Meeting –

       B. Chansois   Children's Christmas Party Committee

           G. Hyl
           M. Kevin
           D. Calaban
           D. Smithson
           B. Brewster
           V. Braun

The third meeting of the Children's Christmas Party Committee will be held in the large Conference Room at 3 p.m. on Monday, 8 November. The agenda will be:

1.   Unfinished business from 22 October meeting

2.   Selection of Hall (D. Calaban)

3.   Purchasing of Gifts (A. Rittman)

4.   Catering Arrangements (B. Chansois)

5.   Decorations - Hall and Tree (B. Brewster)

6.   Entertainment (B. Brewster)

7.   Other Business

R. Davis

Fig. 8–4   Agenda for a Committee Meeting.

appearing to intrude too much in the decision-making. Learn what you should not do from those whose meetings seem to wander from topic to topic before a decision is made, have many "contributors" all speaking at the same time, and last far too long.

The Secretary's Role

Sometimes a stenographer is brought in to act as secretary and record the minutes of a meeting, but more often the chairman appoints one of the participants to take minutes. If his finger happens to point at you, you should know how to go about it.

Recording minutes does not mean writing down everything that is said. Minutes should be brief (otherwise they will not be read), so there is room only to mention the highlights of each topic discussed. Items that must be recorded are: (1) main conclusions reached; (2) decisions made (with, if necessary, the name[s] of the person[s] who made them, or the results of a vote); and (3) what is to be done next and who is to do it. The best way to get this information quickly is to write the agenda topics on a lined sheet of paper, spacing them about two inches vertically. In these spaces jot down the highlights in note form, leaving room to write in more information from memory immediately after the meeting.

The completed minutes should be distributed to everyone present, preferably within 24 hours. They should be a permanent record on which the chairman can base the agenda for the next meeting (if there is to be one), and participants can depend for a reminder of what they are supposed to do. I like the format shown in Figure 8–5, which provides an "action" column to draw participants' attention to their particular responsibilities.

ASSIGNMENTS

Speaking situations you are likely to encounter in industry will develop from projects on which you are working. Hence assignments for this chapter are assumed to grow naturally out of the major writing assignments presented in other chapters.

Technical Briefings

Many of the projects in Chapters 4 to 7 offer opportunities for you to brief a client, management, or other members of your department on the results of a technical investigation. Projects that particularly suit oral reporting are:

H. L. Winman and Associates

PROFESSIONAL CONSULTING ENGINEERS
475 Reston Avenue-Cleveland, Ohio, 44104

CHILDREN'S CHRISTMAS PARTY COMMITTEE

Minutes of Meeting

held 8 November 1971, Main Conference Room

In attendance: R. Davis (Chairman)   A. Rittman
                  B. Chansois (Secretary) D. Calaban
                  G. Hyl   D. Smithson
                  M. Kevin   B. Brewster
                  V. Braun

	MINUTES	ACTION
1.	The meeting opened at 3 p.m.	
2.	**Unfinished Business:** R. Davis advised the committee that he had obtained $800 to spend on the party ($500 from the company and $300 from the social club). This will be divided as follows:	
	2.1 Children's gifts $400	A. Rittman
	2.2 Rental and decoration of hall $100	D. Calaban
	2.3 Entertainment $140	B. Brewster
	2.4 Catering $160	B. Chansois
3.	**Selection of Hall:** The hall used last year cannot be used because of recent fire damage. Discussion of an alternative hall resulted in three possibilities: Viceroy Lodge; Eldwood Club; and Ramona Room. D. Calaban will investigate these and select the most suitable.	D. Calaban
4.	Now that funds are available, the gift selection sub-committee will order gifts. Lists of employees' children have been updated. Two male and two female staff members will be enlisted to help in gift selection. A. Rittman will be responsible for boys' gifts; Miss M. Kevin for girls' gifts.	A. Rittman M. Kevin
5.	Last year's caterers have been contacted and are available on the proposed date. Firm price quotations and final orders for catering will be arranged now that a budget has been established.	B. Chansois

Fig. 8–5   Minutes of a Meeting.

Meetings

Some of the projects in Chapters 4, 5, and 6 also offer excellent opportunities for group participation. If you are assigned one of these projects, try tackling it this way:

1. Install a class manager to coordinate the project (he should be voted into office).
2. Hold a meeting to discuss the project, and assign different individuals or teams to undertake specific aspects. Appoint a secretary to keep minutes, and set up a schedule for the project.
3. Let each team research its part of the project separately.
4. During the research or investigation phase, hold a meeting for the teams to report progress and compare ideas.
5. At the end of the research phase, hold a meeting for the teams to report their initial results.
6. Give the teams time to write their final reports and to prepare technical briefings.
7. Let the class manager and secretary write a management report that outlines the project as a whole and summarizes the teams' general findings.
8. Hold a briefing for the teams to present their findings orally (to individuals defined by your instructor). Let the class manager act as chairman and introduce the speakers.

Projects that particularly suit group participation are:

Choosing the Right Vehicles (Chap. 4, Project 8). At a group meeting ask individual investigators or teams to present their results. Then open the meeting to general discussion. At the end of the discussion, let the group decide by a vote which vehicles are to be purchased.

Fleet Ownership or Rental? (Chap. 5, Project 2). Here all the participants will be manipulating the same information. After individual decisions have been made, hold a meeting to arrive at a group decision.

Is There a Market for Mini-Inns? (Chap. 5, Project 3). When each individual has researched his own area, selected a Mini-Inn site, and written a progress report to Harvey Winman, hold a meeting to select the best sites to include with the feasibility study for Mr. Schmidt. At this meeting ask each participant to describe his site and outline its advantages.

Assign a selection committee to select one-third of the sites.

Let the manager and secretary present the feasibility study orally to Mr. Schmidt.

Conducting a College Travel Survey (Chap. 6, Project 5). Good organization is the key to successful completion of this project. It will need an effective manager, plus well-organized project work groups to carry out the survey efficiently. There will have to be two levels of meetings:

1. Project group meetings.
2. Whole group meetings at which a representative from each project group will present his group's progress.

A Supervisory Training Program for RamSort Corporation (Chap. 6, Project 6). This project requires that you determine the supervisory training methods of a local company and then compare them with those used by other companies. When individual enquiries are complete, call a meeting at which each participant is to present his findings.

Open the meeting to discussion.

Appoint a subcommittee to identify and list the best methods.

A Warehouse for Roper Corporation (Chap. 6, Project 8). Assume that small teams have been dispatched to various parts of your city to research suitable sites. When their research is complete (but before they write their formal reports), hold a meeting to discuss the relative suitability of the various sites. Have each team describe its site, the reason why it was selected, and the type of building the team has designed for it.

Open the meeting to discussion.

Appoint a subcommittee to consider the various sites and then report back to the group at a future meeting with its selection of the three most promising. Have this committee write a report that summarizes all the sites evaluated and recommends the best ones to Mr. Speers.

9

Illustrating Technical Documents

If you open any well-known technical magazine you will notice immediately how illustrations are an integral part of most articles. Some are photographs that display a new product, a process, or the result of some action, others are line drawings that illustrate a new concept, some demonstrate how a test or an experiment was tackled, while another group may consist of charts and graphs that illustrate progress or show technical data in an easy-to-visualize form.

But good illustrations are not limited solely to magazine articles. They serve an equally useful purpose in technical reports, where their primary role is to help readers understand the topic. Interesting illustrations attract a reader's eye and encourage him to read a report. They also break up pages of narrative that lack eye-appeal and may even suggest that the information is dull.

In this chapter I will discuss the types of illustrations seen most often in technical reports, indicate the overlaps that exist between types, and suggest occasions when they can be used most beneficially.

GRAPHS

Graphs are a simple means for showing a change in one function in relation to a change in another. A function used frequently in such comparisons is time. The other function may be temperature, erosion, wear, speed, strength, or any of many factors that vary as time passes.

Suppose, for instance, that technician John Greene wants to find how long it takes a newly painted manometer case to cool down after it comes out of the drying oven. He goes to the paint shop armed with a stopwatch and a special thermometer. When the next manometer case comes out of the oven he starts taking readings at half-minute intervals, and records the results in Table 9–1.

TABLE 9–1—COOLING RATE, MANOMETER CASE MM–7

Time Elapsed (min:sec)	Temperature (deg F)	Time Elapsed (min:sec)	Temperature (deg F)
:30	307	5:30	111
1:00	254	6:00	106
1:30	224	6:30	102
2:00	196	7:00	99
2:30	172	7:30	96
3:00	157	8:00	94
3:30	143	8:30	92
4:00	132	9:00	91
4:30	124	9:30	89
5:00	117	10:00	88

Ambient temperature   73°F   Oven temperature   350°F

These readings are part of a study he is undertaking into the cooling rates of different components manufactured by Robertson Engineering Company. The information will also be used by the production department to establish how long manometer cases must cool before assemblers can start working on them with bare hands (the maximum bare-hand temperature has been established by management/union negotiation to be 100°F).

A quick inspection of this table shows that the temperature of the case drops continuously, that it is within 15° of the ambient temperature after ten minutes (*ambient* means "surrounding environment"), and is down to the bare-hand temperature after seven minutes. A much closer examination determines that the temperature drops rapidly at first, then progressively more slowly as time passes.

Single Curve. John Greene can make the data he has recorded in Table 9–1 much more readily understood if he converts it into the graph in Figure 9–1. Now it is immediately evident that the temperature drops very rapidly at first, then slows down until the rate of change is almost negligible. There is no point in measuring or showing further drops in temperature unless John wants to demonstrate how long it takes the component to cool right down to the ambient temperature (probably 30 minutes or more).

Multiple Curves. In Table 9–2 John compares the temperature readings he has recorded for the cooling manometer case with measurements he has taken under similar conditions for a cover plate and a panel board. This time, however, he simplifies the table slightly by showing the temperatures at one minute intervals.

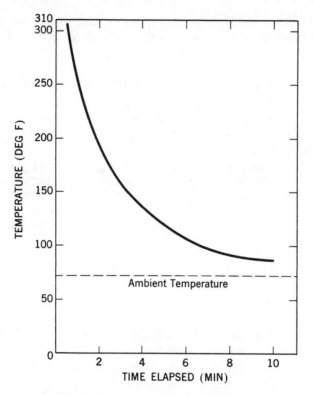

COOLING RATE – MANOMETER CASE MODEL NO. MM-7

Fig. 9–1   Graph with a single curve.

TABLE 9–2–COOLING RATES FOR THREE COMPONENTS

Time Elapsed (minutes)	Temperature (deg F)		
	Cover Plate	Panel Board	Manometer Case
0:30	310	293	307
1	278	207	254
2	234	154	196
3	203	125	157
4	180	108	132
5	161	97	117
6	147	90	106
7	132	85	99
8	121	81	94
9	110	79	91
10	102	77	88

What can we assess from this table? The most obvious conclusion is that in ten minutes the cover plate has cooled down less than the manometer case, and even less than the panel board. We can also see that the initial rate at which the components cooled varied considerably (the panel board, very quickly; the manometer case, fairly quickly; the cover plate, seemingly quite slowly), but it is difficult to assess whether there were any changes in rate of cooling as time progressed.

This can be shown more effectively in a graph, which John Greene has plotted in Figure 9–2. The rapid initial drop in temperature is evident from the initial steepness of the three curves, with each curve flattening out to a slower rate of cooling after two to four minutes. The difference in cooling rates for the three components is much more obvious than in the table.

Graphs will be essential when John Greene presents this data in a future report. If his intent is to present a general description of temperature trends, his narrative can be accompanied only by graphs like these, or by charts. But if he also wants to discuss exact temperature at specific times for each material, then the narrative and graphs will have to be supported by figures similar to those in Table 9–2. Factors that both you and John should consider when preparing graphs, charts, and tables are outlined below.

Constructing a graph usually offers no problems to technical people

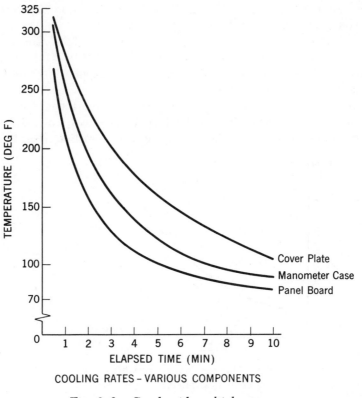

COOLING RATES – VARIOUS COMPONENTS

Fig. 9–2   Graph with multiple curves.

because they recognize a graph as a logical means for conveying statistical information. But constructing a graph that also tells a story is an aspect they may easily overlook. In order to illustrate your reports with graphs designed to emphasize the right information, you must first know the tools you have to work with.

Scales. The two functions to be compared in the graph are entered on two scales: a horizontal scale along the bottom and a vertical scale along the left-hand side. (On large graphs the vertical scale is sometimes repeated on the right-hand side to simplify interpretation.) The scales meet at the bottom left-hand corner, which normally—but not always—is designated as the zero point for both. The curved lines in Figures 9–1 and 9–2 are the actual graphs; they are known as curves even though they may be straight lines, or a series of short, straight lines joining points plotted on the graph.

The functions represented by the two scales are commonly known as the dependent and independent variables, so named because a change in the dependent variable *depends* on a change in the independent variable This can be demonstrated best by an example. If I wanted to show how the fuel consumption of my car increases with speed, I would enter speed as the independent variable along the bottom scale, and fuel consumption as the dependent variable along the left-hand side (see Fig. 9–3). Fuel consumption *depends* on speed (or, if you prefer, speed *influences* fuel consumption); the speed does not depend on the fuel consumption. The same applies to John Greene's temperature measurement graphs: temperature is the dependent variable because it depends on the *time that has elapsed* since the components came out of the oven (the independent variable).

When you construct a graph, the first step is to identify which function should form the horizontal scale and which the vertical scale. Table 9–3 lists some typical situations which show that the same function (e.g. temperature)

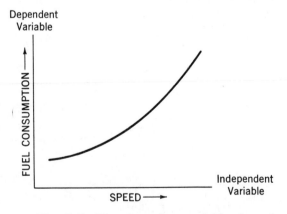

Fig. 9–3   The dependent variable depends on the independent variable.

TABLE 9–3–IDENTIFYING DEPENDENT AND INDEPENDENT VARIABLES		
Graph Illustrates	*Dependent Variable (vertical scale)*	*Independent Variable (horizontal scale)*
1. The effect that frequency has on the gain of a transducer	Gain	Frequency
2. How attendance at a ball game varies with temperature	Attendance	Temperature
3. How much a motor's speed affects the noise it produces	Noise	Speed
4. The changes in temperature brought about by changes in pressure	Temperature	Pressure
5. How much an increase in payload reduces an aircraft's range by limiting the amount of fuel it can carry	Aircraft range (or fuel load)	Payload
6. How much increasing the fuel load of an aircraft to achieve greater range reduces its effective payload	Payload	Fuel load (or aircraft range)
NOTE: A function can be either dependent or independent, depending on its role in the comparison (see temperature in examples 2 and 4, and both functions in examples 5 and 6).		

can be an independent variable in one situation and dependent in another. Selection of the independent variable depends on which function can be more readily identified as influencing the other in the comparison.

The second factor to consider is scale interval. Poorly selected scale intervals, particularly scale intervals that are not balanced between the two variables, can defeat the purpose of a graph by distorting the story it is trying to convey. Suppose John Greene had made the vertical scale interval of his time vs. temperature graph in Figure 9–1 much more compact, but had retained the same spacing for the horizontal scale. The result is shown in Figure 9–4(a). Now the rapid initial decrease in temperature is no longer evident; indeed, the impression conveyed by the curve is that temperature dropped only moderately at first, remained almost constant for the last three minutes, and will never drop to the ambient temperature. The reverse occurs in Figure 9–4(b), which shows the effect of compressing the horizontal scale: now the curve seems to say that temperature plummets downward, and it will be only a minute or two until the ambient temperature is reached. Neither curve creates the correct impression, although technically the graphs are accurate.

I said earlier that both scales normally start at zero, which would be the case when the resulting curve is comfortably balanced in the graph area. If it is crowded against the top or right-hand side, then a zero starting point is unrealistic. See Figure 9–1, in which the curve occupies the top 75% of the graph area. Since it is obvious that no points will ever be plotted below the ambient temperature (which will hover around 72°), the bottom portion of the vertical scale is unnecessary. This can be corrected by starting the vertical scale at a higher value (say 60°, as in Fig. 9–5), or by breaking the left-hand scale to indicate that some scale values have been omitted (Fig. 9–2).

A graph should be easy to read. Multiple-curve graphs should not have

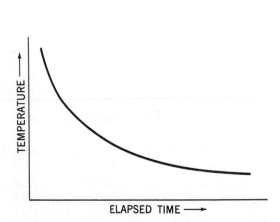
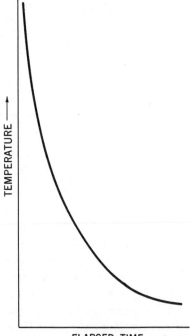

Fig. 9–4 (a) Effect of compressed vertical scale.   Fig. 9–4 (b) Effect of compressed horizontal scale.

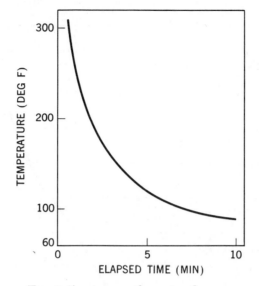

Fig. 9–5   A correctly-centered curve.

too many curves or they will be hard to interpret. If a graph contains more than three curves, it is probably getting too crowded, particularly if the curves cross each other. You can help a reader identify the most important curve by drawing it in more heavily than the others (Fig. 9–6), and can differentiate among curves that cross by using different symbols for each (Fig. 9–7):

Most important curve	—	bold line
Next most important curve	—	light line
Third curve	—	dashes
Least important curve	—	dots

Simplicity. Simplicity is important in graph construction. If a graph illustrates only trends or comparisons, and the reader is not expected to extract specific data from it, then a grid is not necessary. But if the reader will want to extrapolate quantities, a grid must be included. In the graphs we have examined, Figure 9–5 has no grid, Figure 9–1 has an implied grid that only suggests the grid pattern, and Figures 9–6 and 9–7 have full grids so that quantities can be extracted from them. Graphs without grids do not need top and right-hand borders (Fig. 9–3).

Plot points should be omitted and all captions should be horizontal. Captions for the curves should appear at the end of the curve whenever possible (Fig. 9–2), or, alternatively, above or below the curve (Fig. 9–6). They should never be written along the slope of the curve. The only caption that may be entered vertically is the identification for the vertical scale function.

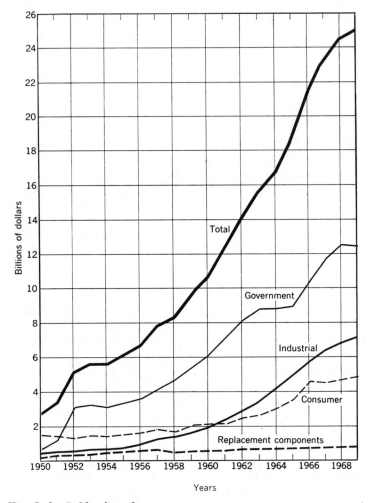

Fig. 9–6   Bolder line draws attention to most important curve of
graph comparing factory sales of electronics by markets. Graphs
normally should carry no more than three curves. [Courtesy IEEE
Spectrum (December 1969).]

CHARTS

Most charts show trends or compare only general quantities. Although
it can be said that many graphs fall within this definition, I choose to separate
them from charts because they have the ability to provide accurate interpretation, whereas charts generally do not. Hence, charts are more often seen in
reports, technical articles, and papers intended for a general readership.

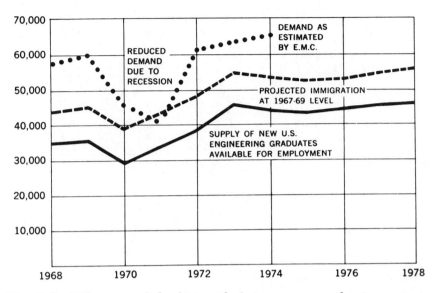

Fig. 9–7 Different symbols distinguish between curves showing current and projected engineering manpower supply and demand. [Source: Engineering Manpower Commission; courtesy Mechanical Engineering (October 1970).]

Bar Charts

You can use bar charts to compare functions that do not necessarily vary continuously. In the graph in Figure 9–1, John Greene plotted a curve to show how temperature decreased continuously with time. He could do this because both functions were varying continuously (time was passing and temperature was decreasing). For the production department, however, he has to prepare a report on how long it takes various components coming from the oven to cool to a safe temperature for bare-hand work. He prepares a bar chart to depict this because he knows the report will be read by both management and union representatives, and some of the readers may need easy-to-interpret data. He also has only one continuous variable to plot: elapsed time. The other variable is noncontinuous because it comprises the various components he has tested. In this case elapsed time is the dependent variable, and the components the independent. The bar chart John constructs is shown in Figure 9–8.

Scales for a bar chart can be made up of such diverse functions as time, age groups, heat resistance, employment categories, percentages of population, types of soil, and quantities (of products manufactured, components sold, oscilloscopes in use, and so on). Charts can be arranged with either vertical or horizontal bars depending on the type of information they portray.

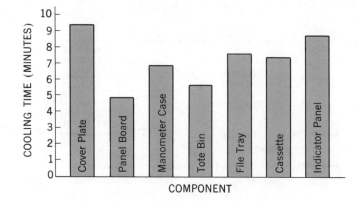

Times Required Before Oven-Dried Components
Reach Bare-Hand Handling Temperature (100°F)

Fig. 9–8   Vertical bar chart with one continuous variable
(cooling time).

The bars normally are separated by spaces the same width as each of the bars.

In a complex bar chart, the bars may be shaded or colored to indicate comparisons within each factor being considered. The horizontal bar chart in Figure 9–9 uses two shades to describe two factors on the one chart, and has a legend to help the reader identify what each shade represents. Sometimes individual bars can be shaded to show proportional content, like these bars showing development times for proposed new products:

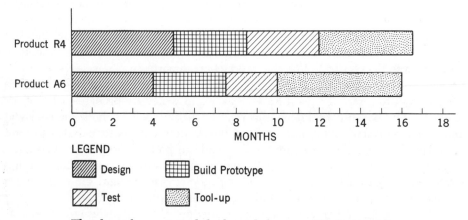

The chart also contained the legend shown on the bottom left.

Horizontal bar charts can be used in an unconventional way when information can be arranged naturally on either side of a zero point, as when

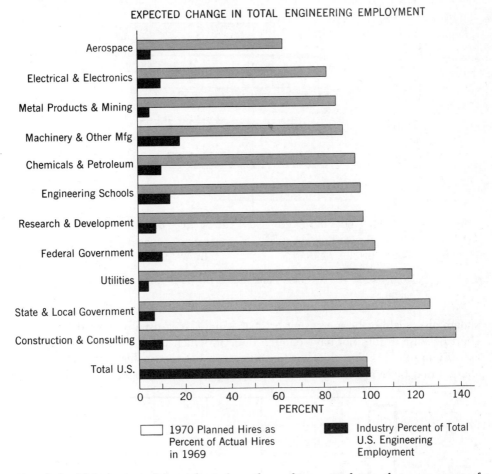

EXPECTED CHANGE IN TOTAL ENGINEERING EMPLOYMENT

Fig. 9–9 This horizontal bar chart does three things: it shows the percentage of engineers employed in major industries; it predicts hirings for the year 1970; and it demonstrates the reduction in employment for engineers in the aerospace industry. [Courtesy Machine Design (November 26, 1970)].

comparing negative and positive quantities, satisfactory and defective products, or passed and failed students. The chart in Figure 9–10 divides products returned for repair into two groups: those that are covered by warranty, and those that are not. Each bar represents 100% of the total number of items repaired in a particular product age group, and is positioned about the zero line depending on the percentage of warranty and nonwarranty repairs.

Histograms

A histogram looks like a bar chart, but functionally it is similar to a graph because it deals with two continuous variables (functions that can be

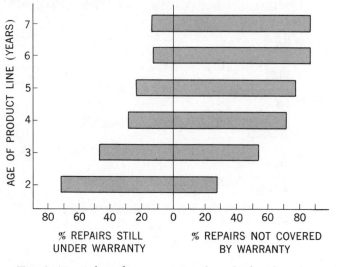

Fig. 9–10   A bar chart constructed on both sides of a zero point.

shown on a scale to be increasing or decreasing). It is usually plotted like a bar chart because it does not have enough data on which to plot a continuous curve (see Figure 9–11). The chief visible difference between a histogram and a bar chart is that there are no spaces between the bars of a histogram.

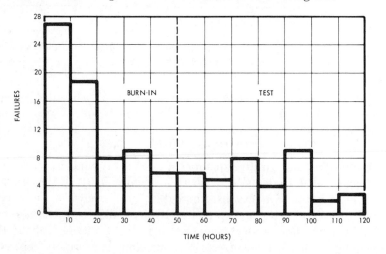

Fig. 9–11   The data for this histogram was obtained from failure records on only 14 units. Considerably more data would have been required to construct a curve. [Courtesy IEEE (Proceedings, 1968 Symposium on Reliability) and the authors (H.S. Minner and H.A. Romero), General Dynamics, Forth Worth, Texas.]

Surface Charts

A surface chart (Figure 9–12) may look like a graph, but it is not. To a technical person its construction may seem so awkward that he might wonder when he would ever need to use one. Yet as a means for conveying information pictorially to nontechnical readers, it can serve a very useful purpose.

Like a graph, a surface chart has two continuous variables that form the scales against which the curves are plotted. But unlike a graph, individual curves cannot be read directly from the scales. The uppermost curve on a surface chart shows the *total* of the data being presented. This curve is achieved as follows:

1. The curve containing the most important or largest quantity of data is drawn in first, in the normal way. This is the Hydro curve in Figure 9–12.
2. The next curve is drawn in above the first curve, using the first curve as a base (i.e. "zero") and adding the second set of data to it. For example, the energy resources shown as being available in 1980 are:
   Hydro: 15,000MW
   Thermal: 7,000MW
   In Figure 9–12, the lower curve for 1980 is plotted at 15,000MW. The 1980 data for the next curve is 7,000MW, which is added to the first set of data so that the second curve indicates a *total* of 22,000MW. (If there is a third set of data, it is added on in the same way.)

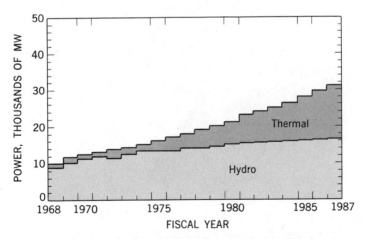

Fig. 9–12   Surface chart adds Thermal data to Hydro data to show total firm energy resources in the Pacific Northwest area of U.S. [Courtesy IEEE Spectrum (December 1968).]

The area between curves is shaded to indicate that the curves represent the boundaries of a cumulative set of data. Normally, the lowest set of data has the darkest shade, and each set above it is progressively lighter.

Pie Charts

The name "pie chart" is an excellent description for this type of chart, because it looks exactly like a whole pie viewed from above with cuts in it ready for giving wedges to people of varying appetites. It is purely a pictorial device for showing approximate divisions of a whole unit. The pie chart in Figure 9–13 depicts the percentage of electronic equipment manufactured in four major product categories.

Readers expect the wedges of a pie chart to add up to a whole unit, such as 100%, $1.00, or 1 (unity). But this may result in a lot of tiny wedges which would be difficult to draw and hard to read. In such cases you may have to combine some of the data into a single wedge and give it a general heading, such as "miscellaneous expenses," "other uses," or "minor effects."

DIAGRAMS

Under this general heading I include any illustration that helps the reader understand the narrative, yet does not fall within the category of graph,

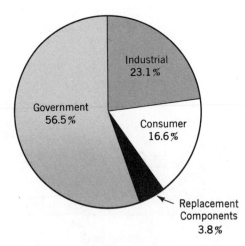

Fig. 9–13   Pie chart shows slices of business done in major product categories. [Courtesy IEEE Spectrum (December 1969).]

chart, or table. It can range from a schematic drawing of a complex circuit to a simple plan of an intersection. I must add one restriction: If included in the narrative part of a report, it must be clear enough to read easily. This means that complex drawings should be placed in an appendix and treated as supporting data.

Diagrams should be simple, easy to follow, and contribute to the story. They can comprise organization charts (see Chap. 2), flow diagrams (Fig. 9–14, and Figs. 4–5 and 5–1), site plans (Fig. 9–15), and sketches.

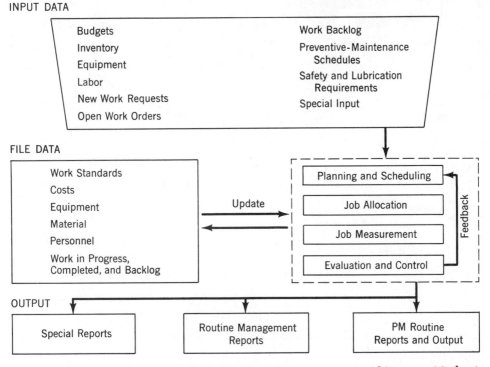

Fig. 9–14   Flow diagram of a Maintenance Information System. [Courtesy Mechanical Engineering (December 1970).]

I also want to mention photographs here. When you are describing something that exists, a photograph can do much to help a reader visualize shape, appearance, complexity, or size. For example, the photographs of Harvey Winman and Wayne Robertson add depth to the narrative description of the two engineering companies in Chapter 2.

The criterion when selecting a photograph is that it be clear and contain no extraneous information that might distract the reader's attention. The photographs in Figure 9-16 show two views of the same equipment. The right-hand view offers only a general impression of the whole tape recording unit.

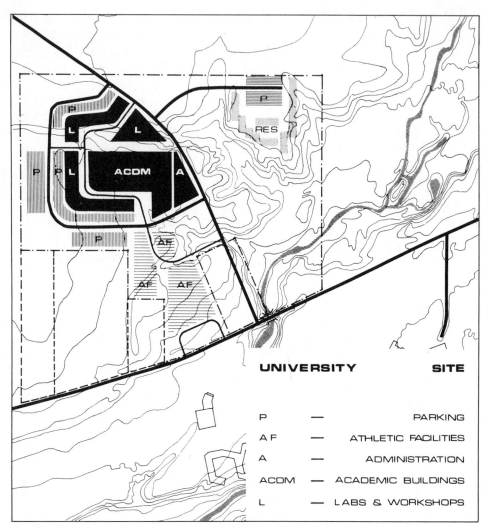

Fig. 9–15   A site plan that illustrates where college facilities are to be located. (Courtesy Smith, Carter, Searle—W.L. Wardrop and Associates Limited, Winnipeg, Canada.)

In the left-hand view, the cover over the tape drive system has been removed and the photographer has zoomed right in to show details of the open-loop tape drive on the transport. The latter photograph is much more useful to users of the equipment.

Photographs are harder to copy and print than drawings. For really clear reproduction, they need professional services often beyond the capability of an in-plant printing department. Sometimes a photograph will lose so much essential detail that it fails to support a report effectively. Before deciding to

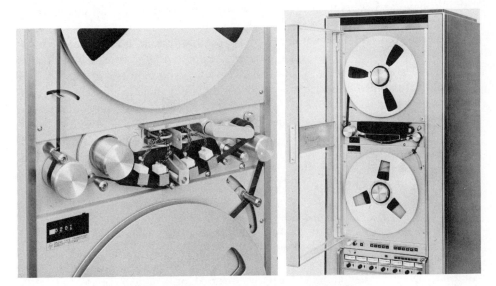

Fig. 9–16 Close-up photograph (left) shows details of tape drive; more distant view (right) shows general appearance of HP-3950 Tape Recorder. (Courtesy Hewlett-Packard Co., Palo Alto, California.)

use photographs, you should find out if your printer can reproduce them clearly and economically. If he cannot handle them easily, you may have to glue individual copies of the photographs into your report (an awkward process), have copies printed professionally, or replace the photographs with good sketches.

TABLES

The criterion for inserting a table into a report is whether the reader will need to refer to it. If he is going to use it as he reads the report, then it should be included in the discussion. If he will be able to understand the report without referring to it, but may want to consult it later either as evidence of statements in the discussion or to extract information, it should be included as an appendix. If the information in the table can be expressed more simply by words, a graph, or a chart, then the table should be omitted.

A table to be inserted as an illustration should be as short as possible so that it can be read easily. If the information compiled during a series of tests is lengthy, the essentials should be summarized and built into a short table (if necessary, the complete table can be included in the appendix). Similarly, it should have as few columns as possible, and each column should contain only data that the reader will need.

From the design viewpoint, I prefer a table that is "open," that is, without lines between the vertical and horizontal columns. (For comparison, Tables 9–1 and 9–2 are open, while Table 9–3 is closed.) The captions at the head of each column should be clear and specific, and should include units of measurement (e.g. decibels, volts, seconds) so that the units need not be repeated throughout the table. This, again, has been done in Tables 9–1 and 9–2.

It is not enough simply to insert a table and assume that the reader will know what to infer from it. The narrative should refer to the table and comment on its reason for being there. This entry draws the reader's attention to a specific area of the table:

> The voltage fluctuations were recorded at ten minute intervals and entered in column 3 of Table 7, which shows that fluctuations were most marked between 8:15 and 11:20 a.m.

Alternatively, a similar comment can be inserted as a note beneath the table.

So far we have assumed that tables contain only technical information, such as the results of tests. As Table 9–3 shows, this is not necessarily true. Frequently, the best way to show comparative data or summaries of analyses is to insert informative abstracts or comments in a table. This has been done on several occasions in this book; see pages 130 and 195 as examples.

MECHANICAL CONSIDERATIONS

Positioning

Not only must the narrative refer to every illustration in a report, but a real effort must be made to keep the illustration on the same page as the narrative it supports. This can become a problem if a description is long. However, a reader who has to keep flipping back and forth between narrative and illustrations will soon tire, and the reason for including the illustrations will be defeated.

When reports are typed on only one side of the paper, full page illustrations can become an embarrassment. The only feasible way to place them conveniently near the narrative is to print them on the back of the preceding page, facing the words they support. But this in turn may pose a printing problem.

A more logical solution is to limit the size of illustrations so that they can be placed beside, above, or below the words, and then to make sure that the stenographer who is typing the report keys them in at the right points.

Plainly, it will not be possible to position every illustration exactly where it is needed. Decisions may have to be made to position the most important technical illustrations at the critical points, and to place others as close as possible to the words they support.

Horizontal full page illustrations may be inserted sideways on a page, but must always be positioned so that they are read from the right (see Fig. 9–17). This holds true whether they are placed on a left- or right-hand page.

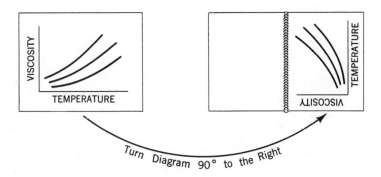

Fig. 9–17   Page-size horizontal drawings should be positioned so they can be read from the right.

When an illustration is too large to fit on a normal page, or is going to be referred to frequently, you should consider printing it on a foldout sheet and inserting it at the back of the report. If the illustration is printed only on the extension panels of the foldout, the page can be left opened out for continual reference while the report is being read. This technique is particularly suitable for circuit diagrams and flow charts.

If the equipment normally used for printing your reports cannot reproduce large foldout sheets, you can prepare the illustrations on tracing paper and make blueline prints of it. An ideal size is 11 × 22 inches, which folds conveniently into a standard 8½ × 11 report (see Fig. 9–18). Such a foldout can be cut from a standard size C (17 × 22) sheet of blueprint paper. Even larger foldouts can be made in this way, but they tend to be unwieldy.

Printing

Always discuss printing methods with the person who will be making copies of your report *before* you start making reproduction copy. Certain reproduction equipment cannot handle some sizes, materials, and colors. For example, heavy blacks and light blues may not reproduce well on some electrostatic copiers, light browns cannot be copied by other types of equipment, and photographs can be reproduced clearly by very few.

(a) Fold-out sheet opened out for reading

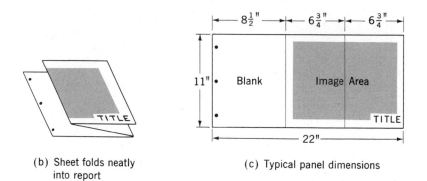

(b) Sheet folds neatly
into report

(c) Typical panel dimensions

Fig. 9–18   Large illustrations can be placed on a fold-out sheet at rear of
report.
   (a)   Fold-out sheet opened out for reading
   (b)   Sheet folds neatly into report
   (c)   Typical panel dimensions

Illustrating a Talk

An illustration for a talk must be utterly simple. Its message must be
so clear that the audience can grasp it in seconds. An illustration that forces
an audience to read and puzzle out a curve detracts from a talk rather than
complements it.

If you want to convert any of the illustrations discussed in this chapter
to a large visual aid or slide, you will have to observe the following limitations:

1. Make each illustration *tell only one story.* Avoid the temptation
   to save preparation time by inserting too much data on one chart.
   Be prepared to make two or three simple charts in place of a
   single complex one.
2. Use bold letters large enough to be read easily by the back row of
   your audience.
3. Use very few words, and separate them with plenty of white
   space.
4. Give the illustration a short title.
5. Insert only the essential points on a graph. Let the curves tell the
   story, rather than be buried in construction detail.
6. Accentuate key figures and curves with a bold or colored pen (but
   remember that from a distance some colors look very similar).

7. Avoid clutter—a simple illustration will draw attention to important facts, whereas a busy one will hide them.

Diagrams that are to illustrate an in-plant briefing may be hand-lettered with a felt pen on large sheets. But those to be used for a technical paper presented before a large audience should be prepared more professionally. In the first instance you are expected to do a workmanlike job at no great expense. In the second, you are conveying an image of yourself and the company you represent; in effect, your diagrams are demonstrating the technical quality of your company's products.

ASSIGNMENTS

Project No. 1

The table below is a record of the average daily water consumption for the City of Montrose, Ohio, over the past calendar year. The City Engineer has to prepare two reports in which he will identify how much water has been consumed by different segments of the community, and analyze when and why variations in consumption occurred. He asks you to prepare a graph, chart, or diagram to accompany each of these reports. They are to comprise:

1. An illustration for a technical report that will be read by engineers and technicians involved in water supply and distribution.
2. An illustration for a water consumption analysis that will be sent to all consumers. (This part of the project should be done in conjunction with the writing assignment outlined in Project No. 9 of Chap. 7.)

CITY OF MONTROSE, OHIO
AVERAGE DAILY WATER CONSUMPTION
(in gallons)

Month	Business & Industry	Private Homes	Schools & Colleges
January	4,256,000	3,608,000	1,910,000
February	4,310,000	3,673,000	2,296,000
March	4,318,000	4,127,000	2,501,000
April	4,325,000	4,980,000	2,507,000
May	4,331,000	5,641,000	2,610,000
June	4,417,000	6,775,000	2,192,000
July	4,484,000	8,926,000	807,000
August	4,491,000	9,681,000	762,000
September	4,369,000	6,810,000	1,418,000
October	4,323,000	5,604,000	2,298,000
November	4,298,000	4,254,000	2,304,000
December	4,243,000	3,672,000	1,641,000

Project No. 2

Prepare an illustration comparing the variations in noise levels at Bluff Heights, Varsity Valley, and Hartland Point (the three sites considered in the formal report, "Evaluation of Sites for Montrose Residential Teachers' College," in Chap. 6). Base your illustration on the data recorded in Appendix "B" to the report (p. 196).

Describe why you selected a particular type of illustration (graph, chart, and so on) to depict this data.

Projects Combined with Other Assignments

Some of the major report writing assignments in Chapters 5 and 6 need good graphic aids if they are to convey their information effectively. Specific projects that can combine illustrating with report writing are:

An Electrostatic Copier for Engineering (Chap. 5, Project 1)
Fleet Ownership or Rental? (Chap. 5, Project 2)
Evaluating Employment Possibilities (Chap. 6, Project 2)
Installing Dial-a-Wash in the U.S. (Chap. 6, Project 4)
Conducting a College Travel Survey (Chap. 6, Project 5)
A Supervisory Training Program for RamSort Corporation (Chap. 6, Project 6)

10

The Technique of Technical Writing

In this chapter I want to concentrate on a few writing techniques that will enable you to convey information both quickly and efficiently. I will not be discussing grammar or "English," because I assume you are already proficient in these areas and can recognize and correct basic writing problems. If you need practice in basic writing I suggest you refer to a textbook such as the *Prentice-Hall Handbook for Writers.* You can also refer to the glossary of terms in Chapter 11 for information on how to resolve many of the problems that crop up in technical writing (how to form abbreviations and compound adjectives, how to spell problem words, when to use numerals or spell out numbers in narrative, and so on).

THE WHOLE DOCUMENT

Throughout this book I have stressed the need to tell the reader immediately what he most wants to know. This means structuring your writing so that the very first paragraph (if possible, the first sentence) satisfies his curiosity. Most executives and many technical readers are busy people who have time to read only essential information. By presenting the most important items first, you can help them decide whether they need to read the whole document, or whether they should hand it to someone more familiar with the topic.

The Pyramid Technique

This method of writing is known as the pyramid technique because it starts at the top with a small morsel of essential information, and then substantiates it with a broad base of details, facts, and evidence. In most letters and short reports it has only two stages: a brief *summary* followed by the *full development*, as shown in Figure 10–1(a). In long reports an additional stage, which I call the *essential details* (see Figure 10–1[b]), is inserted between the summary and the full development.

A reader normally is not consciously aware of the pyramid technique. He simply finds that a document in which it is employed is very easy to follow. In the letter report in Figure 10–2, Stanley Roning summarizes in the first paragraph the two things Wayne Robertson wants to know right away: whether the training course was a success and what results were achieved. He uses the remainder of the letter to fill in background details, to state briefly how the course was run, to report on student participation, and to comment on student reaction.

The pyramid technique requires that the reader know the main elements of the story by the end of the first paragraph. He can then read the full development more intelligently. If he is very busy and the report is long, he may choose to stop reading at the end of the summary and continue with the full development later, or he may scan the full development quickly and pass the report on to someone else for action.

In the full development the writer tells the whole story. He amplifies what he has already stated in the summary by inserting all technical details the reader needs to understand the subject fully. He may develop the subject in whatever order he likes, although much of the time he will find that a past-present-future order is the simplest and most effective. This in turn may present him with a problem, since he will have to maintain continuity between the summary (which frequently ends with a comment on future action) and the first paragraph of the full development (which normally starts by saying how the project began).

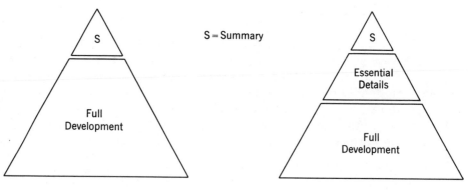

(a) Letters and Short Reports   (b) Long Reports

Fig. 10–1   The Pyramid Technique.

THE RONING GROUP
COMMUNICATION CONSULTANTS
Box 181   Postal Station "C"   Winnipeg 9   Manitoba
Tel. 452-6480

14 January 1972

Mr. Wayne D. Robertson, President
Robertson Engineering Company
600 Deepdale Drive
Toronto, Ontario

Dear Mr. Robertson:

Results of Pilot Report Writing Course

The report writing course we conducted for members of your engineering staff
was completed successfully by 14 of the 16 participants. The average mark
obtained was 63%.

This was a pilot course set up in response to a 13 August 1971 enquiry from
Mr. F. Stokes. At his request, emphasis was placed on giving students
practical experience in writing business letters and technical reports.
Attendance was voluntary, the 16 students being selected at random from 29
applicants.

Best results were achieved by students who recognized their writing problems
before they started the course, and willingly became very actively involved
in the practical work. A few presumably had expected it to be an "information" type of course, and hence were less willing to take part in the
heavy writing program. Our comments on the work done by individual students
are attached.

Course critiques completed by students indicate that the course met their
needs from a letter and report writing viewpoint, but that they felt more
emphasis could have been placed on technical proposals and oral reporting.
Perhaps these topics should be covered in a short follow-up course.

We enjoyed developing and teaching this pilot course for your staff, and
particularly appreciated their enthusiastic participation.

Yours truly,

Stanley G. Roning

Stanley G. Roning
President

SGR:ib
Encl

Fig. 10-2   A Letter Report Written Using the Pyramid Technique.

To make a good transition from the summary *back* to the background information requires skill. It has been done well in Figure 10–2 because the writer begins paragraph two by referring to the training course mentioned in the previous paragraph (he starts the full development by saying: "This was a pilot course. . . ."). Further examples of good transitions can be found in the letter reports in Figure 4–3 and 4–4 (p. 90, 92).

There is no problem in making this transition in the long formal report because the summary and the introduction (which contains the background information) are on on separate pages and are often separated by the table of contents. Thus a technical writer can develop a brief but complete story in the summary without having to consider how he will build a transition to the next paragraph. Indeed, the organization of many modern formal reports will assist him in adapting his writing to the pyramid technique.

In Figure 10–3 the three main sections of the alternative format for the formal report discussed in Chapter 6 are paralleled with the three stages of development demanded by the pyramid technique. The summary of the formal report is the initial stage of the pyramid; the introduction, conclusions, and recommendations are the essential details; and the discussion is the full development. In effect, the formal report written in this way becomes three reports: the brief but informative summary written in general terms that can be understood by any reader; the essential details of the introduction, conclusions, and recommendations, written for the knowledgeable but not necessarily technical reader; and the full development of the discussion, written for the reader who has the interest and technical capability to understand all the evidence the writer has to present.

What does this technique mean to the writer of scientific and engineering documents? Principally, it provides him with the opportunity to cater to more than one level of reader within the same document. This is particularly true of long formal reports, less so of short business letters. But simply adopting the two- or three-stage approach of the pyramid technique is insufficient in itself to ensure good writing. It is a mechanical means to effective writing that

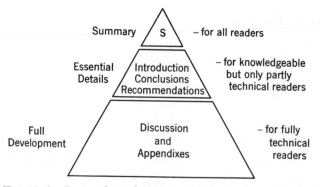

Fig 10–3   Pyramid Technique Applied to the Formal Report.

must be supported by two less tangible but equally important factors: tone and writing style.

Tone

The fact that it is possible to write for several levels of reader in the same document does not imply that you can omit identifying a specific reader. Without this important step you will not be able to set the right tone. The reader you must identify is always the person who will be reading the full development. Thus, you should write primarily for your most technically knowledgeable reader, while directing the essential details and the summary at progressively less technically knowledgeable persons.

Writing Sequence. If you are to set the right tone throughout you must write in reverse order, starting with the full development. Writing a report in the order in which it will be read is difficult, if not impossible. An engineering technician who writes the summary before the full development will use too many adjectives and adverbs, big words when shorter words would be more effective, and dull opening statements such as "This report has been written to describe the investigation into defective MN–1 compasses carried out by H.L. Winman and Associates." He will be writing without having established exactly what he will say in the full development.

The best writing sequence therefore should be the reverse of that shown in the pyramid in Figure 10–3:

Step 1 Write the full development. Direct it to the type of technical reader who will use or analyze your report in depth.

Step 2 Write the essential details (omitted in short informal reports). Make them brief and direct them at a semitechnical reader or person in a supervisory or managerial position. They should comprise:
  a. Introductory information to help the reader understand why the project (and hence the report) was undertaken, and what the terms of reference were.
  b. The results of the project, or the main conclusions that can be drawn from the full development; this should follow naturally from the introductory information and satisfy the terms of reference.
  c. A recommendation (if one is to be made) showing what needs to be done next.

Step 3 Write the summary. Direct it to a nontechnical reader who has absolutely no knowledge of the project or the contents of the report. Make it very brief and make sure it tells:
  a. Why the report was written.
  b. What was found out.
  c. What are the main conclusions and recommendations.

Degree of Formality. Whether your writing should be formal or informal will depend on the situation and your familiarity with the reader. Formal reports should adopt a formal tone. (Note, however, that a formal tone is not stiff or pompous; there is no room for writing that makes the reader feel uncomfortable because he is not as knowledgeable as you are.) A formal report conveys an image of your company's reputation even before the first page is opened; the words between the covers must maintain that image. Business letters should become less formal as their importance decreases. For example, a management-level letter proposing a joint venture on a major defense project would be very formal, whereas letters between two technologists discussing mutual technical problems would be informal. A memorandum report can be quite informal, since normally it is an in-plant document most often written between persons who know each other. But this informality should not be interpreted as the green light for slipshod writing in interpersonal communications.

Varying levels of tone are evident in the following extracts from three separate H.L. Winman and Associates' documents, all written on the same subject.

1. *Extract from a Memorandum.* John Wood's Material Testing Laboratory has compression-tested samples of concrete for Ian Bailey of the Civil Engineering Department. In his memorandum reporting the test results, John writes:

> *informal* I have tested the samples of concrete you took from the
> *tone* sixth floor of Tarryton House and none of them meet the
> 4800 psi you specified. The first failed at 4070 psi; the
> second at 3890 psi; and the third at 4050 psi. Do you want
> me to send these figures over to the architect, or will you
> be doing it?

2. *Extract from a Letter Report.* Ian Bailey conveys this information to the architect in a brief letter report:

> *semiformal* Our tests of three samples taken from the sixth floor of
> *tone* Tarryton House show that the concrete at 52 days still was
> 800 psi below your specification of 4800 psi. We doubt
> whether further curing will increase the strength of this
> concrete more than another 150 psi. We suggest, however,
> that you examine the design specifications before embark-
> ing on an expensive and time-consuming remedy.

3. *Extract from a Formal Report.* The architect rechecked the design specifications and decided that 4200 to 4300 psi still would not satisfy the design requirements. He then requested H.L. Winman and Associates to prepare a formal report that he could present to the general contractor and the concrete supplier. Ian Bailey's report said, in part:

formal   At the request of the Architect we cut three 12 × 6 in.
tone   diameter cores from the sixth floor of Tarryton House 52
days after the floor had been poured. These cores were
subjected to a standard compression test with the following results (detailed calculations are attached at appendix
"A"):

Core No.	Location	Failed at:
1	18 in. W of col 18S	4070 psi
2	33 in. N of col 22E	3890 psi
3	48 in. N of col 46E	4050 psi

The average of 4003 psi for the three cores is 797 psi
below the design specification of 4800 psi. Since further
curing will increase the strength of the concrete by no more
than 150 psi, we recommend that this concrete pour be
rejected.

Although the information conveyed by these three examples is similar, the tone
the writer adopts varies sufficiently to help set the scene for each situation. It
is unlikely that you will have to write the same information in three such different ways, but you should be able to select the right tone for a given situation.

Style

The pyramid technique has no effect on writing style, other than consideration of the reader at each stage. Complexity of subject and technical level
of reader have some effect on style, in that you should use careful, simple
language when describing a very complex topic to a moderately knowledgeable
reader. As the reader's technical level increases, or the complexity of the subject
decreases, you can use longer sentences and words, and more complex sentences.
Here are some suggestions:

1. When presenting low-complexity background information, and descriptions of nontechnical or easy-to-understand processes, write
   in an easygoing style that tells the reader he is encountering information that does not require his total concentration. Use slightly
   longer paragraphs and sentences, and insert a few adjectives and
   adverbs to color the description and make it more interesting.
2. For important or complex data, use short paragraphs and sentences.
   Present one item of information at a time. Develop it carefully
   to make sure it will be fully understood before proceeding to the
   next item. Use simple words. This punchy style will warn the
   reader that the information demands his full attention.
3. When writing instructions or step-by-step descriptions, start with a

narrative-type opening paragraph that introduces the topic and presents any information that the reader should know or would find interesting; then follow it with a series of subparagraphs each describing a separate step. These subparagraphs should conform to the following general rules:

3.1 Each should develop only one item of the process.

3.2 The subparagraphs should be short. Even single sentence subparagraphs are acceptable, particularly for technical instructions.

3.3 They should be parallel in construction; that is, they should generally conform to the same "shape." The importance of parallelism is discussed later in this chapter (p. 331).

3.4 Each step of an instruction should start with an active verb (see p. 240 for suggestions on this topic.)

Paragraph Numbering

The types of industrial documents that most often carry paragraph numbers are military reports, specifications, and technical instructions. Some companies stipulate that all their reports bear paragraph numbers, but they are in the minority. Their reasons for doing so, however, are valid: paragraph numbers are an excellent aid to organization, and they provide a useful means for cross-referencing (both within the report itself and from one report to another).

A paragraph numbering system must be obvious to the reader, simple for you (the writer) to manipulate, and easy for the typist to arrange on the page. The simplest paragraph numbering system starts at 1 and numbers paragraphs consecutively to the end of the document. More complex systems combine numbers, letters, and decimals to allow for subparagraphing. Three possible arrangements appear below.

TYPICAL PARAGRAPH NUMBERING SYSTEMS

METHOD (A)	METHOD (B)	METHOD (C)
1.	A.	1.
2.	1.	2.
3.	2.	3.
3.1	(a)	3.1
3.1.1	(b)	3.2
3.1.2	(1)	a.
3.1.2.1	(2)	b.
	B.	(1)
etc., up to	1.	(2)
6 digits	2.	etc.
	etc.	

Method (A) is a numerals-only decimal system, simple and unambiguous; it is often used by the military services and by specification writers. Although unambiguous, the subparagraphing becomes difficult to follow in long documents, particularly when indentation cannot be used. (Indentation of multiple-digit subparagraphs can quickly result in a very narrow band of information down the right-hand side of the page.)

Method (B) permits indentation and uses a combination of letters and numerals. Extremely simple, it is popular with many report writers even though it can be ambiguous (there can be several paragraph 1's, 2's, etc., one set under A, another under B, and so on).

The third method, (C), combines the decimal and letter-number arrangements for a system with no ambiguity between main paragraphs and subparagraphs. It also permits indentation, even with the rather wide indentations demanded by proportional-spacing typewriters. I limit large paragraphs to the first two levels, and use small paragraphs with the lower case letters, and very short paragraphs or simple points with the numbers in parenthesis:

1. 2.	} main paragraphs
2.1 2.2	} full subparagraphs
a. b.	} short subparagraphs
(1) (2)	} very short subparagraphs } (i.e. one sentence only)

This paragraph numbering system combines well with the heading arrangements illustrated in Figure 10-4. For a partial example of this system used in an informal report, turn to Figure 5-2 on pages 126-30.

Arrangement of Headings

There are three main types of headings you are likely to encounter: center headings, side headings, and paragraph headings. Their use is demonstrated in Figure 10-4.

PARAGRAPHS

The role of the paragraph is complex. It should be able to stand alone, but normally is not expected to. It must contribute to the document of which

<u>MAIN CENTER HEADING</u>

SUBSIDIARY CENTER HEADING

1. The Main Center Heading is always capitalized and underlined. Subsidiary Center
    Headings (if used) are capitalized but not underlined as shown, or may be in lower
    case letters and underlined: <u>Subsidiary Center Heading</u>

<u>Side Heading</u>

2. Side headings are in lower case letters and underlined. They do not require para-
    graph numbers, but frequently are assigned them when a numbering system is used.
    In such a case, the above side heading would be preceded by '2', and this and the
    next paragraph would bear numbers 2.1 and 2.2.

3. The text is typed so that it forms an indented block. It should not run back to
    the left-hand margin (under the paragraph number) since that would make the visible
    evidence of paragraphing and subparagraphing less obvious.

<u>Paragraph Headings</u>

4. <u>Paragraph Heading Preceding a Single Paragraph</u>

    Where only a single paragraph follows a side or paragraph heading, the text is not
    assigned a separate paragraph number and is still typed with the left-hand edge
    level with the normal paragraph position.

5. <u>Paragraph Heading Preceding Several Paragraphs</u>

    5.1 Where several paragraphs follow a heading, the paragraphs are assigned sub-
         paragraph numbers.

    5.2 In this case the text is indented further to the right, as has been done here.

6. <u>Paragraph Heading Built into Paragraph</u>.   In this alternative arrangement the text
    continues immediately after the heading, so that the heading becomes part of the
    paragraph. Subparagraphing can be used quite easily.

    6.1 This would be the first subparagraph.

    6.2 This would be the second subparagraph.

<u>Subparagraph Headings and Subparagraphing</u>

7. The arrangement of headings should be consistent throughout the document. Thus the
    rules that apply to paragraph headings also apply to subparagraph headings and the
    subparagraphs that follow them.

    7.1 <u>Subparagraph Heading</u>

         a. The remaining entries demonstrate progressively further subparagraphing.

         b. These subparagraphs should be short. If there are many, letters "i" and
             "l" should be omitted to avoid confusion with Greek numbers "i" and
             Arabic number "l".

             (1) If even further subparagraphing is necessary (and it should occur
                  only on rare occasions) then this method may be used.

             (2) Entries should be very short, preferably no longer than one
                  sentence.

Fig. 10-4   Combined System of Headings and Paragraph Numbers.

it is a part, yet it must not be obtrusive except when called on to emphasize a specific point. It should convey only one idea, although made up of several sentences each containing a separate thought.

Experienced writers construct effective paragraphs almost subconsciously. They adjust length, tone, and emphasis to suit their topic and the atmosphere they want to create. Literary writers, in particular, knowingly stretch and bend the rules of good paragraph construction to obtain exactly the right impact. But they once had to study and master the techniques of good paragraph construction to enable them to write fluently and creatively. Their writing now is not inhibited by the need to pay attention to minute details. The rhythm of words guides them far more than rules of good paragraph construction.

We are not so fortunate. We have to learn the rules and apply them consciously. But in doing so we must take care not to let our approach become too pedantic, to become so bound by the rules that we write in a stilted manner that is dull, uninteresting, and unrhythmic. So when considering the factors designed to help you construct good paragraphs, remember that they are blocks on which to build your writing, not bars to imprison your creativity.

The three elements essential to good paragraph writing are:

> Unity
> Coherence
> Adequate Development

These elements cannot stand alone. All three must be present if a paragraph is to be useful to the reader.

Unity

For a paragraph to have unity, it must be built entirely around a central idea. This idea is expressed in a topic sentence and developed in supporting sentences. In effect, this permits us to construct paragraphs using the pyramid technique I previously applied to the whole document, with the topic sentence taking the place of the summary and the supporting sentences representing the full development (see Fig. 10–5).

The paragraph below is strongly unified because its topic is clearly expressed and the supporting sentences develop it fully (the numbers that precede each sentence are for reference):

(1) *Joey, when we began our work with him, was a mechanical boy.* (2) He functioned as if by remote control, run by machines of his own powerfully creative fantasy. (3) Not only did he himself believe that he was a machine but, more remarkably, he created this impression in others. (4) Even while he performed actions that are intrinsically human, they never appeared to be

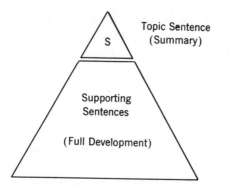

Fig. 10–5 Pyramid Technique Applied
to the Paragraph.

other than machine-started and executed. (5) On the other hand, when the machine was not working we had to concentrate on recollecting his presence, for he seemed not to exist. (6) A human body that functions as if it were a machine and a machine that duplicates human functions are equally fascinating and frightening. (7) Perhaps they are so uncanny because they remind us that the human body can operate without a human spirit, that body can exist without soul. (8) And Joey was a child who had been robbed of his humanity.[1]

This opening paragraph of a technical article demonstrates that technical writing does not have to be dull. Let's analyze it, identify why it has unity, and discover why it holds our interest and makes us want to read on.

The paragraph is summarized in sentence (1), the topic sentence, which sets the scene in nontechnical terms that create interest. Sentence (2) amplifies the topic sentence by using slightly more technical terms and introducing the fact that Joey's mechanical aspects are self-created. Then sentence (3) underscores the strength of Joey's imagination by telling us that he caused adult onlookers to feel that he really was a machine. This is carried further in sentence (4), which relates his machinelike actions to normal activities, and sentence (5), which shows how very real his machinelike qualities were, even in moments of inactivity. Then sentences (6) and (7) introduce the fascination (and an element of apprehension) that Joey's condition caused in persons observing him. Sentence (8) restates the idea expressed in sentence (1), but this time in human as opposed to mechanical terms.

The topic sentence does not always have to be the first sentence in a paragraph; there are even occasions when it does not appear at all and its presence is only implied. But until you are a proficient writer you would be wise to place your topic sentences in the primary position, where both you and your readers can see them. Then, when you have experience to support your actions, you can experiment and try placing topic sentences in alternative positions.

The two examples that follow are paragraphs that do not start with a topic sentence. In this descriptive passage on baseball the topic sentence is at the end of the paragraph (compare it with the similarly structured technical paragraph on p. 238):

> The reason, obviously, is that baseball came up from the sand lots—the small town, the city slum, and the like. It had a rowdy air because rowdies played it. One of the stock tableaux in American sports history is the aggrieved baseball player jawing with the umpire. In all our games, this tableau is unique; it belongs to baseball, from the earliest days it has been an integral part of the game, and even in the carefully policed major leagues today it remains unchanged. *Baseball never developed any of the social niceties.*[2]

In the continuing description of Joey, the "Mechanical Boy," the topic sentence is only implied. If it had been stated it would have read something like this: "Joey's machinelike actions were very realistic."

> For long periods of time, when his "machinery" was idle, he would sit so quietly that he would disappear from the focus of the most conscientious observation. Yet in the next moment he might be "working" and the center of our captivated attention. Many times a day he would turn himself on and shift noisily through a sequence of higher and higher gears until he "exploded," screaming "Crash, crash!" and hurling items from his ever present apparatus—radio tubes, light bulbs, even motors or, lacking these, any handy breakable object. (Joey had an astonishing knack for snatching bulbs and tubes unobserved.) As soon as the object thrown had shattered, he would cease his screaming and wild jumping and retire to mute, motionless nonexistence.[3]

Coherence

Coherence is the ability of a paragraph to hold together as a solid, logical, well-organized block of information with good continuity. To the reader a coherent paragraph is abundantly clear; he can easily follow the writer's line of reasoning, and has no problem in progressing from one sentence to the next.

Most technical people are logical thinkers and should be able to write logical, well-organized paragraphs. But the organization must not be kept a secret; it must be apparent to every reader who encounters their work. Simply

[2]From Bruce Catton, "The Great American Game," *American Heritage*, April 1959. Copyright © 1959 by American Heritage Publishing Co., Inc. Reprinted by permission.
[3]Bettelheim, "Joey, 'A Mechanical Boy,'" *op. cit.*

summarizing a paragraph in the topic sentence and then following it with a series of supporting sentences does not make a coherent paragraph. The sentences must be arranged in an identifiable order, following a pattern that assists the reader in understanding what is being said.

This pattern will depend on the topic and the type of document. Paragraphs describing an event or a process most likely will adopt a sequential pattern; those describing a piece of equipment will probably be patterned on the shape of the equipment or the arrangement of its features. Patterns that can be used for typical writing situations are illustrated in Table 10–1.

Narrative Patterns. You can write narrative-type paragraphs when a sequence of steps or events has to be described. The past-present-future pattern of an occurrence report, such as Hank Williams' accident report in Figure 4–1 (p. 76) and the damage to a videotape recorder described on page 77, are typical examples. In every paragraph the pattern should be clearly evident, as in two of the following three paragraphs:

A coherent paragraph (in chronological order)	The accident occurred when D. Friesen was checking in at the Remick Airlines counter. He placed the company Polaroid camera on the counter while he completed flight boarding procedure. When the passenger ahead of him (Miss Jane Tooke) left the counter, the shoulder strap of her handbag tangled with the carrying strap of the camera and pulled the camera to the floor. D. Friesen examined the camera and discovered a 1½ inch crack across its back. Remick Airlines' representative K. Trane took details of the incident and will be calling you to discuss compensation.
A much less coherent paragraph (containing the same information but not presented in an identifiable pattern)	The accident occurred when D. Friesen was checking in at the Remick Airlines counter. K. Trane, a Remick Airlines representative, took details of the incident and will be calling you to discuss compensation. The damaged camera received a 1½ inch crack across the back. When the passenger ahead of D. Friesen (Miss Jane Tooke) removed her handbag from the counter its shoulder strap tangled with the carrying strap of the company Polaroid camera and pulled it to the floor. D. Friesen had placed the camera on the counter while he completed flight boarding procedure.
A coherent paragraph (tracing events from evidence to conclusion)	A mild shimmy at speeds above 52 mph was noticed about 10 days after the new tires had been installed. A visual check of all four wheels revealed no obvious defects. Rotation of the four wheels to different positions on the vehicle did not eliminate the shimmy but seemed to change its point of origin. To pin down the cause I replaced each

TABLE 10–1—PARAGRAPH PATTERNS USED IN TECHNICAL WRITING

Topics or Subjects		Writing Patterns	Definitions and Examples
Generally Narrative	Event	Chronological order	The sequence in which events occurred; e.g. the steps taken to control flooding, how a new product was developed, how an accident happened
	Occurrence	Logical order	Used when chronological order would be too confusing (when describing a multiple-activity process, or several events that occurred concurrently); requires careful analysis to achieve clear narrative
	Situation / Process	Cause to effect	The factors (causes) leading up to the result (effect) they produce; e.g. dirty air vents and failure to lubricate bearings cause overheating and eventual failure of equipment
	Procedure	Evidence to conclusion	The factors, data, or information (evidence) from which a conclusion can be drawn; e.g. how results obtained from a series of tests help identify the source of a problem
	Method	Comparison	A comparison of factors (price, size, weight, convenience) to show the differences between products, processes, or methods
Generally Descriptive — By Shape	Equipment	Vertical	From top to bottom* of a tall, narrow subject
		Horizontal	From left to right* of a shallow, wide subject
	Scene	Diagonal	From corner to corner (for items arranged across a subject)
	Appearance	Circular	From center to circumference* (for items arranged concentrically)
			Clockwise* (for items arranged around a subject)
Generally Descriptive — By Features	Building	In order of: Size	From smallest item to largest item*
	Location	Importance	From most important to least important item*
	Features	Operation	The sequence in which the items are used when the equipment is operated

* Or vice versa.

317

wheel in turn with the spare wheel, and found that the shimmy disappeared when the spare was in the left front position. The wheel removed from that position was tested and found to have been incorrectly balanced.

Descriptive Patterns. You can write descriptive paragraphs to describe scenes, buildings, equipment, and any subject having physical features. The shape of the subject often dictates the pattern, with a low, horizontal subject calling for a left-to-right pattern, and a tall thin subject calling for a top-to-bottom pattern. Subjects that do not have a clearly defined shape, or on which the features are not arranged regularly, may have to be described in a less obvious pattern, such as the order in which parts are operated, their order of importance, or their arrangement from largest to smallest. For example:

> *Excerpt* The most important control on the bombardier's panel is
> *from para-* the firing button, which when not in use is held in the
> *graph* black retaining clip at the bottom left-hand corner. Next
> *(features in* in importance is the fusing switch at the top right of the
> *order of* panel; when in the "OFF" position it prevents the bombs
> *importance)* from being dropped live. Two safety switches, one im-
> mediately above the firing button retaining clip and the
> other to the right of the bank of selector switches, pre-
> vent the firing button from being withdrawn from its clip
> unless both are in the "UP" position.

Continuity. A fully coherent paragraph must also have smooth transitions between the sentences. Smooth transitions give a sense of continuity that makes the reader feel comfortable. As he finishes one sentence, there is a logical bridge to the next. This can be done by using linking words, and by referring back to what has already been said. In the example I have just quoted there is a natural flow from "The most important . . ." in the first sentence to "Next in impor- tance . . ." in the second; the third sentence then refers back to the firing button to relate the newly introduced safety switches to what has gone before. The transitions are equally good in the first paragraph describing damage to a Polaroid camera, each sentence containing a component that is a development from one of the previous sentences. This is not true of the second paragraph, in which each new sentence introduces a new subject with no reference to what has already been said.

Adequate Development

This element demands good judgment. It means you must identify your reader clearly enough so that you can look at each paragraph from his point of view. Only then can you establish whether your supporting sentences amplify the topic sentence in enough detail to satisfy him.

Simple insertion of additional supporting sentences does not necessarily meet the requirements for adequate development. The supporting sentences must contain just the right amount of pertinent information that is relevant to the particular reader. There must never be too little or too much information. Too little results in fragmented paragraphs that present snippets of information that arouse the reader's interest but do not satisfy his needs. On the other hand, too much information can lead to long, repetitious paragraphs that annoy him because they waste his reading time. Compare the following paragraphs, all describing the result of man's first venture with machinery across the Peel Plateau east of Alaska, intended for a reader interested in the problems of working in the north but who has never seen what the terrain is like.

Inadequate development Trails left by the tractors look like long narrow scars cut in the plateau. Many of them have been there for years. All have been caused by permafrost melting. They will stay like this until the vegetation grows in again.

Adequate development To the visitor viewing this far northern terrain from the air, the trails left by tractors clearing undergrowth for roads across the plateau look like long, narrow scars. Even those that have been there for as long as 28 years are still clearly defined. All have been caused by melting of the permafrost, which started when the surface moss and vegetation were removed and will continue until the vegetation grows in again—perhaps in another 30 years.

Over-development To the conservationist, whose main interest is the protection of the environment, viewing this far northern terrain from the air is a heartrending sight. To him, the trails left by tractors clearing undergrowth for roads across the plateau look like long, narrow scars. The tractors were making way for the first roads to be built by man over an area that until now had been trodden only by Indians indigenous to the area, and the occasional trapper. Some of these trappers had journeyed from Quebec to seek new sources of revenue for their trade. But now, in the very short time span of 28 years, man has defiled the terrain. With his machines he has cut and gouged his way, thoughtlessly creating havoc that will be visible to those that follow for many decades. Those that preceded him for centuries had trodden carefully on the permafrost, leaving no trace of their presence. The new trails, even those that have been there for 28 years, are visible almost as though they had been cut yesterday. And all were caused by melting of the permafrost. . . .

In these examples the descriptive pendulum has swung from one extreme to the other. The first paragraph leaves the reader with questions in his mind: What were the tractors doing? How many years? How soon will the vegetation grow in again? The second develops the topic sentence in just enough detail; it explains why the tractors left semipermanent scars and predicts how long they will remain. The third paragraph, though interesting, is filled with irrelevant information (e.g. where the trappers came from) and repetitive statements that detract from the main theme. It might be suitable for a novel but not for a technical report or description.

The first paragraph of the article on Joey, the "Mechanical Boy" (p. 313) also is adequately developed because it provides just sufficient information to support the author's original intent: to introduce the boy. It confines itself to a brief word picture of Joey's problem (but not of Joey himself) and its effect on the boy and the adults who worked with him. Because its readers would be technical men and women who specialize in other technical fields, the paragraph introduces its subject in general terms that will be understood by and provoke interest in a wide range of readers. The author has not described who Joey is, where he is, and what his surroundings are like. Instead, he has started with a thought provoking topic sentence and used the whole paragraph to develop the image of the "mechanical" child. To have attempted to introduce names, locations, and similar information would have destroyed the impact of this very effective introductory paragraph.

Correct Length

How long should a good paragraph be? My answer to this enigmatic question is simple: It should be no longer than you need to cover a particular topic for a particular reader. To this I must add three conditions:

1. Variety in paragraph length helps the reader. A series of paragraphs of equal length creates the impression of dullness. If the paragraphs are consistently short the reader will feel cheated because he is not presented with enough information. (Consistently short paragraphs create a jack-rabbit effect—all starts and stops with no significant ground covered.) Conversely, a succession of very long paragraphs will make him feel he is swimming in a sea of extraneous information, and he will often find it difficult to identify the main topic.
2. Readers attach importance to a paragraph that is clearly longer or shorter than those surrounding it. A very short paragraph among a a series of generally longer paragraphs attracts attention, just as a longer paragraph among a series of short paragraphs implies that the information it contains is particularly important.

3. Paragraph length needs to be adjusted to suit complexity of topic and technical level of reader. Generally, complex topics demand short paragraphs containing small portions of information. But this should be conditioned by your knowledge of the reader. Even a very complex topic can be covered in long paragraphs for one who is technically competent.

SENTENCES

Although sentences normally form an integral part of a larger unit—the paragraph—they still must be able to stand alone. While helping to develop the whole paragraph, each has to develop a separate thought. In doing so they play an important part in placing emphasis—in stressing points that are important and playing down those that are not.

As I mentioned in the discussion on paragraphs, experienced writers had to learn the elements that contribute to good paragraph writing so that now they can apply them automatically. They also had to learn the elements of good sentence construction, which are:

Unity
Coherence
Emphasis

The first two elements—unity and coherence—perform a function similar to the elements of the same name in a well-written paragraph.

Unity

Although the comparison is not quite so clearly defined, we can still apply the pyramid technique to the sentence in the same way that I applied it to the paragraph and the whole document. In this case the summary is replaced by a primary clause that presents only *one thought*, and the full development by subsidiary clauses and phrases that develop or condition that thought (see Fig. 10–6). Thus a unified sentence, as its name implies, presents and develops only a single thought.

Compare these two sentences:

A unified sentence that expresses one main thought: The Amron Building will make an ideal manufacturing plant because of its convenient location, single-level floor, good access roads, and low rent.

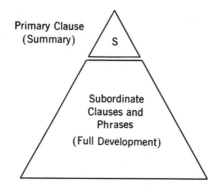

Fig. 10–6   Pyramid Technique Applied
to the Sentence.

A complicated   The copier should never have been placed in the general
sentence that   office, where those using it interrupt the work being done
tries to express   by the stenographic pool, which has been consistently un-
two thoughts: derstaffed since the beginning of the year.

The first sentence has unity because everything it says relates to only one topic: that the Amron Building will make a good manufacturing plant. The second sentence fails to have unity because the reader cannot tell whether he is supposed to be agreeing that the copier should have been placed elsewhere, or sympathizing that there has been a shortage of stenographers. No matter how complex a sentence may be, or how many subordinate clauses and phrases it may have, every clause must either be a development of or actively support only one thought, expressed in the primary clause.

Coherence

Coherence in the sentence is very similar to coherence in the paragraph. Coherent sentences are continuously clear, even though they may have numerous subordinate clauses, so that the message they convey is apparent throughout. Like the paragraph, they need good continuity, which can be obtained by arranging the clauses in logical sequence, by linking them through direct or indirect reference to the primary clause, and by writing them in the same grammatical form. Parallelism, discussed at the end of this chapter, plays an important role here.

The unified sentence discussing the Amron Building has good coherence because its purpose is continuously clear and each of its subordinate clauses links comfortably back to the lead-in statement: "because of its. . . ." The clauses are written in parallel form (i.e. they have the same grammatical pattern), which is the best way to carry the reader smoothly from point to point. The first sentence below lacks coherence because there is no logic to the arrangement or form of the subordinate clauses. Compare it with the second sentence, which despite its length is still coherent because it continuously develops the thought of "late" and "damaged" expressed in the primary clause.

An incoherent sentence:	The Amron Building will make an ideal manufacturing plant because of its convenient location, which also should have good access roads, the advantage of its one-level floor, and it commands a low rent. (34 incoherent words)
A coherent sentence:	Many of the 61 samples shipped in December either arrived late or were damaged in transit, even though they were shipped one week earlier than usual to avoid the Christmas mail tie-up, and were packed in polyurethane as an extra precaution against rough handling. (45 coherent words)

Emphasis

Whereas unity and coherence ensure that a sentence is clear and uncluttered, properly placed emphasis helps the reader identify its important parts. By arranging a sentence effectively you can help the reader attach importance to the whole sentence, to a clause or phrase, or even to a single word.

Emphasis on the Whole Sentence. You can give a sentence more emphasis than sentences that precede and follow it by manipulating its length or by stressing or repeating certain words. You can also imply to the reader that all the clauses within a sentence are equally important (without affecting its relationship with surrounding sentences) by balancing its parts.

Although a writer should strive for variety in the length of sentences, there will be occasions when he will want to emphasize a specific sentence through its length. Readers will attach importance to a short sentence placed among several long sentences, or to a long sentence among predominantly short ones. They will also detect a sense of urgency in a series of short sentences that carry them quickly from point to point. This technique is used effectively by storytellers:

> The prisoner huddled against the wall, alone in the dark. He listened intently. He could hear the guards, stomping and muttering. Cursing the cold, probably.

In technical writing we have little occasion to write very short sentences, except perhaps to impart urgency to a warning of a potentially dangerous situation:

Dangerously high voltages are present on exposed terminals. Before
opening the doors:
1. Set the master control switch to "OFF."
2. Hang the red "NO!" flag on the operator's panel.
Never cheat the interlocks.

Neither must we err in the opposite direction and write overly long
sentences that are confusing. The rule that I applied to paragraph writing—
adjust the length to suit complexity of topic and technical level of reader—applies equally to sentence writing.

Repetition of key words is another effective technique for making a
sentence strong and emphatic, but it must be used sparingly. Its sudden use will
catch a reader's attention; repeated utilization will seem awkward and contrived.
For instance, a product manager who wanted to say "no" in very strong, definite
terms might write:

> I know it's clever and I know it's interesting; I know it's cheap and
> I know it's reliable; but I also know it's not the product for our
> market.

How much stronger this is than the more straightforward:

> I know it's clever, interesting, cheap, and reliable; but it's not the
> product for our market.

The somewhat exaggerated first example also demonstrates how parallelism can
give a sentence continuity. The reader, recognizing that the clauses have a similar
shape, proceeds easily from one to the next.

Similarity of shape can also signify that all parts of a sentence are of
equal importance. Clauses separated by a coordinate conjunction (mainly and,
or, but, and sometimes a comma) tell a reader that they have equal emphasis.
The following sentences are "balanced" in this way:

> Eight test instruments were used for the rehabilitation project, *and*
> were supplied free of charge by the Dere Instrument Company.

> The gas pipeline will be 300 miles shorter than the oil pipeline,
> *but* will have to cross much more difficult terrain.

> The upper knob adjusts the instrument in the vertical plane, *and* the
> lower knob adjusts it in the horizontal plane.

Emphasis on Part of a Sentence. While coordination means giving equal
weight to all parts of a sentence, subordination means emphasizing a specific
part and deemphasizing all other parts. It is effected by placing the most important information in the primary clause, and placing less important informa-

tion in subordinate clauses. In each of the following sentences the main thought is italicized to identify the primary clause:

> When the technician momentarily released his grip, *the control slipped out of reach.*

> *The bridge over the underpass was built on a compacted gravel base,* partly to save time, partly to save money, and partly because materials were available on site.

> *He lost control of the vehicle* when the wasp stung him.

> When the wasp stung him, *he lost control of the vehicle.*

The two initial sentences of the paragraph introducing these examples use subordination and coordination. In the first sentence the last 11 words are emphasized (from "subordination means" to "all other parts") and the first part of the sentence is deemphasized; thus the sentence has subordination. In the second sentence, the word "and" indicates that the parts on either side of it are of equal importance; thus the sentence has coordination.

Emphasis on Specific Words. Where we place words in a sentence has a direct bearing on their emphasis. The first and last words in a sentence are automatically emphasized. Notice how unimportant words have robbed the following sentences of emphasis by stealing the impact-bearing positions:

Emphasis misplaced:	Such matters as equipment calibration will be handled by the standards laboratory however.
Emphasis restored:	Equipment calibration will, however, be handled by the standards laboratory.
Emphasis misplaced:	Without exception change the oil every six days at least.
Emphasis restored:	Change the oil at least every six days.

The verbs we use have a powerful influence on emphasis. Strong verbs attract the reader's attention, whereas weak verbs tend to divert it. Active verbs are strong because they tell *who did what.* Passive verbs are weak because they merely pass along information; they describe "what was done by whom." The sentences below are written using both active and passive verbs. Note that the versions written in the active voice are consistently shorter:

Passive Verbs	*Active Verbs*
Elapsed time is indicated by a pointer.	A pointer indicates elapsed time.
The project was completed by the installation crew on May 2.	The installation crew completed the project on May 2.

Passive Verbs	*Active Verbs*
It is suggested that meter readings be recorded hourly.	I suggest you record meter readings hourly.
The samples were passed first to quality control for inspection, and then to the shipping department where they were packed in polyurethane.	Quality control inspected the samples and then the shipping department packed them in polyurethane.

There are occasions when you will have to use the passive voice because you are reporting an event without knowing who took the action. For instance, if you want to state that during a test the strain gage was read by technician Kevin McCaughan at ten minute intervals, you can write in the active voice:

Kevin McCaughan read the strain gage at ten minute intervals.

If you want to deemphasize Kevin's role, you can simply omit his name:

A technician read the strain gage at ten minute intervals.

But if several technicians took part in the test, or you were not present when the test took place but have to report what was done, you will have to write in the passive voice:

The strain gage was read at ten minute intervals.

Unfortunately, many scientists and engineers in industry still believe that everything should be written in the passive voice. They shun the active voice because they feel it is too strong and does not fit their professional character. With the passage of time, this outdated belief is slowly being eroded.

WORDS

Word selection is important. The right word in the right place at the right moment can greatly influence a reader. A heavy, ponderous word will slow him down; an overused expression will make him doubt your sincerity; a complex word that he does not recognize will annoy him; and a weak or vague word will make him think of you as indefinite. But the right word—short, clear, specific, and necessary—will help him understand your message quickly and easily.

Words That Tell a Story

Words should convey images. We have many strong, descriptive words in our individual vocabularies, but most of the time we are too lazy to employ them, preferring to use the same old routine words because they spring easily to mind. We write "put" when we would do better to write "place," "position," "insert," "drop," "slide," or any one of numerous descriptive verbs that describe the action better. Compare these examples of vague and descriptive words:

Vague Words	*Descriptive Words*
He *got it out* using the MX extraction tool.	He *pulled* it out He *extracted* it He *unlocked* it He *unscrewed* it
While the crew was in town they *picked up* spare parts	they *bought* they *purchased* they *ordered* they *borrowed* they *stole*
The project will *take a long time*	*require 300 man-hours* *employ two men for six weeks* *last four months*

Story writers use descriptive words to convey active images to their readers. The excerpt on page 315 effectively demonstrates this. When Bruce Catton writes: ". . . the aggrieved baseball player *jawing* with the umpire," the reader can *see* the jaw going up and down much more clearly than if Catton had written "arguing" in its place. Our concern with technical writing does not mean that we should never seek colorful words to help the reader *see* the topic. (But a word of warning: Reserve most of these colorful, descriptive words for your verbs and nouns, rather than for adverbs and adjectives.) One descriptive word that defines size, shape, color, smell, taste, and so on, is much more valuable than a dozen words that only generalize.

Analogies can offer a useful means for describing an unfamiliar item in terms a nontechnical reader will recognize:

A resistor is a piece of ceramic-covered carbon about the size of a cribbage peg, with a two inch length of wire protruding from each end.

Specific words tell the reader that you are a definite, purposeful individual. Vague generalities imply that you are unsure of yourself. If you write:

It is considered that a fair percentage of the samples received from one of our suppliers during the preceding month contained a contaminant.

you are giving the reader four opportunities to wonder whether you really know much about the topic:

1. "It is considered"   Who has voiced this opinion?
2. "a fair percentage"   How many?
3. "one of our suppliers"   Who? One in how many?
4. "contained a contaminant"   What contaminant? In how
                                   strong a concentration?

All these generalities can be avoided in a shorter, more specific sentence:

We estimate that 60% of the samples received from RamSort Chemicals last June were contaminated with 0.5% to 0.8% mercuric chloride.

This statement tells the reader that you know exactly what you are talking about. As a technical person, you should never create any other impression.

Long vs. Short Words

Writing that is full of long words when shorter words could carry the same message makes a writer sound either pompous or insecure. They act as a barrier between him and the reader. Sometimes writers use big words to hide lack of knowledge, or because they started writing without defining clearly what they wanted to say. There are many long scientific words that we have to use—we should surround them with short words whenever possible so that our writing does not become ponderous and overly complex.

There are also many words and expressions that we use solely because we see other writers use them. Vance Richards, H.L. Winman and Associates' top architect, writes as flamboyantly as he dresses, and his draftsmen tend to copy him. Thus intermediate draftsman Jim Kingsley models his writing on Vance's, and has replaced the simple, direct style he brought with him as a new employee with wordy writing like this:

Vague Pursuant to the client's original suggestion, Mr. Richards
and was of the opinion that the structure planned for the
Wordy client would be most suitable for immediate erection on
the site until recently occupied by the old established
costume manufacturer known as Garrick Garments. In
accordance with the client's anticipated approval of this
site, Mr. Richards has taken great pains to design a multi-
level building that can be considered to use the property
to an optimum extent.

"Waffle" is the word I apply to such cumbersome writing. Full of hackneyed expressions ("pursuant to," "in accordance with") and words of low information content ("of the opinion that," "for immediate erection on," "can be considered to use," "the optimum extent"), it uses 76 words to state what can be said as clearly in 54 direct words:

> *Clear* Mr. Richards believes that the type of building that best
> *and* suits the client's ideas should be at the corner of Main
> *Direct* and Willis, on the site previously occupied by Garrick
> Garments. On the assumption that the client will approve
> this site, Mr. Richards has designed a multi-level building
> that fully develops the property.

Low Information Content Words

Words of low information content (LIC) contribute little or nothing to the facts conveyed by a sentence. They are untidy lodgers. Remove them and the sentence appears neater and says just as much. The problem is that practically everyone inserts LIC words into sentences, and we become so accustomed to them that we do not notice how they fill up space without adding any information.

The partial list of LIC words and phrases in Table 10–2 indicates the types of expressions that we must try to delete from our writing. They are difficult to identify because they often sound like good but rather vague prose. In the following sentences the LIC words have been italicized; they should either be deleted or replaced as indicatd by the parenthesized notations at the end of each sentence.

> The control is actuated by *means of* No. 3 valve. (delete)
> Adjust the control *as necessary* to obtain maximum deflection. (delete)
> Tests were run for *a period of* three weeks. (delete)
> If the project drops behind schedule *it will be necessary to* bring in extra help. (*I, you, he, we,* or *they will* bring in extra help)
> By Wednesday we had a backlog of 632 units, *and for this reason* we adopted a two-shift operation. (*so*)
> A new store will be opened *in an area* where market research has *given an indication that there actually is a* need for more retail outlets. (delete; *indicated the*)

Hackneyed expressions are similar to LIC words and phrases, except that their presence is more obvious and their effect can be more damaging. Whereas LIC words are intangible in that they impart a sense of vagueness, hackneyed expressions can make a writer sound pompous, garrulous, insincere, or artificial. Referring to oneself as "the writer," starting and ending letters with

TABLE 10–2
EXAMPLES OF LOW INFORMATION CONTENT (LIC)
WORDS AND PHRASES

The LIC words and phrases in this partial list are followed by an expression in parenthesis (to illustrate a better way to write the phrase) or by an (X), which means that it should be dropped entirely.

actually (X)	in connection with (about)
a majority of (most)	in fact (X)
a number of (many; several)	in order to (to)
as a means of (for; to)	in such a manner as to (to)
as a result (so)	in terms of (in; for)
as necessary (X)	in the course of (during)
at the rate of (at)	in the direction of (toward)
at the same time as (while)	in the event that (if)
bring to a conclusion (conclude)	in the form of (as)
by means of (by)	in the neighborhood of; in the
by use of (by)	vicinity of (about; approxi-
communicate with (talk to;	mately; near)
telephone; write to)	involves the use of (employs;
connected together (connected)	uses)
contact (talk to; telephone;	involve the necessity of (de-
write to)	mand; require)
due to the fact that (because)	is designed to be (is)
during the time that (while)	it can be seen that (thus; so)
end result (result)	it is considered desirable (I or
exhibit a tendency (tend)	we want to)
for a period of (for)	it will be necessary to (I, you,
for the purpose of (for; to)	or we must)
for this reason (because)	of considerable magnitude (large)
in an area where (where)	on account of (because)
in an effort to (to)	prior to (before)
in close proximity to (close to;	subsequent to (after)
near)	with the aid of (with)

NOTE: Many of these phrases start and end with little connecting words such as *as, at, by, for, in, is, it, of, to,* and *with.* This knowledge can help you to identify LIC words and phrases in your writing.

such overworked phrases as "We are in acknowledgement of" and "please feel free to call me," and using semilegal jargon ('the aforementioned correspondent," "in the matter of," "having reference to") are all typical hackneyed expressions. Further examples are listed in Table 10–3.

Presenting the right word in the correct form is an important aspect of technical writing. Comments on how to abbreviate words, how to form compound adjectives, whether or not you should split an infinitive (generally, you may), and when you should spell out numbers or present them as numerals, are contained in a glossary of usage in Chapter 11.

TABLE 10–3
TYPICAL HACKNEYED EXPRESSIONS

and/or	in the matter of
as per	last but not least
attached hereto	please feel free to
enclosed herewith	pursuant to your request
for your information	regarding the matter of
(as an introductory phrase)	this will acknowledge
if and when	we are pleased to advise
in reference to	we wish to state
in short supply	with reference to
in the foreseeable future	you are hereby advised

PARALLELISM

Parallelism in writing means "similarity of shape." It is applied loosely to whole documents, more firmly to paragraphs, sentences, and lists, and tightly to grammatical form. It is important in all forms of writing because it creates a sense of balance. Readers generally do not notice that parallelism is present in a good piece of writing—they only know that it reads smoothly. But they are certainly aware that something is wrong when parallelism is lacking. Good parallelism makes a reader feel comfortable, so that even in long or complex sentences he never loses his way. Its ability to help the reader through difficult passages makes parallelism particularly applicable to technical writing.

The Grammatical Aspects

If you keep your verb forms similar throughout a sentence in which all parts are of equal importance (i.e. in which the sentence has coordination, or is balanced), you will have taken a major step toward preserving parallelism. For example, if I write "unable to predict" in the early part of a balanced sentence, I should write "able to convince" rather than "successful in convincing" in a latter part:

Parallelism violated: Mr. Johnson was *unable to predict* the job completion date, but was *successful in convincing* management that the job was under control.

Parallelism maintained (A) Mr. Johnson was *unable to predict* the job completion date, but was *able to convince* management that the job was under control.

More direct Mr. Johnson *predicted* no completion date, but *convinced*
alternative (B) management that the job was under control.
(parallelism
maintained)

The rhythm of the words is much more evident in the latter two sentences, particularly the alternative sentence in which the parallelism has been stressed by converting the still awkward ". . . unable . . . able . . ." of sentence (A) to the positively parallel "predicted . . . convinced . . ." of sentence (B). Two other examples follow:

Parallelism Pete Hansk likes surveying airports and to study new con-
violated: struction techniques.

Parallelism Pete Hansk likes to survey airports and to study new con-
maintained: struction techniques.
or
Pete Hansk likes surveying airports and studying new construction techniques.

Parallelism His hobbies are designing solid-state circuits and hi-fi
violated: component construction.

Parallelism His hobbies are designing solid-state circuits and con-
maintained: structing hi-fi components.
or
His hobbies are solid-state circuit design and hi-fi component construction.

Parallelism is particularly important when joining sentence parts with the coordinating conjunctions *and, or* and *but* (as in the examples above), sometimes " , ", and when using correlatives, which are:

either . . . or
neither . . . nor
not only . . . but also

Each part of a correlative must be followed by an expression with the same grammatical form. That is, if *either* is followed by a verb, then *or* must also be followed by the same form of verb:

Parallelism You may either repair the test set or it may be replaced
violated: under the warranty agreement.

Parallelism You may either repair the test set or replace it under the
maintained: warranty agreement.

Similarly, if one part of a correlative is followed by a phrase, then the second part must be followed by the same form of phrase:

Parallelism violated: The instrument not only requires mechanical repair, but also it will have to be realigned electrically.

Parallelism maintained: The instrument requires not only mechanical repair but also electrical realignment.

(Whenever I demonstrate parallelism after the correlative "not only . . . but also," I feel that I have constructed a clumsy sentence. An astute reader will have noticed that it would be much simpler to write: "The instrument requires both mechanical repair and electrical realignment." This particular correlative can successfully emphasize a point, but does look awkward when the sentence in which it appears is taken out of context.)

Application to Technical Writing

Although parallelism is a useful means for maintaining continuity in general writing, it is in technical writing that it seems to have special application. Parallelism can clarify difficult passages and give rhythm to what otherwise might be dull, uninteresting material. When building sentences that have a series of clauses, you can help the reader see the connection between elements by molding them in the same shape throughout the sentence. There is complete loss of continuity in this sentence because it lacks parallelism:

Parallelism violated: In our first list we inadvertently omitted the 7 lathes in room B101, 5 milling machines in room B117, and from the next room, B118, we also forgot to include 16 shapers.

When the sentence is written so that all the items have a similar shape, the clarity is restored:

Parallelism restored: In our first list we inadvertently omitted 7 lathes in room B101, 5 milling machines in room B117, and 16 shapers in room B118.

Emphasis can also be achieved by presenting clauses in logical order. If they are arranged climactically (in ascending order of importance), or in order of increasing length, the reader's interest will gain momentum right through the sentence:

No one ever offered to help the navigator struggle out to the aircraft, a parachute hunched over his shoulder, a roll of plotting charts

tucked under his arm, a sextant and an astrocompass dangling from
one hand, and his precious navigation bag heavy with instruments,
log tables, radar charts, and the route plan for the ten hour flight
clenched firmly in the other.

Within the paragraph parallelism has to be applied more subtly. If it
is too obvious, the similarity in construction can be dull and repetitive. The
verbs often are the key: keep them generally in the same mood and they will
help bind the paragraph into one cohesive unit (this is closely tied in with
coherence, discussed under "Paragraphs"). This paragraph has good parallelism:

> One of these structures is the student union, a rallying point for
> snacking, dalliance and amusement. From morning until night it re-
> sounds to the blare of the jukebox, the clink of coffee cups, the clatter
> of bowling pins, the click of billiard balls, the slap of playing cards,
> the gentle creak of lounge chairs and, in the plushier ones, the splash
> of languid bodies in tepid swimming pools. There's likely to be an
> informal dance here every Friday and Saturday night. They have a
> ball—banquet or name-band dance—about every weekend in the ball-
> room.[4]

Similarity of shape is most obvious in the second sentence, with its
rhythmic:

> *blare* of the jukebox
> *clink* of coffee cups
> *clatter* of bowling pins
> *click* of billiard balls
> *slap* of playing cards
> *creak* of lounge chairs
> *splash* of languid bodies in
> tepid swimming pools

Here we have words that paint strong images. They not only have the same
grammatical form but also are bound together because they relate to the same
sense: hearing. The longest clause is at the end of the sentence, where the reader
can relax a moment with the "languid bodies in tepid swimming pools."

Of course, parallelism can be of particular effect in descriptive writing
like this, where the author wants to convey exciting images. In technical writing
the subjects are more mundane, giving us less opportunity for imaginative
imagery. Nevertheless, you should try to use parallelism to carry your reader
smoothly through your paragraphs, as in this description of part of a surveyor's
transit:

[4]From Jerome Ellison, "Are we Making a Playground out of College?" *The Saturday Evening Post*, March 7, 1959. Reprinted by permission.

Two sets of clamps and tangent screws are used to adjust the leveling head. The upper clamp fastens the upper and lower plates together, while the upper tangent screw permits a small differential movement between them. The lower clamp fastens the lower plate to the socket, while the lower tangent screw turns the plate through a small angle. When the upper and lower plates are clamped together they can be moved freely as a unit; but when both the upper and lower clamps are tightened the plates cannot be moved in any plane.

Application to Subparagraphing

Subparagraphing, which seldom occurs in literature, is a useful technical writing technique for separating a series of steps or describing different operations that follow from or belong to the main topic. Parallelism comes into play in subparagraphing as a means for reminding the reader that he is following a series of events or steps; it can also act as a reminder to the writer that he has to adopt a similar grammatical approach for each step. This paragraph has been subdivided to describe a series of tests (only the initial words are included for each test):

Three tests were conducted to isolate the fault:
    In the first test a matrix was imposed upon the face of the cathode ray tube and. . . .
    The second test consisted of voltage measurements taken at. . . .
    A continuity tester was connected to the unit for test 3 and. . . .

The last subparagraph is not parallel with the first two. To be parallel it must adopt the same approach as the others (i.e. first mention the test number and then say what was done):

For the third test, a continuity tester was connected to the unit and. . . .

To be truly parallel the subparagraphs should start in *exactly* the same way:

In the first test a matrix was. . . . .
In the second test a series of. . . .
In the third test a continuity tester was. . . .

But this would be too repetitive. The slight variations in the original version make the parallelism more palatable.

If more than three tests have to be described, a different approach is necessary. To continue with "A fourth test showed . . .," "For the fifth test . . .," and so on, would be dull and unimaginative. A better method is to insert a number in front of each paragraph:

> Seven tests were conducted to isolate the fault:
> 1. A matrix was imposed upon the face of the cathode ray tube and. . . .
> 2. Voltage measurements were taken at. . . .
> 3. A continuity tester was connected to the unit and. . . .

Parallelism has been retained, and we now have a more emphatic tone that lends itself to technical reporting.

A third version employs active verbs together with parallelism to build a strong, emphatic description that can be written in either the first or third person:

> In laboratory tests to isolate the fault we:
> 1. *Imposed* a matrix upon the face of the cathode ray tube and. . . .
> 2. *Measured* voltage at. . . .
> 3. *Connected* a continuity tester to the unit and. . . .

To change from the first person to the third person, only the lead-in sentence has to be rewritten:

> In tests conducted to isolate the fault, the laboratory:
> 1. *Imposed.* . . .
> 2. *Measured.* . . .
> 3. *Connected* . . .

Another area of technical writing where parallelism plays an important role is parts listing. This very specific application is discussed in Chapter 7.

ASSIGNMENTS

Exercise 1. Identify words and phrases of low information content in the following sentences. In some cases you will be able to delete the unnecessary words; in others you will have to revise a phrase or rewrite the sentence.

1. Measure the current with the aid of an ammeter connected to test points 9 and 14.
2. At operating speed the flywheel turns at the rate of 1200 revolutions per minute.
3. Lake Wabagoon is located in close proximity to Montrose, Ohio.
4. We have examined your concrete batcher and consider it will be

necessary to effect repairs that will cost in the region of $647.49.

5. The new output phase involves the use of cathode followers in cascade.
6. We plan tentatively to conduct a preliminary heat run on March 16.
7. An automatic switch-off valve is a feature that prevents damage to the equipment should the temperature rise above 180F.
8. The end result was in the form of a mixture that required further analysis.
9. We are pleased to be able to report that your receiver has been tested and aligned and found to meet all the manufacturer's specifications.
10. We wish to state that if you have any further queries regarding this matter, please feel free to contact Mr. A.B. Jones at any time.
11. The training program has been planned in such a manner as to reduce waiting time to a minimum.
12. The purchase of a new machine would result in an increase in operating expenditure in the approximate amount of $1800.00.
13. For your information, the new laboratory is now ready to receive the first shipment of test samples.
14. If pressure drops below 70 pounds per square inch it will be necessary to switch over to the standby unit.
15. A check in the amount of $37.42 is attached, in respect of laboratory services performed during the preceding month.
16. In an effort to increase production, it was found necessary by the plant superintendent to hire four additional assemblers.
17. Since all the spare parts needed for the overhaul have been received, work will be able to start immediately. Orders for additional spare parts are neither envisaged nor necessary.
18. A stability test did in fact prove that damaged instruments were a contributing factor in reducing measurement accuracy.
19. The angle is measured in terms of degrees and minutes of arc, and is recorded on a roll chart by means of a special type of pen.
20. We are in agreement with the committee's decision to make an effort to encourage greater student participation in activities conducted by the community.

Exercise 2. The following short passages lack compactness, simplicity, or clearness. Improve them by deleting unnecessary words, or by partial or complete rewriting.

1. Mr. Wasalusky has worked very efficiently as company janitor for seven years and for this reason it is recommended that he be considered for an increase in remuneration.
2. Personnel desirous of availing themselves of the opportunity of joining the bowling league are requested to contact J. Soames.

3. Connect the signal generator to the microphone terminals on the junction box with shielded wire.

4. High grass and brush around the storage tanks impeded the technicians' progress and should be cleared before they grow too dense.

5. It is entirely within the realms of possibility that the inspector is not cognizant of the specification.

6. The radioactive source must be maintained in an area where unauthorized personnel cannot gain access.

7. We apologize for the prolonged delay in settling your account, and wish to advise that your request for payment has again been forwarded to the Accounts Department with a request that a check be issued to you at the earliest possible opportunity.

8. Our project has now dropped rather far behind schedule, although we plan to make up as much as possible of this loss within a reasonable length of time.

9. On completion of the inspection, we found no evidence to support the view that negligence had occurred.

10. In an effort to consolidate purchases of consumable goods by different departments, the chief buyer will screen all purchase requisitions for the same with a view to reducing the quantity of orders for similar items.

11. An extensive evaluation of the field operations department has effectively demonstrated beyond a shadow of doubt that it has a mammoth administrative organization that is grossly out of proportion to the productivity of the department.

12. All the components supplied for the installation were examined and many were found to be defective or missing.

13. This instrument is relatively inexpensive, although it could be improved by the addition of a muting control, and will undoubtedly appeal to the majority of "ham" operators.

14. A replacement klystron was installed in the transmitter and appeared to operate satisfactorily for two to three hours. During the lunch hour there was a distinct flash, followed by what seemed like intermittent arcing, which signaled its early demise.

15. With regard to your valued enquiry of January 21, we can supply four type J locks at $12.75 and eight type MR control units at $16.24, and will be glad to ship same immediately on receipt of your valued order.

16. Fracture occurred at a force well below the minimum stipulated by the government specification (73,000 lb).

17. If it is intended that the instruments are to be monitored hourly, then it will be necessary to arrange for the maintenance technician to record the meter readings on form 168.

18. As well as presenting a monthly outage analysis, each field technician also gave a brief report on the serviceability status of the monitoring equipment at his pumping station.

19. In an effort to reduce the number of failures of operational equip-

ment, it was decided to perform mechanical tests of all components before assembly. This was done for a period of one month to assess whether the failure rate would in fact decrease. At the end of the month a 43% decrease in failures had been recorded.

20. Although the new contract was authorized by the government prior to the end of the old contract, it was not received until subsequent to a layoff of 320 men had occurred.

Exercise 3. The following sentences offer choices between words that sound similar or are frequently misused. Select the correct word in each case.

1. Vic Braun's panel truck was (stationary/stationery) 47 yards north of the intersection.
2. From Dr. Hilbury's remarks, the staff (implied/inferred) that they were expected to work harder.
3. The BRILLIANCE control (affects/effects) the intensity of the image on the screen.
4. Ground work could not be started this week (as/because/for) the soil was still frozen.
5. The (amount/number) of books on numerical control held by the library has increased significantly since last year's count.
6. This year's consumption was slightly (fewer/less) than in previous years, probably because the maintenance crew conducted (fewer/less) inspections.
7. It is a lot (farther/further) from Cleveland to Montrose, Ohio, than from Duluth to Reece, Minnesota.
8. No outside work was done last week because of the inclement (climate/weather).
9. Please (accept/except) my apologies for the delay in replying to your request for technical information.
10. Mr. Winman's suggestion that the company purchase an executive jet (appears/seems) to be feasible.
11. During May the laboratory processed 67 of the 103 samples received from the Roper Corporation. The (balance/remainder) will be processed in June.
12. Instrument serviceability increased by 20% two months after the preventative/preventive) maintenance program was started.
13. The power house boilers are fired up early in October, and operate (continually/continuously) until they are shut down for annual overhaul in mid-May.
14. Andy Rittman reported from the construction (cite/sight/site) that progress was being hampered by (defective/deficient) excavation equipment that had been subject to many breakdowns.
15. Before continuing with step 23, check that the sheet is free (from/of) surface defects.
16. We concluded that there was little difference (among/between) the four pantographs evaluated.

17. The frequency and loudness of the noise decreased as the flywheel (deaccelerated/decelerated).
18. The line supervisor (complemented/complimented) the production crew on the quality of their work.
19. To obtain an objective appraisal of his draft report, Hank Williams had it read by (a disinterested/an uninterested) technician who was not familiar with the topic.
20. The (principal/principle) reason I have included this exercise is that it will introduce you to the glossary of technical usage.

Exercise 4. Rewrite the following sentences to make them more emphatic (i.e. change from passive voice to active voice).

1. When a check was made of the aircraft's compasses, it was discovered by the calibration crew that a $+3°$ installation error existed.
2. It was recommended by the architect that a sulfate-resistant concrete should be used by the installation contractor.
3. At a project group meeting held on 28 September, it was unanimously agreed that no further action should be taken until after Mr. Dawes had returned from New York.
4. Just before fracture of the specimen occurred, the needle on the gage was noted to fluctuate wildly.
5. Because of low cloud and drizzle, Remick Airlines Flight 176 was instructed by the duty Air Traffic Controller to overfly Montrose Municipal Airport and to make a landing at Syracuse.
6. Accommodation for the night was arranged at airline expense for passengers who were stranded.
7. It was noted in the report that the cause of the accident was due to a bearing that was defective.
8. Although there had been a decrease in noise level of 12.7 dB for three weeks, complaints of "uncomfortable" noise were still being received by the Industrial Relations Department.
9. A new book on construction materials was received by Mr. Winman in the morning's mail, which the Materials Testing Laboratory was asked to evaluate.
10. It is our considered opinion that 167 yards of pavement are in need of rebuilding. Further, it is recommended that the remaining 421 yards of pavement be resurfaced.

Exercise 5. Improve the parallelism in the following sentences.

1. I recommend that we purchase the Smoothset blueprint machine because it is fast, quiet, and is not expensive.
2. The technicians were given training in organizing technical data and in how to present their conclusions.

3. Three of the applicants were given promotions, and transfers were arranged for the other four applicants.
4. We have found that the new system has four disadvantages:
    Too costly to operate
    It causes delays.
    Fails to use any of the existing equipment.
    It permits only one in-process examination.
5. If you purchase only one set the price will be $350.00, whereas the price will be $275.00 per set if you purchase six or more sets.
6. Although relatively inaccessible, the Kettle Generating Station is frequently visited by engineers interested in the technicalities of the project and artists.
7. A night crew was started to accelerate construction, rather than as a hindrance to progress by causing further problems.
8. The analysis was needed not only to determine permafrost regression, but also as a help in planning the proposed pipeline route.

Exercise 6. Check that the numbers in the following sentences have been expressed in the proper form (refer to Article 4 of Chapter 11 for rules for writing numbers in narrative). Rewrite those that have been written incorrectly.

1. The run-up will start at 10 o'clock on Tuesday morning.
2. As stated in chapter four, page 127, para 14.2, a machining tolerance of .003 in. is specified.
3. The survey covered 10857 males and 9881 females.
4. The heat tests must be conducted during week 3 and the results must be tabulated by the end of the 4th week.
5. 100 of the units are to be manufactured with a 1–3/4 in. diameter. 287 units are to be manufactured with a 3.25 in. diameter.
6. Twenty-six students wrote test number eight, with approximately twenty percent obtaining a mark of better than eighty percent.
7. For the first phase of test 14 2 spring balances will be required, plus 7 10 pound weights. In phase 2, ¼ pound of mercury and 7½ gallons of oil will be needed. These items will be used for ½ a day.
8. In reply to your letter of October 21st, and your subsequent Telex of third November, our price quotation for four (4) vertical drafting tables is six hundred and forty-seven dollars each. If you decide to purchase six (6) or more units, the price will be twenty percent less.

Exercise 7. Abbreviate the terms shown in *italics* in the following sentences. In some cases you will also have to express numerals in the proper form. (Rules for forming abbreviations appear in Article 2 of Chapter 11.)

1. Please supply *twenty-four pounds* of bitumastic compound and *three* rolls of *thirty-six inch wide* roofing material.
2. By driving at a constant 56 *miles an hour* we achieved a fuel consumption of 25.4 *miles per gallon.*
3. A noise level of 52.6 *decibels* was recorded at 825 *hertz.*
4. Rotate the control *clockwise* until a *continuous wave* signal is heard in the headset.
5. The transformer is rated for operation up to *eighty-five degrees centigrade.*
6. The municipality's 24 *inch inside diameter* storm sewers are fed into a 109 *inch diameter* culvert.
7. A meeting has been called for 10:30 *in the morning on Tuesday the fifth of September 1972.*
8. The heater produces *fifty-six thousand British thermal units* of heat per hour.
9. The base of the antenna is 726.3 *feet above mean sea level.*
10. Lake Wabagoon is 3.8 *miles northeast* of Montrose, Ohio.
11. The sun's elevation at *zero-eight hours, seventeen minutes, and twenty-six seconds Greenwich mean time* was calculated to be *twenty-three degrees, four minutes, and fourteen seconds* of arc.
12. We are shipping our voltmeter *serial number* 4257 to you for recalibration.
13. Television station DMON of Montrose, Ohio has been allocated a transmitting frequency in the *ultrahigh frequency* band (*three hundred megahertz to three gigahertz*).
14. The United States gallon of 231 *cubic inches* is roughly one-fifth smaller than the British 277.42 *cubic inch imperial gallon.*
15. Production from *number three* well averages *four hundred barrels per day.*
16. Strip *one inch* of insulation from both ends of *thirty-nine inches of number twenty-two American Wire Gage* hookup wire.

Exercise 8. Some of the expressions in the following sentences need to be compounded into one word or joined by hyphens. Indicate where such linking should occur. (Rules for compounding appear in Article 1 of Chapter 11.)

1. The time tables will be ready by month end.
2. An independently controlled survey failed to produce the desired end result.
3. This photo multiplier is home made.
4. Although still under age, his qualifications were greater than those of his co workers.
5. His post graduate studies were concerned mostly with high altitude radar research.
6. A rack mounted pre amplifier was installed in each of the odd numbered bays.
7. We have air expressed the radio isotopes to you.
8. A quick once over failed to reveal why the inter lock failed to

operate. More detailed fault finding traced the cause to a loose lock washer.

9. The time base was displayed on a cathode ray oscilloscope.
10. Break down of the air conditioner resulted in a seven hour time loss.
11. An air supported radome protected the height finder radar set from wind and rain.
12. The proposed change over to the new product line means that we must phase in a new training scheme.
13. A log periodic antenna was installed in place of the non directional antenna.
14. Phase out of the moving target indicator will occur early in the new year.

Exercise 9. Reorganize the following information and rewrite it as a single paragraph.

a. A battery is composed of one or more cells connected in series.
b. Within a battery each cell is a source of chemical energy. This chemical energy is converted into electrical energy—but only when needed.
c. There are two types of cells.
d. One type of cell is the primary cell, which can supply electricity until its chemical energy is exhausted, then you have to throw it away.
e. Some cells can be recharged after use. In this type, an electrical charge can be used to return the chemicals within the cell to their original state (i. e. prior to discharge). This type of cell is known as the secondary cell.

Exercise 10. Rewrite this one-paragraph notice to make it clearer and more personal.

PROCEDURE RE EXPENSE CLAIMS

Expense claims must be handed to the Accounts Section before 10 a.m. on Wednesday for payment on Friday. Personnel failing to hand in their forms at the proper time, may do so at any time until 4:30 on Thursday but must wait until Monday for payment. Under no account will a late claim be paid in the same week that it was filed. Claims handed in after Thursday will be processed with the following week's claims and will be paid on the next Friday.

Exercise 11. Write a single paragraph describing the following situation.

a. Your company has a lot of telephone switching equipment to maintain. Most is old, but some is new.

b. New telephone switching equipment is under warranty for one year.

c. The manufacturer's warranty specifies new equipment must be lubricated with Vaprol.

d. You currently use Vaprol for all telephone switching equipment.

e. Tests by the Engineering Department show Gra-Lub to be a better lubricant.

f. Gra-Lub has a graphite base; Vaprol has not.

g. You want to adopt Gra-Lub for all telephone switching equipment.

h. You don't want to stock and use two types of lubricant for the same equipment.

i. You have written to the manufacturer to request authorization to use Gra-Lub on equipment that is still under warranty.

11

Glossary of Technical Usage

A standard glossary of usage contains rules for combining words into compound terms, for forming abbreviations, for capitalizing, and for spelling unusual or difficult words. This glossary also offers suggestions for handling many of the technical terms peculiar to industry. Hence, it is oriented toward the technical rather than the literary writer.

The information, arranged alphabetically, is preceded by five brief supplementary articles on specific aspects that would be too lengthy to include in the glossary. These are:

Article 1—Combining Words into Compound Terms
Article 2—Abbreviating Technical and Nontechnical Terms
Article 3—Capitalization and Punctuation
Article 4—Writing Numbers in Narrative
Article 5—Summary of Numerical Prefixes and Symbols

The glossary also contains a selection of words most likely to be misused or misspelled. These include:

* Words that are similar and frequently confused with one another; e.g. *imply* and *infer*, *diplex* and *duplex*, *principal* and *principle*.
* Common minor errors of grammar, such as *comprised of* (should be *comprises*), *most unique* (unique cannot be compared), *liaise* (an unnatural verb formed from *liaison*).
* Words that are particularly prone to misspelling; e.g. *desiccant*, *oriented, immitance.*
* Words for which there may be more than one "correct" spelling; e.g. *programer* and *programmer.*

Where differences exist in the spelling of certain words, and two versions are in general use (e.g. *symposiums* and *symposia*), both are entered in the glossary and a preference shown for one of them.

Definitions have been included where they will help you select the correct word for a given purpose, or to differentiate between similar words having different meanings. These definitions are intentionally brief and intended to act only as a guide; for more comprehensive definitions you should consult an authoritative dictionary (I recommend Webster's New Collegiate Dictionary).

All entries in the glossary are in lower case letters. Capital letters indicate that they are recommended for the specific word, phrase, or abbreviation in which they are used. Similarly, periods have been eliminated except where they form part of a specific entry. For example, the abbreviation for "inch" is *in.*, and the period that follows it has been inserted intentionally to indicate that it is not the word "in," but an abbreviation.

Finally, I would prefer you to think of this glossary as a guide rather than a collection of hard and fast rules. Our language is continually changing, and what was fashionable yesterday may seem pedantic today and a cliché tomorrow. I expect that in some cases your views will differ from mine. Where they do, I hope that the comments and suggestions I offer will help you to choose the right expression, word, or abbreviation, and that you will be able to do so both consistently and logically.

ARTICLES

Article 1—Combining Words into Compound Terms

One of the biggest problems facing technical writers (and particularly the typists who have to record their work) is to know whether multiword expressions should be compounded fully, joined by hyphens, or allowed to stand as two or more separate words. For example, should you write:

> cross check, cross-check, or crosscheck?
> counterclockwise, counter-clockwise, or counter clockwise?
> change over, change-over, or changeover?

The tendency today is to compound a multiword expression into a single term. But this bare statement cannot be applied as a general rule because there are too many variations, some of which appear in the glossary.

Most multiword expressions are compound adjectives. When two words combine to form an adjective they are either joined by a hyphen or compounded to form one word. They are usually joined by a hyphen if they are formed from a noun-adjective expression:

Noun-Adjective	As a Compound Adjective
vacuum tube	vacuum-tube voltmeter
cathode ray	cathode-ray tube
high frequency	high-frequency oscillator

But when one of the combining words is a verb, they often combine into a one word adjective. Under these conditions they normally will compound into a single-word noun:

Two Words	As a Noun	As an Adjective
lock out	lockout	lockout voltage
shake down	shakedown	shakedown test
cross over	crossover	crossover network

Three or more words that combine to form an adjective in most cases are joined by hyphens. For example, *lock test pulse* becomes *lock-test-pulse generator*. Occasionally, however, they are compounded into a single term, as in *counterelectromotive force*. Specific examples are identified in the glossary.

Obviously, these "rules" cannot be taken at full face value because there are occasions when they do not apply. A useful guide for doubtful combinations is the *Government Printing Office Style Manual*[1], which contains a list of over 19,000 compound terms.

Article 2—Abbreviating Technical and Nontechnical Terms

You may abbreviate any term you like, and in any form you like, providing that you indicate clearly to the reader how you intend to abbreviate it. This can be done by stating the term in full, then showing the abbreviation in parentheses to indicate that from now on you intend to use the abbreviation. A typical example follows:

> Always spell out single digit numbers (sdn). The only time that sdn
> are not spelled out is when they are being inserted in a column of
> figures.

When forming abbreviations of your own, take care not to form a new abbreviation when a standard one already exists. For example, if you did not know that there is a commonly accepted abbreviation for pound (weight), you might be tempted to use *pd*. This would not sit well with your readers, who might resent replacement of their old friend *lb*.

There are three basic rules that you should observe when forming abbreviations:

---

[1] *United States Government Printing Office Style Manual*, Superintendent of Documents, Washington, D.C., 20402.

1. *Use lower case letters:*
   foot   — ft
   pound   — lb
   centimeter   — cm
   approximately   — approx

2. *Omit all periods* unless the abbreviation forms another word, then insert a period after it:
   horsepower   — hp
   cubic feet per second   — cfs
   cathode-ray tube   — crt
   miles per hour   — mph
   inch   — in.
   singular   — sing.

3. Write plural abbreviations in the same form as the singular abbreviation:
   inches   — in.
   pounds   — lb
   gallons — gal
   hours   — hr

Of course there are exceptions which have grown as part of the language. Through continued use these unnatural abbreviations have been generally accepted as the correct form, and now cannot easily be displaced. A few examples follow:

for example   — e.g.
  (exempli grati)
that is   — i.e.
  (id est)
morning   — a.m.
  (ante meridiem)
inside diameter   — ID
number(s)   — No.

Other unusual abbreviations exist in specific technical disciplines. The glossary offers a partial list of standard abbreviations and some technical abbreviations. Consult a good dictionary for general guidance, or refer to the list of standard abbreviations compiled by one of the technical societies in your discipline for specific technical terms.

Article 3—Capitalization and Punctuation

Lower case letters (these words are in lower case) should be used as much as possible in technical writing. Too many capital letters cause an untidy appearance and reduce the effectiveness of capital letters inserted for emphasis.

The glossary therefore recommends lower case letters except where usage has resulted in general adoption of a capitalized form (as in *No.*, *GCA*, and the Brinell and Rockwell hardness tests); for proper nouns; and for expressions coined from proper nouns which have not yet been accepted as common terms. (Note also that although we still use capital letters when referring to Doppler's principle, we write of doppler radar in lower case letters.)

The tendency today is to underpunctuate technical writing. This does not mean you may omit punctuation with a blasé "when in doubt, leave out" attitude. But you need not assiduously punctuate every clause and subclause. The intent is to obtain smooth reading, the criterion to make the message clear. Hence, you should insert punctuation where it is needed rather than rigorously divide the information into tight little formal compartments.

Article 4—Writing Numbers in Narrative

The conventions that dictate whether a number should be written out or expressed in figures differ between ordinary writing and technical writing. In technical writing you are much more likely to express numbers in figures, as the rules listed below indicate.

These rules are intended mainly as a guide. There will be many times when you will have to make a decision between two rules that conflict. Your decision should then be based on three criteria:

Which method will be most readable?
Which method will be simplest to type?
Which method did I use previously, under similar circumstances?

Good judgment and a desire to be consistent will help you to select the best method each time.

The basic rule for writing numbers in technical narrative is:

* Spell out single digit numbers (one to nine inclusive).
* Use figures for multiple digit numbers (10 and above).

Exceptions to this rule follow:

Always Use Figures:
1. When writing specific technical information, such as test results, dimensions, tolerances, temperatures, statistics, and quotations from tabular data.
2. When writing any number that precedes a unit of measurement: 3 in.; 7 lb; 121.5 MHz.
3. When writing a series of both large and small numbers in one

passage: During the week ending 27 May we tested 7 transmitters, 49 receivers, and 38 power supplies.

4. When referring to section, chapter, page, figure (illustration), and
table numbers: Chapter 7; Figure 4.
5. For numbers that contain fractions or decimals: 7¼; 7.25.
6. For percentages: 3% gain; 11% sales tax.
7. For years, dates, and times (including hours): At 3 p.m. on January 9, 1972; 0817, 20 Feb 72.
8. For sums of money: $2000; $28.50; 20 dollars; $0.27 (preferred)
or 27 cents.
9. For ages of persons.

Always Spell Out:
1. Round numbers that are generalizations (unless spelling them out
would result in a cumbersome quantity): about five hundred;
approximately forty thousand.
2. Fractions that stand alone: repairs were made in less than three-
quarters of an hour.
3. Numbers that start a sentence. (Better still, rewrite the sentence
so that the number is not at the beginning.)

Additional rules are:

* Spell out one of the numbers when two numbers are written consecutively and are not separated by punctuation: 36 fifty-watt
amplifiers or thirty-six 50-watt amplifiers. (Generally, spell out
whichever number will result in the simplest or shortest expression.)
* Insert a zero before the decimal point of numbers less than one:
0.75; 0.0037.
* Use decimals rather than fractions (they are easier to type), except
when writing numbers that are customarily written as fractions.
* Insert commas in large numbers containing five or more digits:
1,275,000; 27,291; 4056. (Insert a comma in four digit numbers
only when they appear as part of a column of numbers.)
* Write numbers that denote position in a sequence as 1st, 2nd, 3rd,
4th . . . 31st . . . 42nd . . . 103rd . . . 124th. . . .

Article 5—Summary of Numerical Prefixes and Symbols

The table below summarizes the numerical prefixes and abbreviations
to be found in the glossary.

Multiple and Submultiple		Prefix	Symbol
1,000,000,000,000	$= 10^{12}$	tera	T
1,000,000,000	$= 10^{9}$	giga	G
1,000,000	$= 10^{6}$	mega	M
1,000	$= 10^{3}$	kilo	k

$$100 = 10^2 \quad \text{hecto} \quad \text{h}$$
$$10 = 10 \quad \text{deca} \quad \text{da}$$
$$0.1 = 10^{-1} \quad \text{deci} \quad \text{d}$$
$$0.01 = 10^{-2} \quad \text{centi} \quad \text{c}$$
$$0.001 = 10^{-3} \quad \text{milli} \quad \text{m}$$
$$0.000.001 = 10^{-6} \quad \text{micro} \quad \mu$$
$$0.000.000.001 = 10^{-9} \quad \text{nano} \quad \text{n}$$
$$0.000.000.000.001 = 10^{-12} \quad \text{pico} \quad \text{p}$$
$$0.000.000.000.000.001 = 10^{-15} \quad \text{femto} \quad \text{f}$$
$$0.000.000.000.000.000.001 = 10^{-18} \quad \text{atto} \quad \text{a}$$

THE GLOSSARY

General abbreviations used throughout the glossary are:

abbr	abbreviate(d); abbreviation
def	definition
lc	lower case
pl	plural
pref	prefer, preferred, preference

A

a; an use an before words that commence with a silent *h* or a vowel; use *a* when the *h* is sounded or if the vowel is sounded as *w* or *y*; *an hour* but *a hotel, an onion* but *a European*

abbreviate; abbreviation throughout this glossary, abbr is used to denote either of these two words

aberration

ab initio def: from the beginning

abrasion

abscissa

absolute abbr: **abs**

absorb(ent); adsorb(ent) absorb means to swallow up completely (as a sponge absorbs moisture); adsorb means to hold on the surface, as if by adhesion

accelerate; accelerator; accelerometer

accept; except accept means to receive (normally willingly), as in *he accepted the company's offer of employment*; except generally means exclude, as in *the night crew completed all the repairs except rewiring of the control panel*

access(ible)

accessory; accessories abbr: **accy**

accommodate; accommodation

account abbr: **acct**

accumulate; accumulator

achieve means to conclude successfully, probably after considerable effort; avoid using *achieve* when the intended meaning is simply to reach or to get

acknowledg(e)ment *acknowledgment* pref

acre

across not *accross*

actuator

A.D. def: Anno Domini

adapter; adaptor *adapter* pref

adaption; adaptation *adaption* pref

addendum pl: *addenda*

adhere to never use *adhere by*

ad hoc def: set up for one occasion

adjective (compound) two or more words that combine to form an adjective are either joined by a hyphen or compounded into a single word; see Article 1

adsorb(ent)   see *absorb*

advice; advise   use advice as a noun and advise as a verb; e.g. *the engineer's advice was sound; the technician advised the driver to take an alternative route*

aerial   see *antenna*

aero-   a prefix meaning of the air; it combines to form one word: *aerodynamics, aeronautical*; in some instances it has been replaced by *air*, as in *airplane, aircraft*

affect; effect   affect is used only as a verb, never as a noun; it means to produce an effect upon or to influence (as in *the potential difference affects the transit time*); effect can be used either as a verb or as a noun; as a verb it means to cause or to accomplish (as in *to effect a change* or *give effect to*); as a noun it means the consequences or result of an occurrence (as in *the detrimental effect upon the environment*), or refers to property, such as *personal effects*

after-   as a prefix, after- combines to form one word: *afteracceleration, afterburner, afterglow, afterheat, afterimage*

agenda   although actually plural, agenda is generally treated as singular; e.g. *the agenda is complete*

aggravate   the correct definition of aggravate is to increase or intensify (worsen) a situation; it should not be used when the meaning is simply to annoy

aggregate

aging; ageing   *aging* pref

agree to; agree with   to be correct, you should *agree to* a suggestion or proposal, but *agree with* another person

airborne

air-condition(ed);(er);(ing)

air horsepower   abbr: ahp

airline; air line   an airline provides aviation services; an air line is a line or pipe that carries air

alkali; alkaline   the plural is *alkalis* (pref) or *alkalies*

allot(ted)

all ready; already   all ready means that all (everyone or everything) is ready; already means by this time; e.g. *the samples are all ready to be tested; the samples have already been tested*

all right   def: everything is satisfactory; never use *alright*

all together; altogether   all together means all collectively, as a group; altogether means completely, entirely; e.g. *the samples have been gathered all together, ready for testing; the samples are altogether useless*

alphanumeric   def: in alphabetical, then numerical sequence

alternating current   abbr: **ac**

alternate; alternative   alternate(ly) means by turns or turn and turn about; e.g. *the inspector alternated among the four construction sites*; alternative(ly) offers a choice between only two things; e.g. *the alternative was to return the samples*; it is incorrect to say *there are three alternatives*

alternator

altitude   abbr: **alt**

a.m.   def: before noon (*anti meridiem*)

amateur

ambient   abbr: **amb**

ambiguous; ambiguity

American Wire Gage   abbr: **AWG**

among; between   use among when referring to three or more items; use between when referring to only two; *amongst* is seldom used

amount; number   use amount to refer to general quantity, as in *the amount of time taken as sick leave has decreased*; use number to refer to items that can be counted, as in *the number of technicians assigned to the project was 30% greater than anticipated*

ampere(s)   abbr: **A** (pref) or **amp**

ampere-hour(s)   abbr: **Ah** (pref) or **amp-hr** (more common)

amplitude modulation   abbr: **AM**

an   see **a**

analog; analogous

analyse(r); analyze(r)   *analyze(r)* pref

AND-gate

and/or   avoid using this term; in most cases it can be replaced by either *and* or *or*

angstrom   abbr. Å

anion   def: negative ion

anneal(ing)

annihilation

antarctic   see comment for **arctic**

ante- a prefix that means before; combines to form one word: *antecedent, anteroom*

ante meridiem def: before noon; abbr: **a.m.**; can also be written as *antemeridian*, but never as *antimeridian* (which see)

antenna the proper plural in the technical sense is *antennas*; although *antennae* is sometimes seen, its use should be limited to zoology; *antenna* has generally replaced the obsolescent *aerial*

anti- a prefix meaning opposite or contradictory to; generally combines to form one word: *antiaircraft, antiastigmatism, anticapacitance, anticoincidence, antisymmetric*; if combining word starts with *i* or is a proper noun, insert a hyphen: *anti-icing, anti-American*

antimeridian def: the opposite meridian (of longitude); e.g. the antimeridian of 96° 30′W is 83° 30′E

anybody; any body anybody means any person; any body means any object; e.g. *anybody can attend; discard the batch if you find any body containing foreign matter*

anyone; any one anyone means any person; any one means any single item; e.g. *you may take anyone with you; you may take any one of the samples*

anyway; any way anyway means in any case or in any event; any way means in any manner; e.g. *the results may not be as good as you expect, but we want to see them anyway; the work may be done in any way you wish*

anxious although anxious really implies anxiety, current usage permits it to be used when the meaning is simply keen or eager

apparent(ly)

appear(s); seem(s) use appears to describe a condition that can be seen: *the equipment appears to be new*; use seems to describe a condition that cannot be seen: *he seems to be clever*

appendix def: the part of a report that contains supporting data; the plural is *appendixes* (pref) or appendices

appreciate means to value or to cherish; should not be used as a synonym for *understand*, as in *we appreciate your difficulty in finding spare parts*

approximate(ly) abbr: **approx**

arbitrary

arc; arced; arcing

architect

arctic capitalize when referring to a specific area: *beyond the Arctic Circle*; otherwise use lc letters: *in the arctic*

areal def: having area

around def: on all sides, surrounding; should not be used as a synonym for *about*

arrester; arrestor *arrester* pref

as avoid using when the intended meaning is *since* or *because*; to write *he could not open his desk as he left his keys at home* is incorrect (replace *as* with *because*)

as per avoid using this hackneyed expression

assure means to state with confidence that something has been or will be made certain; it is sometimes confused with *ensure* and *insure*, which it does not replace; see **ensure**

assembly; assemblies abbr: **assy**

asymmetrical

asynchronous

athletic not *atheletic*

atmosphere abbr: **atm**

atomic weight abb: **at. wt**

attenuator

atto def: 10^{-18}; abbr: **a**

audible; audibility

audio frequency abbr: **af** (pref) or **a-f**

audiovisual

aural means that which is heard; it should not be confused with *oral*, which means that which is spoken

auto- a prefix meaning self; combines to form one word: *autoalarm, autoconduction, autoionization, autotransformer*

automatic frequency control abbr: **afc** (pref) or **AFC**

automatic volume control abbr: **avc** (pref) or **AVC**

auxiliary abbr: **aux**

average see **mean**

ax; axe ax pref; the plural is *axes*

axis the plural also is *axes*

azimuth abbr: **az**

B

balance; remainder use balance to describe a state of equilibrium (as in *discontinuous permafrost is frozen soil delicately balanced between the frozen and unfrozen state*), or as an accounting term; use remainder when the meaning is the rest of, as in *the remainder of the shipment will be delivered next week*

bandwidth

barometer abbr: **bar.**

barrel(l)ed; barrel(l)ing the single *l* is pref; the abbr of *barrel(s)* is **bbl**

barretter

bases this is the plural of both *base* and *basis*

B.C. def: Before Christ

because; for use because when the clause it introduces identifies the cause of a result, as in *he could not open his desk because he left his keys at home*; use for when the clause introduces something less tangible, as in *he failed to complete the project on schedule, for reasons he preferred not to divulge*

begin(ning)

benefit; benefited; benefiting

beside; besides beside means alongside, at the side of; *besides* means as well as

between see **among**

bi- a prefix meaning two or twice; combines to form one word: *biangular, bidirectional, bifilar, bilateral, bimetallic, bizonal*

biannual(ly); biennial(ly) biannual(ly) means twice a year; biennial(ly) means every two years; the *bi* of *bimonthly* and *biweekly* means every two

bias; biased; biasing

billion electron volts although the pref abbr is **GeV**, *beV* and *bev* are more commonly used

Bill of Materials abbr: **BOM**

bimonthly def: every two months

binaural

bioelectronics

bionics def: application of biological techniques to electronic design

birdseye (view)

biweekly def: every two weeks

board feet abbr: **fbm** (derived from *feet board measure*)

boiling point abbr: **bp**

bona fide def: in good faith, authentic, genuine

borderline

brake horsepower; brake horsepower-hour abbr: **bhp, bhp-hr**

brand-new

break- when used as a prefix to form a compound noun or adjective, break combines into one word; *breakfast, breakdown, breakup*; in the verb form it retains its single word identity: *it was time to break up the meeting*

bridging

Brinell hardness number abbr: **Bhn**

British thermal unit abbr: **Btu**

buoy; buoyant

burned; burnt *burned* pref

bur(r) *burr* pref

bypass

by-product

C

calendar; calender a calendar is the arrangement of the days in a year; calender is the finish on paper, cloth, and so on

caliber; calibre *caliber* pref

calking; caulking *calking* pref

cal(l)iper *caliper* pref

calorie abbr: **cal**

calorimeter; colorimeter a calorimeter measures quantity of heat; a colorimeter measures color

cancel(l)ed; cancel(l)ing *canceled, canceling* pref, but always *cancellation*

candela def: unit of luminous intensity (replaces candle); abbr: **cd**; recommended

candela (*continued*)
abbr for candela per square foot and square meter are **cd/ft²** and **cd/m²**

candlepower; candlehour(s) abbr: **cp, c-hr**; although *candle* is still commonly used, it is being replaced by *candela*

candoluminescence

capacitor

capacity for never use *capacity to* or *capacity of*

capillary

capital letters abbr: **caps.**; use capital letters as little as possible (see Article 3)

carburet(t)or *carburetor* pref; a third, seldom used spelling is *carburetter*

caseharden

caster; castor pref spelling is *caster* when the meaning is to swivel freely; castor is used when the reference is to castor oil, etc.

catalog(ue) catalog pref, but *catalogued* and *cataloguing* always retain the *u*

catalyst

category; categories; categorical

cathode-ray tube abbr: **crt** (pref) or **CRT** (commonly used)

cation def: positive ion

-ceed; -cede; -sede only one word ends in -sede: *supersede*; only three words end in -ceed: *exceed, proceed, succeed*; all others end in -cede: e.g. *precede, concede*

centerline abbr: **₵** (pref) or **CL**

Celsius abbr: **C**; see **temperature**

center-to-center abbr: **c-c**

centi- def: 10^{-2}; as a prefix combines to form one word: *centiampere, centigram*; abbr: **c**; other abbr:

centigram	cg
centiliter	cl
centimeter	cm
centimeter-gram-second	cgs

centigrade abbr: **C**; see **temperature**

centri- a prefix meaning center; combines to form one word: *centrifugal, centripetal*

chamfer

changeable

channel(l)ed; channel(l)ing single *l* pref

chapter abbr: **chap.**

chassis both singular and plural are spelled the same

chisel(l)ed; chisel(l)ing single *l* pref

chrominance

cipher

circuit abbr: **cct**

cite def: to quote; see **site**

climate avoid confusing climate with *weather*; *climate* is the average type of weather, determined over a number of years, experienced at a particular place; *weather* is the state of the atmospheric conditions at a specific place at a specific time

clockwise (turn) abbr: **CW**

co- as a prefix, co- generally means jointly or together; it usually combines to form one word; *coexist, coequal, cooperate, coordinate, coplanar* (*co-worker* is an exception); it is also used as the abbr for *complement of* (an arc or angle), as in *codeclination, colatitude*

coalesce

coarse; course coarse means rough in texture or of poor quality; course implies movement or passage of time; e.g. *a coarse granular material; the technical writing course*

coaxial abbr: **coax.**

coefficient abbr: **coef**

cologarithm abbr: **colog**

colon when a colon is inserted in the middle of a sentence to introduce an example or short statement, the first word following the colon is not capitalized; when a colon is used at the end of a sentence to introduce subparagraphs that follow, the first word of each subparagraph should be capitalized unless all the subparagraphs are very short (i.e. only one sentence long); a hyphen should not be inserted after the colon

colorimeter see **calorimeter**

column abbr: **col.**

combustible

comma a comma normally need not be used immediately before *and, but,* and *or,* but may be inserted if to do so will increase understanding or avoid ambiguity; also see Article 3

commence in technical writing, replace commence with the more direct *begin* or *start*

commit; commitment; committed; committing

committee

compare; comparable; comparative use *compared to* when suggesting a general likeness; use *compared with* when making a definite comparison

compatible; compatibility

complement; compliment complement means the balance required to make up a full quantity or a complete set; to compliment means to praise; e.g. *in a right angle, the complement of 60° is 30°; Mr. Perchanski complimented him for writing a good report*

composed of; comprising; consists of all three terms mean "made up of" (specific items); if any one of these terms is followed by a list of items, it implies that the list is complete; if the list is not complete, the term should be replaced by *includes* or *including*

compound terms two or more words that combine to form a compound term are joined by a hyphen or are written as one word, depending on accepted usage and whether they form a verb, noun, or adjective; see Article 1

comprise; comprised; comprising to write *comprised of* is incorrect, because the verb comprise includes the preposition *of*; also see **composed of**

conform use *conform to* when the meaning is to abide by; use *conform with* when the meaning is to agree with

conscience

conscious

consensus means a general agreement of opinion; hence to write *consensus of opinion* is incorrect; an example of correct usage is: *the consensus was that a further series of tests would be necessary*

consistent with never use *consistent of*

consists of; consisting of see **composed of**

contact should not be used as a verb when *write, visit, speak,* or *telephone* better describes the action to be taken

continual; continuous continual(ly) means

continual; continuous (*continued*) happens frequently but not all the time; e.g. *the generator is continually being overloaded* (is frequently overloaded); continuous(ly) means goes on and on without stopping; e.g. *the noise level is continuously above 100 dB* (it never drops below 100dB)

continue(d) abbr: **cont**

continuous wave abbr: **cw**

contra-rotating

contrast when used as a verb, contrast is followed by *with*; when used as a noun, it may be followed by either *to* or *with* (*with* pref)

conversant with never use *conversant of*

corollary

correspond *to correspond to* suggests a resemblance; *to correspond with* means to communicate in writing

cosecant abbr: **csc** (pref) or **cosec**

cosine abbr: **cos**

cotangent abbr: **cot**

counter- a prefix meaning opposite or reciprocal; combines to form one word: *counteract, counterbalance, counterflow, counterweight*

counterclockwise (turn) abbr: **CCW**

counterelectromotive force abbr: **cemf**; also known as *back emf*

counts per minute abbr: **cpm**

course see **coarse**

criterion the plural is *criteria*

criticism; criticize

cross-refer(ence) abbr: **x-ref**

cryogenic

crystal abbr: **xtal**

crystalline; crystallize

cubic abbr: **cu**; other abbr:

cubic centimeter(s)	cm^3 (pref);	cu cm; cc
cubic foot (feet)	ft^3 (pref); cu ft	
cubic feet per minute	cfm (pref); ft^3/min	
cubic feet per second	cfs (pref); ft^3/sec	
cubic inch(es)	in.^{3} (pref); cu in.	
cubic meter(s)	m^3 (pref); cu m	
cubic micron(s)	μ^3 (pref); cu μ; cu mu	

cubic (*continued*)

cubic millimeter(s)   mm³ (pref); **cu mm**
cubic yard(s)   yd³ (pref); **cu yd**

curriculum the plural is *curriculums* (pref) or *curricula*

cursor

cycles per minute abbr: **cpm**

cycles per second abbr: **cps**; although still occasionally used, this term has been replaced by **hertz** (which see)

cylinder abbr: **cyl**

D

daraf def: the unit of elastance

data def: gathered facts; data is plural (derived from the singular *datum*, which is rarely used); although it is sometimes used as a singular noun, the pref usage is plural; e.g. *now that all the data have been received, the analysis can begin*

date(s) avoid vague statements such as "last month" and "next year" because they soon become indefinite; write as a specific date, using day (in numerals), month (spelled out), and year (in numerals); e.g. *January 27, 1972* or *27 January 1972* (the latter form has no punctuation); to abbreviate, reduce month to first three letters and year to last two digits; *Jan 27, 72* or *27 Jan 72*; never abbr year without also abbreviating month, and vice versa

days days of the week are capitalized: *Monday, Tuesday*

de- a prefix that generally combines to form one word; *deaccentuate, deactivate, decentralize, decode, deemphasize, deenergize, deice, derate, destagger*; an exception is *de-ionize*

debug(ging)

decelerate def: to slow down; never use *deaccelerate*

decibel abbr: **dB**; the abbr for decibel referred to 1 mw is **dBm**

decimals for values less than unity, place a zero before the decimal point; e.g. *0.17, 0.0017*

decimate def: to reduce by one-tenth; it does not mean reduce to fragments

decimeter abbr: **dm**

declination abbr: **dec**

defective; deficient defective means unserviceable or damaged (i.e. generally lacking in quality); deficient means lacking in

defective; deficient (*continued*)
quantity (it is derived from *deficit*), and in the military sense incomplete; e.g. *a short circuit resulted in a defective transmitter; the installation was completed on schedule except for a deficient rotary coupler*

defense; defence *defense* pref

defer; deferred; deferring; deference

definite; definitive definite means exact, precise; definitive means conclusive, fully evolved; e.g. *a definite price* is a firm price; *a definitive statement* concerns a topic that has been thoroughly considered and evaluated

degree(s) abbr: **deg** (pref in narrative) or ° (following numerals); see **temperature**

demarcation

demi- a little-used prefix meaning half (generally replaced by **semi-**); combines to form one word: *demivolt*

dependent; dependant *dependent* pref; *dependant* is rarely used

deprecate; depreciate deprecate means to disapprove of; depreciate means to reduce the value of; e.g. *the use of "as per" in technical writing is deprecated; the vehicles depreciated by 50% the first year, and 20% the second year*

depth

desiccant

desirable

despatch see **dispatch**

develop not *develope*

device; devise the noun is device, the verb is devise; e.g. *a unique device; he devised a new circuit*

dext(e)rous *dexterous* pref

diagram; diagrammatic

dial(l)ed; dial(l)ing single *l* pref

dialog(ue) *dialogue* pref

diaphragm

diazo

dielectric

diesel; diesel-electric

differ use *differ from* to demonstrate a difference; use *differ with* to describe a difference of opinion

different *different from* is preferred, although *different to* is sometimes used; never use *different than*

diffraction

diffusion

dilemma means to be faced with a choice between two unhappy alternatives; should not be used as a synonym for *difficulty*

diplex; duplex *diplex operation* means the simultaneous transmission of two signals using a single feature, e.g. an antenna; *duplex operation* means that both ends can transmit and receive simultaneously

direct current abbr: **dc**

directly def: immediately; do not use when the meaning is as soon as

disassemble never use *dissemble* when the meaning is to take apart

disassociate see **dissociate**

disc; disk both are correct and commonly used; I recommend *disc*

discernible

discreet; discrete discreet means prudent or discerning, as in *his answer was discreet*;

discreet; discrete (*continued*)
discrete means individually distinctive and separate, as in *discrete channels; discretion* is formed from *discreet*, not from *discrete*

disinterested; uninterested disinterested means unbiased, impartial; uninterested simply means not interested

dispatch; despatch *dispatch* pref

disseminate

dissimilar

dissipate

dissociate; disassociate *dissociate* pref; disassociate is seldom used

distil(l) *distil* pref; but always *distilled, distillate, distillation*

donut; doughnut for electronics/nucleonics use, *donut* preferred

doppler capitalize only when referring to the Doppler principle

dozen abbr: **doz**

drawing(s) abbr: **dwg**

drier; dryer the adjective is always *drier*; the pref noun is *dryer*; e.g. *this material is drier; place the others back in the dryer*

drop; droppable; dropped; dropping

due to an overused expression; *because of* pref

duo- a prefix meaning two; combines to form one word: *duocone, duodiode, duophase*

duplex see **diplex**

E

each abbr: **ea**

east capitalize only if east is part of the name of a specific location, as in *East Africa*; otherwise use lc letters: *the east coast of the U.S.*; abbr: **E**; the abbr for *east-west* (control, movement) is **E-W**; *eastbound* and *eastward* are written as one word

eccentric

echo; echoes

economic; economical use economical to describe economy (of funds, effort, time); use economic when writing about econom-

economic; economical (*continued*)
ics; e.g. *an economical operation* (it did not cost much); *an economic disaster* (refers to economics)

effect see **affect**

efficacy; efficiency efficacy means effectiveness, ability to do the job intended; efficiency is a measurement of capability, the ratio of work done to energy expended; e.g. *we hired a consultant to assess the efficacy of our training methods; the powerhouse is to have a high-efficiency boiler*

e.g. abbr for *exempli gratia* (the Latin of *for example*); avoid confusing with **i.e.**; no comma is necessary after *e.g.* (see how it has been used in the comments for **efficacy** and **economic**)

electric; electrical use electric when the meaning is produces or carries electricity; use electrical when the meaning is related to the generation or carrying of electricity; abbr: **elec**

electro- a prefix generally meaning pertaining to electricity; it normally combines to form one word: *electroacoustic, electroanalysis, electrodeposition, electromechanical, electroplate;* if the combining word starts with *o*, insert a hyphen: *electro-optics, electro-osmosis*

electromagnetic units abbr: **emu**

electromotive force abbr: **emf**

electronic(s) use electronic as an adjective, electronics as a noun; e.g. *electronic countermeasures; your career in electronics*

electron volt(s) abbr: **eV** (pref) or **ev**

electrostatic units abbr: **esu**

elevation abbr: **el**

ellipse

embedded

emigrate; immigrate emigrate means to go away from; immigrate means to come into

emit; emitter, emittance; emission; emissivity

enamel(l)ed; enamel(l)ing single *l* pref

encase; incase *encase* pref

encipher

enclose; inclose *enclose* pref; *inclose* is used mainly as a legal term

endorse, indorse *endorse* pref

enquire *inquire* pref

enrol; enroll both are correct, but *enroll* pref; universal usage prefers *ll* for *enrolled* and *enrolling*, but only a single *l* for *enrolment*

en route def: on the road, on the way; never use *on route*

ensure; insure; assure use ensure when the meaning is to make certain of, as in *use the new oscilloscope to ensure accurate calibration*; use insure when the meaning is to protect against financial loss, as in *we insured all our drivers*; use assure when

ensure; insure; assure (*continued*) the meaning is to state with confidence that something has been or will be made certain, as in *he assured the meeting that production would increase by 8%*

entrust; intrust *entrust* pref

envelop; envelope envelop is a verb that means to surround or cover completely; envelope is a noun that means a wrapper or a covering

environment

equal *equality* and *equalize* have only one *l*; *equally* always has *ll*; *equaled* and *equaling* preferably have only one *l*, but sometimes are seen with *ll*

equi- a prefix that means equality; combines to form one word: *equiphase, equipotential, equisignal*

equip; equipped; equipping; equipment

equivalent abbr: **equiv**

erase; erasable

erratic

erratum (singular); **errata** (plural)

especially see **specially**

et al. def: and others

et cetera abbr: **etc**; def: and so forth, and so on; use with care in technical writing: *etc.* can create an impression of vagueness or unsureness; e.g. *the transmitters, etc, were tested* is much less definite than either *the transmitter, modulator, and power supply were tested*, or (if to restate all the equipment is too repetitious) *the transmitting equipment was tested*

extremely high frequency abbr: **ehf**

everybody; every body everybody means every person, or all the persons; every body means every single body; e.g. *everybody was present; every body was examined for gunpowder scars*

everyone; every one everyone means every person, or all the persons; every one means every single item; e.g. *everyone is insured; every one had to be tested in a saline solution*

exaggerate

except def: to exclude; see **accept**

exhaust

extracurricular

extraordinary

F

Fahrenheit   abbr: **F**; see **temperature**

fail-safe

farad   abbr: **F**

farther; further   farther means greater distance, as in *he traveled farther than the other salesmen*; further means a continuation of (as an adjective) or to advance (as a verb), as in *the promotion was a further step in his career plan*, and to *further his education, he took a part-time course in industrial drafting*

fasten(er)

faultfinder; faultfinding

feet; foot   abbr: **ft**; other abbr:

feet per minute	fpm
feet per second	fps
feet board measure	fbm
(board feet)	
foot-candle(s)	fc (pref); **ft-c**
foot-Lambert(s)	fL (pref); **ft-L**
foot-pound(s)	fp (pref); **ft-lb**
foot-pound-second	fps
(system)	

feasible

femto   def: 10^{-15}; abbr: **f**; other abbr:

femtoampere(s)	**fA**
femtovolt(s)	**fV**

ferri-   a prefix meaning contains iron in the ferric state; combines to form one word: *ferricyanide, ferrimagnetic*

ferro-   a prefix meaning contains iron in the ferrous state; combines to form one word: *ferroelectric, ferromagnetic, ferrometer*

ferrule; ferule   a ferrule is a metal cap or lid; a ferule is a ruler

fewer; less   use fewer to refer to items that can be counted, as in *fewer technicians than we predicted have been assigned to the project*; use less to refer to general quantities, as in *there was less water available than predicted*

fiber; fibrous; Fiberglas   *Fiberglas* is a trade name

figure numbers   in text, spell out the word *Figure* in full, or abbr it to *Fig.*; e.g. *the circuit diagram in Figure 26* and *for details, see Fig. 7*; use the abbreviated form

figure numbers (*continued*)
beneath an illustration; always use numerals for the figure number

final; finally; finalize

first   to write *the first two* . . . (or three, etc.) is better than to write *the two first* . . .; never use *firstly*

fix   in technical usage, fix means to firm up or establish as a permanent fact; avoid using it when the meaning is repair

flammable   def: easily ignited; see **inflammable**

fluid   abbr: **fl**; the abbr for fluid ounce is **fl oz**

fluorescence

focus; focused; focuses; focusing   the plural is *focuses* (pref) or *foci*

foot   see **feet**

for   see **because**

forecast   this spelling applies to both present and past tenses

foresee

forestall

foreword; forward   a foreword is a preface or preamble to a book; forward means onward; e.g. *the scope is defined in the foreword to the book; he requested that we bring the meeting date forward*

for example   abbr: **e.g.** (which see)

former; first   use former to refer to the first of only two things; use first if there are more than two

formula   the plural is *formulas* (pref) or *formulae*

forty   def: 40; it is not spelled *fourty*

fractions   when writing fractions that are less than unity, spell them out in descriptive narrative, but use figures for technical details; e.g. *by the end of the heat run, nine-tenths of the installation had been completed; a flat case 15¼ in. square by 7/8 in. deep*; use decimals rather than fractions, except when a quantity is normally stated as a fraction (such as *3/8 in. plywood*)

free from   use *free from* rather than *free of*; e.g. *he is free from prejudice*

free on board abbr: **fob** (pref) or **f.o.b.** (commonly used)

frequency abbr: **freq**

frequency modulation abbr: **FM**

fulfil(l) the pref spelling is *fulfill, fulfilled, fulfilling,* and *fulfillment; fulfil* and *fulfilment* can also be spelled with a single *l.*

funnel(l)ed; funnel(l)ing single *l* pref

further see **farther**

fuse as a verb, means join together or weld; as a noun, means a circuit protection device

fuze def: a detonation initiation device

G

gage; gauge both spellings are correct; gage is recommended because it is less likely to be misspelled; *gaging* or *gauging* do not retain the *e*

gallon gallons differ between U.S. (231 in.³) and Britain (277.42 in.³); Canada uses British imperial gallon; abbr: **gal**; other abbr:

gallons per day	**gpd**
gallons per hour	**gph**
gallons per minute	**gpm**
gallons per second	**gps**

gang; ganged; ganging

gaseous

gauge see **gage**

geiger (counter)

gelatin(e) *gelatin* pref

geo- a prefix meaning of the earth; combines to form one word: *geocentric, geodesic, goemagnetic, geophysics*

giga def: 10⁹; abbr: **G**; other abbr:
gigahertz	**GHz**
gigawatt(s)	**GW**
gigacycle(s) per second	**Gc** (pref); **Gc/s**

giga (*continued*)
(latter obsolescent, replaced by *gigahertz*)

gimbal

glue; gluing

glycerin(e) *glycerin* pref

gotten never use this expression; use *have got* or simply *have*

government capitalize when referring to a specific government either directly or by implication; use lc if the meaning is government generally; e.g. *the U.S. Government; the Government specifications; no government would sanction such restrictions*

gram abbr: **g**; abbr for gram-calorie is **g-cal**

Greenwich mean time abbr: **GMT**

grill(e) def: a loudspeaker covering, or a grating; *grille* pref

ground (electrical) abbr: **gnd**

guage wrongly spelled; the correct spelling is *gauge*

guarantee never **guaranty**

guideline(s)

gyroscope abbr: **gyro**

H

half; halved; halves; halving as a prefix, half combines erratically; some common compounds are: *half-hour(ly), half-life, half-month(ly),* halftone, half-wave; for others, consult your dictionary

handful; handfuls

hangar; hanger a hangar is a large building for housing aircraft; a hanger is a supporting bracket

haversine abbr: **hav**

headforemost

heat- as a prefix, heat combines erratically; some typical compounds are: *heat-resistant, heat-run,* heatsink, *heat-treat*; for others, consult your dictionary

heavy-duty

hectare abbr: **ha**

height (not *heighth*) abbr: **ht**

heightfinder; heightfinding

helix the plural is *helices* (pref) or *helixes*

hemi- a prefix meaning half; combines to form one word: *hemisphere, hemitropic* henry(s) abbr: **H**

here- when used as a prefix, here- combines to form one word: *hereafter, hereby, herein, herewith*

hertz def: a term used to denote a unit of frequency measurement (similar to *cycle per second*, which it replaces); abbr: **Hz**

heterodyne

heterogeneous; homogeneous heterogeneous means of the opposite kind; *homogeneous* means of the same kind

high fidelity abbr: **hi-fi**

high frequency abbr: **hf**

high-pressure (as an adjective) abbr: **h-p**

high voltage abbr: **hv** (pref) or **HV**

hinge; hinged; hinging

homogeneous see **heterogeneous**

horizontal abbr: **hor**

horsepower abbr: **hp**; the abbr for horsepower-hour is **hp-hr**

hour(s) abbr: **hr**

hundred abbr: **C**

hundredweight def: 112 lb; abbr: **cwt**

hydro- a prefix meaning of water; combines to form one word: *hydroacoustic, hydroelectric, hydromagnetic, hydrometer*

hyper- a prefix meaning over; combines to form one word: *hyperacidity, hypercritical*

hyperbola the plural is *hyperbolas* (pref) or *hyperbolae*

hyperbole def: an exaggerated statement

hyperbolic cosine; sine; tangent abbr: **cosh, sinh, tanh**

hyphen in compound terms you may omit hyphens unless they need to be inserted to avoid ambiguity or to conform to accepted usage; e.g. *preemptive* is preferred without a hyphen, but *photo-offset* and *re-cover* (when the meaning is *to cover again*) both require one; refer to individual entries and Article 1

hypothesis the plural is *hypotheses*

I

ibid. def: Latin abbr for *ibidem*, meaning in the same place; use only for footnoting (see Chap. 6)

i.e. abbr for *id est* (the Latin of *that is*); avoid confusing with **e.g.**; no comma is necessary after *i.e.* and *e.g.* (see how *e.g.* has been used in the comment for *hyphen*)

if and when avoid using this hackneyed expression; use either *if* or *when* alone

ignition abbr: **ign**

illegible

im- see **in-**

imbalance this term should be restricted for use in accounting and medical terminology; use *unbalance* in other technical fields

immigrate see **emigrate**

immittance

impeller

impermeable

impinge; impinging

imply; infer speakers and writers can imply something; listeners and readers infer it from what they hear or read; e.g. *in his closing remarks, Mr. Smith implied that*

imply; infer (*continued*)
  further studies were in order; the technician inferred from the report that his work was better than expected

impracticable; impractical impracticable means not feasible; impractical means not practical; a less preferred alternative for impractical is *unpractical*

in; into in is a passive word; into implies action; e.g. *ride in the car; step into the car*

in-; im-; un- all three prefixes mean not; all combine to form one word: e.g. *ineligible, impermeable, unintelligible*; if you are not sure whether you should use in-, im-, or un-, use *not*

inaccessible

inaccuracy

inadvisable, unadvisable *inadvisable* pref

inasmuch as

incandescence

incase *encase* pref

inch(es) abbr: **in.**; other abbr:
  inches per second **ips** (pref); **in./s**
  inch-pound(s)   **in.-lb**

incidentally in most cases the word *incidentally* is unnecessary; it can almost always be deleted from technical writing

inclose see **enclose**

includes; including abbr: **incl**; when followed by a list of items, *includes* implies that the list is not complete; if the list is complete, use *comprises* or *consists of* (which see)

index the preferred plural is *indexes*, except in mathematics (where *indices* is common)

indicated horsepower abbr: **ihp**; the abbr for indicated horsepower-hour is **ihp-hr**

indifferent to never use *indifferent of*

indispensable

indorse *endorse* pref

inessential; unessential both are correct; *unessential* pref

infer see **imply**

inflammable def: easily ignited (derived from *inflame*); *flammable* may be a better choice, to prevent readers from mistakenly thinking the *in* of *inflammable* means not

infrared

ingenious; ingenuous ingenious means clever, innovative; ingenuous means innocent, naive; *ingenuity* is a noun derived from ingenious

inoculate

inquire; enquire *inquire* pref; also *inquiry*

inside diameter abbr: **ID**

in situ def: in the normal position

insofar as

instal(l) *install, installed, installer, installing, installation,* and *installment* pref; a single l is equally acceptable for *instal and instalment*

instantaneous

instrument

insure the pref def is to protect against financial loss; can also mean make certain of; see **ensure**

integer

integral, integrate, integrator

intelligence quotient abbr: **IQ**

intelligible

inter- a prefix meaning among or between; normally combines to form one word: *interact, intercarrier, interdigital, interface, intermodulation*

intermediate frequency abbr: **if.** (pref) or **i-f**

intermediate-pressure (as an adjective) abbr: **i-p**

intermittent

internal abbr: **int**

interrupt

into see **in**

intra- a prefix meaning within; normally combines to form one word: *intranuclear*; if combining word starts with *a*, insert a hyphen: *intra-atomic*

intrigue; intriguing

intrust *entrust* pref

irrational

irregardless never use this expression; use *regardless*

irrelevant frequently misspelled as *irrevelant*

irreversible

iso- a prefix meaning the same, of equal size; normally combines to form one word: *isoelectronic, isometric, isotropic*; if combining word starts with *o*, insert a hyphen: *iso-octane*

its; it's its means belonging to; it's is an abbr for it is; e.g. *the transmitter and its modulator; if the fault is not present in the remote equipment, then it's most likely to be found in master control*; in technical writing *it's* should seldom be used: replace with *it is*

J

joule abbr: **J** (pref) or **j**
judg(e)ment *judgment* pref

juxtaposition def: side by side

K

kilo   def: 10^3; abbr: **k**; other abbreviations:

kilocalorie(s)	**kcal**
kilocycles per second	**kc** (pref);
(obsolescent, replaced	**kc/s**
by *kilohertz*)	
kiloelectron volt(s)	**kcV**
kilogram(s)	**kg**
kilogram-calorie(s)	**kg-cal**
kilograms per cubic meter	**kg/m³**
kilohertz	**kHz**

kilo (*continued*)

kilohm(s)	**kΩ; kohm**
kiloliter(s)	**kl**
kilometer(s)	**km**
kilovolt(s)	**kV**
kilovolt-ampere(s)	**kVA**
kilovolt-ampere(s), reactive	**kVAr**
kilowatt(s)	**kW**
kilowatthour(s)	**kWh** (pref);
	kw-hr

L

label(l)ed; label(l)ing   single *l* pref

laboratory   abbr: **lab**

lacquer

lambert   abbr: **L**

lampholder

last, latest, latter   last means final; latest means most recent; latter refers to the second of only two things (if more than two, use *last*); it is better to write *the last two* . . . (or three, etc.) than *the two last* . . .

lath; lathe   a lath is a strip of wood; a lathe is a piece of machine shop equiment

latitude   abbr: **lat** or **Φ**

learned; learnt   *learned* pref

least common multiple   abbr: **lcm**

left-hand(ed)   abbr: **LH**

lend; loan   use lend as a verb, loan as a noun; to write or say "loan me your iron" is wrong, but "*lend* me your iron" is correct

lengthy   not *lengthly*

less   see **fewer**

letter of intent; letter of transmittal   the plural is *letters of intent*

level(l)ed; level(l)ing   single *l* pref

liable to   means under obligation to; avoid using as a synonym for *apt to* or *likely to*

liaison   liaison is a noun; it is sometimes used uncomfortably as a verb: e.g. *liaise*

libel(l)ed; libel(l)ous   single *l* pref

licence; license   *license* pref: *licence* is sometimes used as a noun

lightening; lightning   lightening means to make lighter; lightning is an atmospheric discharge of electricity

linear   abbr: **lin**; the abbr for lineal foot is **lin ft**

lines of communication   not *line of communications*

liquefy

liquid   abbr: **liq**

liter   abbr: **l**

loan   see **lend**

loath; loathe   loath means reluctant; *loathe* means to dislike intensely

locus   the plural is *loci*

locknut; lockwasher

logarithm   abbr: common—**log**; natural—**ln**

logbook

logistic(s)   use *logistic* as an adjective, *logistics* as a noun; e.g. *logistic control*; the *logistics of the move*

longitude   abbr: **long.** or **λ**

long-play(ing) (record)   abbr: **LP**

lose; loose   lose is a verb that refers to a loss; loose is an adjective or a noun that means free or not secured; e.g. *three loose nuts caused us to lose a wheel*

louver; louvre   *louver* pref

low frequency   abbr: **lf**

low-pressure (as an adjective)   abbr: **l-p**

lubricate; lubrication   abbr: **lub**

lumen   abbr: **lm**; other abbr:

lumen-hour(s)	**lm-hr** (pref); **lhr**
lumens per square foot	**lm/ft²**
lumens per square meter	**lm/m²**

lumen (*continued*)

lumens per watt   lm/W (pref);
                         lpw

luminance; luminescence; luminosity; luminous

lux   abbr: lx

M

macro-   a prefix meaning very large; combines to form one word: e.g. *macroscopic*

magneto   the plural is *magnetos*; as a prefix, it normally combines to form one word: *magnetoelectronics, magnetohydrodynamics, magnetostriction*; if combining word starts with *o* or *io*, insert a hyphen: *magneto-otpics, magneto-ionization*

magneton; magnetron   a magneton is a unit of magnetic moment; a magnetron is a vacuum tube controlled by an external magnetic field

maintain; maintenance

majority   use majority mainly to refer to a number, as in *a majority of 27*; avoid using it as a synonym for many or most: do not write *the majority of technicians* when the intended meaning is *most technicians*

malfunction

malleable

maneuver; manoeuvre   *maneuver, maneuvered, maneuvering* pref

man-hour(s)

manufacturer   abbr: mfr

marshal(l)ed;   marshal(l)ing;   marshal(l)er   single *l* pref

marvel(l)ed;   marvel(l)ing;   marvel(l)ous   single *l* pref

material; materiel   material is the substance or goods out of which an item is made; when used in the plural, it describes items of a like kind, such as *writing materials*; materiel are all the equipment and supplies necessary to support a project or undertaking (a term commonly used in military operational support)

matrix   the plural is matrices

maximum   the plural is *maximums* (pref) or *maxima*; abbr: max

meager; meagre   *meager* pref

mean; median   the mean is the average of a number of quantities; the *median* is the midpoint of a sequence of numbers; e.g.

mean, median (*continued*)
     in the sequence of five numbers 1, 2, 3, 7, 8, the mean is 4.2 and the median is 3

mean effective pressure   abbr: mep

mean sea level   abbr: msl (pref) or MSL

medium   when medium is used to mean substances, liquids, materials, or the means for accomplishing something (such as advertising), the plural is *media*; in all other senses the plural is *mediums*

mega   def: 10^6; abbr: M; other abbr:

megacycle(s) per second	Mc (pref); Mc/s (obsolescent, replaced by *megahertz*)
megaelectronvolt(s)	MeV
megahertz	MHz
megavolt(s)	MV
megawatt(s)	MW
megohm(s)	$M\Omega$; Mohm

memorandum   the plural is *memorandums* (pref) or *memoranda*; abbr: memo (singular) or memos (plural)

metal   a single *l* is pref for *metaled* and *metaling* (although *ll* also is acceptable); *metallic* always has *ll*

meteorology; metrology   meterology pertains to the weather; metrology pertains to weights, measures, and calibration

meter; metre   *meter* pref; def: metric unit of length; abbr: m

micro   def: 10^{-6}; abbr: μ or u; other abbreviations:

microampere(s)	μA; uA
microfarad(s)	μF; uF
microgram(s)	μg; ug
microhenry(s)	μH; uH
micromho(s)	μmho; umho
microsecond(s)	μs (pref); μsec; usec
microvolt(s)	μV; uV
microwatt(s)	μW; uW

as a prefix meaning very small, micro- normally combines to form one word: *microammeter, micrometer, microswitch, microwave; micromicro-*(10^{-12}) has been replaced by pico (which see)

microphone abbr: **MIC** (pref) or **mike** (slang)

mid- a prefix that means in the middle of; generally combines into one word: *midday, midpoint, midweek*; if used with a proper noun, insert a hyphen; mid-Atlantic

mile the word mile is generally understood to mean a statute mile of 5280 ft, so the statement *I drove 326 miles* implies statute miles; when referring to the *nautical mile* (6080 ft), always identify it as such: *the flight distance was 4210 nautical miles* (or *4210 nmi*); abbr:

statute mile(s)	mi
nautical mile(s)	nmi (pref); **n.m.**
miles per gallon	mpg
miles per hour	mph

mileage; milage *mileage* pref

milli def: 10^{-3}; abbr: **m**; other abbr:

milliampere(s)	**mA**
millifarad(s)	**mF**
milligram(s)	**mg**
millihenry(s)	**mH**
millilambert(s)	**mL**
milliliter(s)	**ml**
millimeter(s)	**mm**
millimho(s)	**mmho**
milliohm(s)	**mΩ; mohm**
millisecond(s)	**ms** (pref); **msec**
milliroentgen(s)	**mR**
millivolt(s)	**mV**
milliwatt(s)	**mW**

as a prefix, *milli* combines to form one word: *milliammeter, millibar, milligram, millimicron*

miniature; miniaturization

minimum the plural is *minimums* (pref) or *minima*; abbr: **min**

minority use minority mainly to refer to a number: *a minority by* 2; avoid using it as a synonym for several or a few; to write *a minority of the technicians* is incorrect when the intended meaning is *a few technicians*

minute abbr:

time	**min**
angular measure	. . .'

mis- a prefix meaning wrong(ly) or bad(ly); combines to form one word: *misalign, misfired, mismatched, or misshapen*

miscellaneous

misspelled; misspelt *misspelled* pref

miter(ed); mitre(d) *miter(ed)* pref

mnemonic

model(l)ed; model(l)er; model(l)ing single *l* pref

mold; mould *mold* pref

mono- a prefix meaning one or single; combines to form one word: *monopulse; monorail, monoscope*

monotonous

months the months of the year are always capitalized: *January, February,* etc; if abbr, use only the first three letters: *Jan, Feb,* etc; the abbr for *month is* mo

mortice; mortise *mortise* pref

mosaic

most never use as a short form for *almost;* e.g. to say *most everyone is here* is incorrect

movable; moveable *movable* pref

multi- a prefix meaning many; combines to form one word: *multiaddress, multicavity, multielectrode, multistage*

municipal; municipality

N

nano def: 10^{-9}; abbr: **n**; other abbr:

nanoampere(s)	**nA**
nanofarad(s)	**nF**
nanosecond(s)	**ns** (pref); **nsec**

naphtha(lene)

nationwide

nautical mile def: 6080 ft; abbr: **nmi** (pref) or **n.m.**; see mile

N.B. means note well, and is the abbr for *nota bene*

NC abbr for *normally closed* (contacts)

nebula the plural is *nebulas* (pref) or *nebulae*

negative abbr: **neg**

negligible

nevertheless

next   it is better to write *the next two* (or next three, etc) than *the two next* (etc)

night   never use *nite*; write *nighttime* as one word

nineteen; ninety; ninth   all three are frequently misspelled

NO   abbr for *normally open* (contacts)

No   abbr for **number** (which see)

noise-cancel(l)ing   single *l* pref

nomenclature

nomogram; nomograph   *nomogram* pref

non-   as a prefix meaning not or negative, non- normally combines to form one word: *nonconductor, nondirectional, nonnegotiable, nonlinear, nonstop*; if combining word is a proper noun, insert a hyphen: *non-American*; avoid forming a new word with non- when a similar word that serves the same purpose already exists (i.e. you should not form *nonaudible* because *inaudible* already exists)

none   when the meaning is not one, treat as singular; when the meaning is not any, treat as plural; e.g. *none* (not one) *was satisfactory; none* (not any) *of the receivers were repaired*

nonplus(s)ed   *nonplused* pref

norm   def: the average or normal (situation or condition)

normalize

normally closed; normally open (contacts) abbr: **NC, NO**

normal to   def: at right angles to

north   abbr: **N**; other abbr:
  northeast   **NE**
  northwest   **NW**
  north-south   **N-S**
  (control, movement)
  northbound and *northward* are written as one word; for rule on capitalization, see **east**

not applicable   abbr: **N/A**

note well   abbr: **N.B.** (derived from *nota bene*)

NOT-gate

noticeable

not to exceed   an overworked phrase that should be used only in specifications; in all other cases use *not more than*

nth (harmonic, etc.)

nucleus   the plural is *nuclei*

null

number   although no. would appear to be the most logical abbr for number (and is **pref**), **No.** is much more common (the symbol # is not an abbr for number); the abbr *no.* or *No.* must always be followed by a quantity in numerals; it is incorrect to write *we have received a No. of shipments*; for the difference in usage between *amount* and *number,* see **amount**

numbers (in narrative)   as a general **rule**, spell out up to and including nine, and use numerals for 10 and above; for specific rules, see Article 4

O

oblique

oblivious   def: unaware or forgetful; oblivious should be followed by *of*, not *to*; e.g. *absorbed in his work, he was oblivious of the disturbance caused by the installation crew*

obstacle

obtain; secure   use obtain when the meaning is simply to get; use secure when the meaning is to make safe or to take possession of (possibly after some difficulty); e.g. *we obtained four additional samples; we secured space in the prime display area*

occasional(ly)

occur; occurred; occurrence; occurring

off-   as a prefix either combines into one word, or a hyphen is inserted; typical examples are *offset, off-center(ed), off-scale, off-the-shelf*

ohm   abbr: Ω or spell out; abbr for ohm-centimeter(s) is **ohm-cm**; *ohmmeter* has *mm*

oil-filled

O.K.; okay   these are slang expressions that should never appear in technical writing omit; omitted; omitting

omni- a prefix meaning all or in all ways; combines to form one word: *omnibearing, omnidirectional, omnirange*

op. cit. def: Latin abbr for *opere citato*, meaning work cited; use only for footnoting (see Chap. 6)

operate; operator; operable   not *operatable*

optimum   the plural is *optima* (pref), and sometimes *optimums*

oral   def: spoken; see **aural**

orbit; orbital; orbited; orbiting

OR-gate

orientation   this is the noun; the verb form is *orient, oriented, orienting*

orifice

origin; original; originally

oscillate

oscilloscope   slang abbr: **scope**

ounce(s)   abbr: **oz**; other abbreviations:
ounce-foot   **oz-ft**
ounce-inch   **oz-in.**

out-of-phase

outside diameter   abbr: **OD**

over- as a prefix meaning above or beyond, normally combines to form one word: *overbunching, overcurrent, overdriven, overexcited, overrun*; avoid using over as a synonym for more than, particularly when referring to quantities: e.g. *more than 17 were serviceable* is better than *over 17 were serviceable*

overage   means either too many or too old

overall   an overworked word; as an adjective it often gives unnecessary additional emphasis (as in *overall impression*) and should be deleted; avoid using as a synonym for *altogether, average, general,* or *total*

oxidize   *oxidation* is better than *oxidization*

oxyacetylene

P

page; pages   abbr: **p**; **pp**

pair(s)   abbr: **pr**

panel(l)ed; panel(l)ing   single *l* pref

paperwork

parabola(s); parabolic; paraboloid

parallax

paragraph(s)   abbr: **para**

parallel; paralleled; paralleling; parallelism; parallelogram   both *parallel to* and *parallel with* are correct

parenthesis   this is the singular form; the plural is *parentheses*

particles

partly; partially   use partly when the meaning is a part of or in part; use partially when the meaning is to a certain extent, or when preference or bias is implied, e.g. *you are partly correct; a partially trained technician*; if still in doubt, use *partly*

parts per million   abbr: **ppm**

passed; past   as a general rule, used passed as a verb and past an an adjective or a noun; e.g. *the test equipment has been passed by quality control; past experience*

passed; past (*continued*)
has demonstrated a tendency to fail at low temperature; in the past . . .

pencil(l)ed; pencil(l)ing   single *l* pref

pendulum   the plural is *pendulums*

people; persons   use persons when quantities are involved; e.g. *three persons were interviewed*; otherwise, use people; *all the people present*

per   in technical writing it is acceptable to use per to mean either by or a(n), as in *per diem* (by the day) and *miles per hour*; in literary writing, take care not to use per in place of *a* or *an*; avoid using *as per* in all writing

per cent; percentage   the abbr for per cent is %; avoid using the expression *a percentage of* as a synonym for *a part of* or *a small part*

perform   not *preform* (when the meaning is to take part)

permeable; permeameter; permeance

permissible

permittivity

perpendicular abbr: **perp**

perseverance

persistent; persistence

personal; personnel personal means concerning one person; personnel means the members of a group, or the staff; e.g. *a personal affair*; the *personnel in the powerhouse*; for *person(s)* see **people**

phase in the nonelectric sense, phase means a stage of transmission or development; it should not be used as a synonym for aspect; it is used correctly in *the second phase called for a detailed cost breakdown*

phenolic

phenomenon the plural is *phenomena*

photo- as a prefix, normally combines to form one word: *photoelectric, photogrammetry, photoionization, photomultiplier*; if combining word starts with *o* insert a hyphen: *photo-offset*

pico def: 10⁻¹²; abbr: **p**; other abbreviations:

picoampere(s) **pA**
picofarad(s) **pF**
picosecond(s) **ps** (pref); **psec**
picowatt(s) **pW**

always combines to form one word

piezoelectric; piezo-oscillator

pint abbr: **pt**

pipeline

plateau the plural is *plateaus* (pref) or *plateaux*

plug; plugged; plugging

plumbbob

p.m. def: after noon (post meridiem)

pneumatic

polarize; polarization; polarizing

poly- a prefix meaning many; combines to form one word: *polydirectional, polyethylene, polyphase*

polyvinyl chloride abbr: **pvc**

positive abbr: **pos**

post- a prefix meaning after or behind; combines to form one word: *postacceleration, postgraduate*

post meridiem def: after noon; abbr: **p.m.** can also be written as *postmeridian* (less pref)

postpaid

potentiometer abb: **pot.**

pound(s) (weight) abbr: **lb**; other abbreviations:

pound-foot	**lb-ft**
pound-inch(es)	**lb-in.**
pounds per square foot	**psf** (pref); **lb/ft²**
pounds per square inch	**psi** (pref); **lb/in²**
pounds per square inch absolute	**psia**

power factor abbr: **pf** or spell out

powerhouse; power line; powerpack

practicable; practical these words have similar meanings but different applications that sometimes are hard to identify; practicable means feasible to do, as in *it was difficult to find a practicable solution* (one that could reasonably be implemented); practical means handy, suitable, able to be carried out in practice, as in *a practical solution would be to combine the two departments*

practice; practise the noun always is *practice*; the verb is *practice* (pref), but can also be *practise*

pre- a prefix meaning before or prior; normally combines to form one word: *preamplifier, predetermined, preemphasis, preignite, preset*; if combining word is a proper noun, insert a hyphen: *pre-Nixon*

precede; proceed precede means go before; proceed generally means carry on or continue; e.g. *the dinner was preceded by a brief business meeting; after dinner, we proceeded with the annual presentation of awards*; see **proceed**

precedence; precedent precedence means priority (of position, time, etc), as in *the pressure test has precedence* (it must be done first); a precedent is an example that is or will be followed by others, as in *we may set a precedent if we grant his request* (others will expect similar treatment)

prefer; preferred; preference; preferable avoid overstating preferable (which states its meaning quite clearly on its own) by using it to make a comparison; e.g. it is wrong to write *more preferable* or *highly preferable*

preform means to form beforehand; frequently written as a misspelled version of *perform* (to carry out, to act)

presently use presently only to mean soon or shortly; never use it to mean *now* (use *at present* instead)

pressure-sensitive

pretense; pretence *pretense* pref

preventive; preventative *preventive* pref; e.g. *preventive maintenance has reduced equipment outages*

previous def: earlier, that which went before; avoid writing *previous to* (see **prior**)

principal; principle as a noun, principal means (1) the first man in importance, the leader; or (2) a sum of money on which interest is paid; e.g. *one of the firm's principals is Mr. H. Winman; the invested principal of $10,000 earned $650 in interest last year*; as an adjective, it means *most important* or *chief*; principle means a strong guiding rule, a code of conduct, a fundamental or primary source (of information, etc); e.g. *his principles prevented him from taking advantage of the error*

prior; previous use only as adjectives meaning earlier, as in *he had a prior appointment*, or *a previous commitment prevented Mr. Perchanski from attending the meeting*; write *before* rather than *prior to* or *previous to*

privilege

proceed; proceedings; procedure use *proceed*

proceed; proceedings; procedure *(continued)* *to* when the meaning is to start something new; use *proceed with* when the meaning is to continue something that was started previously

program(me) use of program and programme varies widely according to personal preference; Mr. Winman and I prefer *program, programed, programing* and *programer; program(m)er-analyst* contains a hyphen

prohibit use *prohibit from*; never *prohibit to*

prominent; prominence

propel; propelled; propelling; propellant (noun); **propellent** (adjective)

prophecy; prophesy use prophecy only as a noun, prophesy only as a verb

proportion avoid writing *a proportion of* or *a large proportion of* when *some, many,* or a specific quantity would be simpler or more direct

proposition in its proper sense, proposition means a suggestion put forward for argument; it should not be used as a synonym for *plan, project,* or *proposal*

pro rata def: assign proportionally; sometimes used in the verb form as *prorate: I want you to prorate the cost over two years' operations*

proved; proven use proven only as an adjective or in the legal sense; otherwise use proved, e.g. *he has been proven guilty; he proved his case*

purge; purging

Q

quality control abbr: **QC**

quantity; quantitative the abbr of quantity is **qty**

quart abbr: **qt**

quasi- a prefix meaning seemingly or almost; insert a hyphen between the prefix and the combining word: *quasi-active, quasi-bistable, quasi-linear*

question mark insert a question mark after a direct question: *how many booklets will you require?*; omit the question mark when the question posed is really a demand; e.g. *may I have your decision by noon on Monday*

questionnaire

quiescent

R

rack-mounted

racon def: a radar beacon

radian abbr: **rad**

radio- as a prefix, radio- combines to form

radio- (*continued*)

one word: *radioactive, radiobiology, radioisotope, radioluminescence*; if combining word starts with *o* omit one of the *o's*: *radiology, radiopaque*; in other instances *radio* may be either combined or treated as a separate word, depending on accepted usage; typical examples are *radio compass, radio countermeasures, radio direction-finder, radio frequency* (as a noun), *radio-frequency* (as an adjective), *radio range, radiosonde, radiotelephone*

radio frequency   abbr: **rf**

radius   the plural is *radii*

radix   the plural is *radices* (pref) or *radixes*

range; ranging; rangefinder; range marker

rare; rarity; rarefy; rarefaction

ratemeter

re   def: a Latin word meaning in the case of; avoid using re in technical writing, particularly as an abbr for *regarding, concerning, with reference to*

re-   a prefix meaning to do again, to repeat; it normally combines to form one word: *reactivate, rediscover, reemphasize, reentrant, reignition, rerun, reset*; if the compound term forms an existing word that has a different meaning, insert a hyphen to identify it as compound; e.g. *re-cover* (to cover again) cannot be combined into a single word because it would be confused with *recover* (to get back, to regain)

reaction   use reaction to describe chemical or mechanical processes, not as a synonym for *opinion* or *impression*

reactive kilovolt-ampere; reactive volt-ampere   see **kilo** or **volt**

readability

receive; receiver; receiving

rechargeable

recommend

reconnaissance

recur; recurrence   these are the correct spellings; never use *reoccur(ence)*

reducible

reenforce; reinforce   reenforce means to enforce again, reinforce means to strengthen; e.g. *Rick Davis reenforced his original instructions by circulating a second memo-*

reenforce; reinforce (*continued*)
randum; the Artmo Building required 34,750 tons of reinforced concrete

reference   it is better to write *with reference to* than *in reference to*

referendum   plural is *referendums* (pref); a less pref alternative is *referenda*

reoccur(rence)   never use; see **recur**

repairable; reparable   both words mean in need of repair and capable of being repaired; reparable also implies that the cost to repair the item has been taken into account, and that it is economically worthwhile to effect repairs

repellant; repellent   use repellant as a noun, repellent as an adjective; **repeller**

replaceable

reservoir

reset; resetting; resettability

resin; rosin   these words have become almost synoymous, with a preference for *resin*; I recommend using resin to describe a gluey substance used in adhesives, and rosin to describe a solder flux-core

respective(ly)   this overworked word is not really needed in sentences that differentiate between two or more items; e.g. it should be deleted from sentences such as: *pins 4, 5, and 7 are marked R, S, and V respectively*

retro-   a prefix meaning to take place before, or backward; normally combines to form one word: *retroactive, retrofit, retrogression*; if combining word starts with *o*, insert a hyphen: *retro-operative*

reverse; reverser; reversal; reversible

revolutions per minute; revolutions per second   abbr: **rpm; rps**

rheostat

rhombus   the plural is *rhombuses* (pref) or *rhombi*

rhythm

ricochet; ricocheted; ricocheting

right-handed(ed)   abbr: **RH**

rivet; riveted; riveting

roentgen   abbr: **R**

role; roll   a role is a person's function or

role; roll (*continued*)
the part that he plays (in an organization, project, or play); a roll, as a technical noun, is a cylinder; as a verb, it means to rotate; e.g. *the technicians' role was to make the samples roll toward the magnet*

root mean square   abbr: **rms**

rosin   see **resin**

rotate; rotator; rotatable; rotary

ruggedize

rustproof; rust-resistant

S

salable; saleable   *salable* pref

same   avoid using same as a pronoun; e.g. to write *we have repaired your receiver and tested same* is incorrect; the correct version is *we have repaired and tested your receiver*

sapphire

satellite

saturate; saturation; saturable

sawtooth; saw-toothed

scalar; scaler   scalar is a quantity that has magnitude only; scaler is a measuring device

scarce; scarcity

sceptic(al)   see **skeptic(al)**

schedule

schematic   although really an adjective (as in *schematic diagram*), in technical terminology schematic can be used as a noun (meaning *a schematic drawing*)

screwdriver; screw-driven

seamweld

seasonal; seasonable   seasonal means affected by or dependent on the season; seasonable means appropriate or suited to the time of year; e.g. *a seasonal activity; seasonable weather*

seasons   the seasons are not capitalized: *spring, summer, autumn* or *fall, winter*

secant   abbr: **sec**

second   abbr:
time   **sec**
angular measure   **. . .**"

secure   see **obtain**

-sede   *supersede* is the only word to end with -sede; others end with *-cede* or *-ceed* (which see)

seem(s)   see **appear(s)**

self-   insert a hyphen when used as a prefix to form a compound term: *self-absorption, self-bias, self-excited, self-locking, self-resetting*; but there are exceptions: *selfless, selfsame*

semi-   a prefix meaning half; normally combines to form one word: *semiactive, semiannually* every six months), *semiconductor, semimonthly* (half-monthly), *semiremote,* semiweekly half weekly); if combining word starts with *i*, insert a hyphen: *semi-idle, semi-immersed*

separate; separable; separator

serial number   abbr: **ser no.** or **S/N**

series-parallel

serrated

serviceable; serviceman

servo-   as a prefix, servo- combines to form one word: *servoamplifier, servocontrol, servosystem*; as a noun, servo is an abbreviation for *servomotor* or *servomechanism*

sewage; sewerage   sewage is waste matter; sewerage is the drainage system that carries away the waste matter

short-   as a prefix, most often combines with a hyphen: *short-circuit; short-form* (report); *short-handed; short-lived*; in some cases it combines into one word: *shorthand* (writing); *shortcoming; shortsighted*

shrivel(l)ed; shrivel(l)ing   single *l* pref

sight   def: having the ability to see; see **site**

signal(l)ed ;signal(l)ing; singal(l)ler   single *l* pref

signal-to-noise (ratio)

silverplate; silver-plate   use silverplate as a noun, silver-plate as a verb or adjective

similar   not similiar

sine   abbr: **sin**

singe; singeing   the *e* must be retained to avoid confusion with *singing* singlehanded

siphon   not syphon

sirup; syrup   syrup pref

site; sight; cite   three words that often are misspelled; a site is a location, as in *the construction site*; sight implies the ability to see, as in *mud up to the axles became a familiar sight*; cite means quote, as in *I cite the May 17 progress report as a typical example of good writing*

siz(e)able   *sizable* pref

skeptic(al); sceptic(al)   *skeptic(al)* pref

skil(l)ful   skillful pref; note that general usage dictates that *ll* is pref, even though it is contrary to the preference in most *l* and *ll* situations in this glossary

slip-   mostly combines into a single word; examples are *slippage, slipshod, slipstream*; but *slip ring(s)*

smelled; smelt   *smelled* pref

smo(u)lder   *smolder* pref

someone; some one   someone is correct when the meaning is any one person; some one is seldom used

some time; sometimes   some time means an indefinite time, as in *some time ago*; sometimes means occasionally, as in *he sometimes works until after midnight*

sound   *sound-absorbent; sound-absorbing; sound-powered; soundproof; sound track; sound wave*

south   abbr: **S**; other abbr:
    southeast   **SE**
    southwest   **SW**
    southbound and *southward* are written as one word; for rule on capitalization, see **east**

spare(s)   as a noun, frequently means spare part(s)

specially   *specially* is better than *especially*; *specialize* has only one *l*

specific gravity   abbr: **sp gr**

specific heat   abbr: **sp ht**

spectrum   the plural is *spectra*

spectro-   as a prefix, combines to form one word: *spectrometer, spectroscope*; if combining word starts with *o*, omit one *o*: *spectrology*

spelled; spelt   *spelled* pref

spilled; spilt   *spilled* pref

split infinitive   to split an infinitive is permissible when not to split it would result in awkward construction or ambiguity, or require extensive rewriting

spoiled; spoilt   *spoiled* pref

spotweld

square   abbr: **sq** ;other abbr:
    square foot   ft^2 (pref); **sq ft**
    square inch   in.^{2} (pref); **sq in.**
    square meter   m^2 (pref); **sq m**
    square yard   yd^2 (pref); **sq yd**

standby; standoff; standstill   these terms combine into one word when used as a noun or an adjective

standing-wave ratio   abbr: **swr**

state-of-the-art

stationary; stationery   stationary means not moving, as in *the vehicle was stationary when the accident occurred*; stationery refers to writing materials, as in *the main item in the October stationery requisition was an order for one thousand writing pads*

statute mile   def: 5280 ft; see **mile**

stencil(l)ed; stencil(l)ing   single *l* pref

stereo-   as a prefix, combines to form one word: *stereometric, stereoscopic; stereo* can be used alone as a noun meaning multichannel system

stimulus   the plural is *stimuli*

stock; stockholder; stocklist; stock market; stockpile

stop-   *stopgap; stopnut; stopover* (when used as a noun or adjective); *stoppage; stop watch*

strato-   a prefix that combines to form one word: *stratocumulus, stratosphere*

stratum   the plural is *strata*

structural

stylus   the plural is *styluses* (pref) or *styli*

sub-   a prefix generally meaning below, beneath, under; combines to form one word: *subassembly, subcarrier, subcommittee, subnormal, subpoint*

subparagraph   abbr: **subpara**; the abbr for *subsubparagraph* is **subsubpara**

sufficient in technical writing, *enough* is a better word than *sufficient*

sulfur; sulphur *sulfur* pref; as a prefix, **sulf-** combines to form one word: *sulfanilamide*

super- a prefix meaning greater or over; combines to form one word: *superconductivity, superregeneration*

superhigh frequency abbr: **shf**

superimpose; superpose superimpose means to place or impose one thing on top of another; superpose means to lay or place exactly on top of, so as to be coincident with

supersede see **-sede**

supra- a prefix meaning above; normally combines to form one word: *supramolecular*; if combining word starts with *a*, insert a hyphen: *supra-auditory*

surveillance

susceptible

switch- *switchboard; switchbox; switchgear*

swivel(l)ed; swivel(l)ing single *l* pref

syllabus the plural is *syllabuses* (pref) or *syllabi*

symmetry; symmetrical

symposium the plural is *symposia* (pref) or *symposiums*

synchro- as a prefix combines to form one word: *synchromesh, synchronize, synchroscope; synchro* can also be used alone as a noun meaning synchronous motor

synonymous use *synonymous with*, not *synonymous to*

synopsis the plural is *synopses*

synthetic

syphon *siphon* pref

systemwide

T

tablespoon use *tablespoonfuls* rather than *tablespoonsful*; abbr: **tbsp**

tailless def: without a tail

take- *takeoff; takeover; takeup*; as nouns and adjectives these terms all combine into a single word

tangent abbr: **tan**

tax- tax-exempt; taxpaid; taxpayer

teaspoon use *teaspoonfuls* rather than *teaspoonsful*; abbr: **tsp**

tele- a prefix that means at a distance; combines to form one word: *teleammeter, telemetry, telephony, teletype(writer)*

telecom; telecon telecom is the abbr for *telecommunication(s)*; telecon is the abbr for *telephone conversation*

television abbr: **TV**

temperature abbr: **temp**; combinations are *temperature-compensating* and *temperature-controlled*; when recording temperatures, the abbr for *degree* (deg or °) may be omitted: e.g. *an operating temperature of 85C; the water boils at 212F*

tempered

template; templet both spellings are correct; *template* pref; def: a pattern or guide

tensile strength abbr: **ts**

tera def: 10^{12}; abbr: **T**

terminus the plural is *termini* (pref) or *terminuses*; note that the preference for this plural is contrary to the preferred plurals for most other *-us* endings

that is abbr: **i.e.**

their; there; they're the first two of these words are frequently misspelled, more through carelessness than as an outright error; their is a plural possessive roughly meaning belonging to, as in *the staff took their holidays earlier than normal*; there means in that place; e.g. *there were 18 desks in the room*, or *put it there*; they're is a contraction of *they are* and should not appear in technical or business writing

there- as a prefix combines to form one word: *thereafter, thereby, therein, thereuopn*

therefor(e) *therefore* pref

thermo- a prefix generally meaning heat; combines to form one word: *thermoammeter, thermocouple, thermoelectric, thermoplastic*

thesis the plural is *theses*

thousand abbr:
    thousand foot-pound(s) **kip-ft**
    thousand pound(s)   **kip**

three- when three is used as a prefix, a hyphen normally is inserted between the combining words: *three-dimensional, three-phase, three-ply, three-wire*; exceptions are *threefold* and *threesome*

threshold

through never use *thru*

tieing; tying *tying* pref

timber; timbre timber is wood; timbre means tonal quality

time always write time in numerals, preferably using a 24 hour clock; e.g. *0817* or *8:17* a.m.; *1530* or *3:30* p.m.; never use the term "o'clock" in technical writing; write 3 p.m. rather than *3 o'clock*

time- typical combinations are *time base; timecard; timeclock; time constant; timeconsuming, time lag; timesaving, timetable; timewasting*

tinplate; tin-plate use *tinplate* as a noun, *tin-plate* as a verb or adjective

to; too; two frequently misspelled, most often through carelessness; to is a preposition that can mean in the direction of, against, before, or until; too means as well; two is the quantity "2"

today; tonight; tomorrow never use tonite

tolerance abbr: **tol**

ton the U.S. ton is 2000 lb and is known as a *short ton*; the British ton is 2240 lb and is known as a *long ton*; the metric ton is 1000 kg (2204.6 lb); other terms: *tonmile* and *tonnage*

top- *top-heavy; top-loaded; top-up*

torque

total(l)ed; total(l)ing single *l* pref

toward(s) *toward* pref

traceable

trans- a prefix meaning over, across, or through; it normally combines to form one word: *transadmittance, transcontinental, transship*; if combining word is a proper noun, insert a hyphen: *trans-Canada* (an exception is *transatlantic*, which through common usage has dropped the capital A and combined into one word)

transceiver def: a transmitter-receiver

transfer; transferred; transferring; transference

transverse; traverse transverse means to lie across; traverse means to track horizontally

travel(l)ed; travel(l)ing; travel(l)er single *l* pref

tri- a prefix meaning three or every third; combines to form one word: *triangulation, tricolor, trilateral, tristimulus, triweekly*

trouble-free; troubleshoot(ing)

tune; tunable; tune-up

tunnel(l)ed; tunnel(l)ing single *l* pref

turbo- a prefix meaning turbine-powered; combines to form one word: *turboelectric, turboprop*

turbulence

turn *turnaround* and *turnover* form one word when used as adjectives or nouns; *turnstile* and *turntable* always form one word; *turns-ratio* is hyphenated

two- when used as a prefix to form a compound term, a hyphen normally is inserted: *two-address, two-phase, two-ply, two-position, two-wire*; an exception is *twofold*

type- as a prefix normally combines into one word: *typeface; typeset(ting); typewriter; typewriting*

U

ultimatum the plural is *utlimatums* (pref) or *ultimata* (seldom used)

ultra- a prefix meaning exceedingly; normally combines to form one word: *ultrasonic, ultraviolet*; if combining word starts with a, insert a hyphen: *ultra-audible, ultra-audion*

ultrahigh frequency abbr: **uhf**

un- a prefix that generally means not or

un- (*continued*)

negative; it normally combines to form one word: *uncontrolled, undamped, unethical, unnecessary*; if combining word is a proper noun, insert a hyphen: *un-American*; similarly, if term combines to form an existing word that has a different meaning, insert a hyphen: e.g. *un-ionized* (meaning not ionized) cannot be combined into a single

un- (*continued*)
word because it could be confused with unionized (to belong to a union); if uncertain whether to use *un-*, *in-*, *or* *im-*, try using *not*

unadvisable; inadvisable both are correct; *inadvisable* pref

unbalance; imbalance for technical writing, *unbalance* pref

unbias(s)ed *unbiased* pref

under- a prefix meaning below or lower; combines to form one word: *underbunching, undercurrent, underexposed, underrated, undershoot, undersigned*

underage means a shortage or deficit, or too young

unequal(l)ed *unequaled* pref; see **equal**

unessential; inessential *unessential* pref

uni- a prefix meaning single or one only; combines to form one word: *uniaxal,*

uni- (*continued*)
unidirectional, unifilar, univalent

uninterested def: not interested; avoid confusing with *disinterested* (which see)

unique def: the one and only, without equal, incomparable; use with great care and never in any sense where a comparison is implied e.g. you cannot write *this is the most unique design;* rewrite as *this design is unique,* or (if a comparison must be made) *this is the most unusual design.*

unparalleled

unpractical *impractical* pref

unserviceable abbr: **u/s**

unstable *instability* is better than *unstability*

up- as a prefix combines to form one word: *update, upend, upgrade, uprange, upswing*

upper case def: capital letters; abbr: **uc**

uppermost

up-to-date

use; usable; usage; using; useful

V

vacuum

valve-grind(ing)

variance write *at variance with;* never write *at variance from*

vari- as a prefix meaning varied, combines into one word: *varicolored, variform*

varimeter; varmeter def: a meter for measuring reactive power; *varimeter* pref

vender; vendor *vendor* pref

versed sine abbr: **vers**

versus def: against; abbr: **vs**

vertex def: top; the plural is *vertexes* (pref) or *vertices;* avoid confusing with *vortex*

very high frequency abbr: **vhf**

vice versa def: in reverse order

video- *videocast; videotape; video mapper*

video frequency abbr: **vf**

viewfinder; viewpoint

visor; vizor *visor* pref

viz def: namely; this term is seldom used in technical writing

volt(s) abbr: **V;** other abbr:

volt-ampere(s)	**VA**
volt-ampere(s), reactive	**VAr**
volts, alternating current	**Vac**
volts, direct current	**Vdc**
volts, direct current, working	**Vdcw**

as a noun, combines into one word: *voltammeter, voltohmyst*

volume abbr: **vol**

vortex def: spiral; the plural is *vortexes* (pref) or *vortices;* avoid confusing with *vertex*

VU-meter

W

walkie-talkie

war- as a prefix, combines to form one word: *warfare, wartime*

warranty

waste; wastage

water *water-cool(ed); waterflow; water level; waterline; waterproof; water-soluble; watertight*

watt(s) abbr: **W;** the abbr for *watt-hour(s)* is **Wh** (pref) or **W-hr;** as a prefix, *watt-* forms *watthourmeter* and *wattmeter*

wave- normally combines to form one word: *waveband, waveform, wavefront, waveguide, wavemeter, waveshape*; exceptions are *wave angle* and *wave-swept*

wavelength abbr: λ

weather use weather only as a noun; i.e. never write of *weather conditions*; avoid confusing with *climate* (which see)

weatherproof

week(s) abbr: **wk**

weight abbr: **wt**

west abbr: **W**; *westbound* and *westward* are written as one word; for rule on capitalization, see **east**

where- as a prefix, combines to form one word: *whereas, wherefore, wherein*; when

where- (*continued*) combining word starts with *e*, omit one *e*: *wherever*

while; whilst while pref

whoever

wide; width abbr: **wd**

wideband; widespread

wirecutter(s); wire-cutting; wirewound

withheld; withhold

words per minute abbr: **wpm**

work- *workbench; workflow; workload; workshop*

working volts, dc abbr: **Vdcw**

worldwide

writeoff; writeup both combine into one word when used as noun or adjective

X

x- *x-axis; X-band; x-particle; x-radiation; x-ray*

X-Y recorder

Y

y- *Y-antenna; y-axis; Y-connected; Y-network; Y-signal*

yards(s) abb: **yd**

yardstick

year(s) abbr: **yr**; typical combinations are *year-end* and *year-round*

your; you're your means belonging to or originating from you, as in *I have examined your prototype analyzer*; you're is a contraction of *you are* and should not appear in technical or business writing

Z

Z-axis

zero the plural is *zeros* (pref) or *zeroes*; typical combinations are *zero-access,*

zero-adjust, zero-beat, zero-hour, zero level, zero-set, zero reader

Index